Springer Asia Pacific Mathematics Series

Volume 4

The Springer Asia Pacific Mathematics Series promotes high-quality scientific research connected to Asia-Pacific countries. Monographs, edited volumes, lecture notes, and textbooks will cover topics related to pure and applied mathematics as well as interdisciplinary research.

* * *

All material will be rigorously peer-reviewed and should be related to mathematical research conducted at or with institutions in the APAC region. This series is open to subseries, which may be topical or exclusively supplied by a single research institution.

* * *

Jingli Ren • Yiwen Tao

Modeling in Life Sciences and Ecology

Machine Learning and Dynamical Systems

Springer

Science Press
Beijing

Jingli Ren
School of Mathematics and Statistics
Zhengzhou University
Zhengzhou, China

Yiwen Tao
School of Mathematics and Statistics
Zhengzhou University
Zhengzhou, China

ISSN 3091-2555 ISSN 3091-2563 (electronic)
Springer Asia Pacific Mathematics Series
ISBN 978-981-95-1037-5 ISBN 978-981-95-1038-2 (eBook)
https://doi.org/10.1007/978-981-95-1038-2

Jointly published with Science Press
The print edition is not for sale in China (Mainland). Customers from China (Mainland) please order the print book from: Science Press.

This Springer imprint is published by the registered company Springer Nature Singapore Pte Ltd.
The registered company address is: 152 Beach Road, #21-01/04 Gateway East, Singapore 189721, Singapore

Preface

The natural world, encompassing its associated life sciences and ecological processes, reveals a conspicuous level of complexity. Within this complexity lies a multitude of interconnected elements, presenting a formidable challenge for direct comprehension and prediction. The primary mission of scientists is to unravel this intricacy, ultimately facilitating the anticipation of events and phenomena within the natural realm. This predictive ability holds paramount significance, bearing implications for problem-solving, decision-making, and the advancement of human knowledge and well-being. To attain this overarching objective, scientists employ various modeling methodologies. These approaches serve the purpose of illuminating essential facets of complex phenomena, particularly the critical processes that shape their temporal and spatial dynamics. Furthermore, they underscore the importance of interconnected descriptions to enhance our comprehension of the intricate interplay between these processes.

In the contemporary landscape of life sciences and ecological research, machine learning and dynamical systems serve as indispensable tools. The choice of modeling methodologies is intricately linked to several critical factors, including the inherent nature of the research questions, the accessibility and relevance of the available data, and the depth of domain-specific expertise possessed by the researchers. Dynamical systems modeling, marked by its pronounced emphasis on elucidating the underlying physical mechanisms and systematic regularities governing a given system, often hinges upon a foundation of a priori knowledge to construct models that faithfully capture the intrinsic dynamics. Conversely, machine learning modeling, deeply rooted in its data-driven paradigm, harnesses a panoply of algorithms to autonomously distill discernible patterns and glean valuable insights from the vast reservoirs of available data.

This book commences by delving into the fundamental concepts of dynamical systems and machine learning modeling, elucidating the workflow of these two modeling approaches. While primarily tailored as an introductory textbook for both undergraduate and graduate students, its broader aim is to captivate the interest of seasoned ecologists and life scientists, beckoning them to explore the realm of modeling. The introduction and development of each section adhere to a practical problem-driven approach, aiming to address real-world issues. The focus is on addressing how to establish and evolve appropriate models based on practical problems or data. What are the fundamental principles for refining the models? The book encompasses an extensive array of models, including plankton ecosystem models, predator-prey models, and glucose regulatory models, survival and incidence prediction models, mortality risk prediction models, drug-target protein interaction prediction models, and attribution models. The methods covered in this book include feature engineering, model evaluation, optimization, interpretation, as well as statistical and algorithmic theories. Additionally, it delves into the stability, bifurcation theory of time-delayed, continuous, and discrete dynamical systems.

During the course of this book endeavor, the author received invaluable support and guidance from esteemed experts, especially Prof. Lei Guo (Academician of Chinese Academy of Sciences), Prof. Song Jiang (Academician of Chinese Academy of Sciences), Prof. Junzhi Cui (Academician of Chinese Academy of Engineering), and Prof. Junbiao Chang (Academician of Chinese Academy of Sciences). The author extends heartfelt appreciation to them for their contributions.

Furthermore, the authors acknowledge the support provided by the National Natural Science Foundation of China (Nos. U23A2065, 52071298, and 12201577). Simultaneously, they thank the students in their team, Xinjia Zhao, BengBeng Wang, Jinjin Wu, Man Zhu, and Ruijia Sun, for their contributions to this book.

Due to limitations in knowledge and expertise, this book may have certain shortcomings and imperfections. The authors welcome constructive criticism and feedback from readers.

Zhengzhou, China
September 2023

Jingli Ren

Contents

Chapter 1
Introduction to Dynamical Systems

A dynamical system encompasses a collection of interconnected entities or variables that evolve over time or in correspondence with other independent factors. These entities could represent physical objects, biological components, economic factors, or any other elements that interact and influence each other's behavior. The dynamics of such a system are captured by a set of mathematical equations that describe the relationships and interactions between the variables.

The fundamental concept behind studying dynamical systems is to unravel the intricate interplay and evolution of the entities involved. By comprehending the underlying mathematical structure, researchers can gain insights into how the system's behavior changes as time progresses or as external factors vary. This analysis often involves investigating equilibrium points, oscillations, stability conditions, and the emergence of complex behaviors like chaos.

1.1 Definition of a Dynamical System

Some basic terminology will be introduced before formally defining a dynamical system.

1.1.1 State Space

The complete array of feasible states within a system finds its representation through the points within a designated set denoted as X. This set is aptly termed the state space of the system. It is imperative to acknowledge that the identification of a point $x \in X$ should encompass more than just detailing the present "location" of the system. It should also encompass the ability to predict its subsequent

J. Ren, Y. Tao, *Modeling in Life Sciences and Ecology*, Springer Asia Pacific Mathematics Series 4, https://doi.org/10.1007/978-981-95-1038-2_1

development. Various disciplines present us with suitable state spaces, which are frequently referred to as phase spaces, a terminology rooted in the legacy of classical mechanics.

1.1.2 Time

The progression of a dynamical system entails a modification in its state over a span of time $t \in T$, where T denotes a set of numerical values. Our examination encompasses two classifications of dynamical systems: those reliant on continuous (real) time ($T = \mathbb{R}^1$) and those contingent upon discrete (integer) time ($T = \mathbb{Z}$). The former are referred to as continuous-time dynamical systems, while the latter are identified as discrete-time dynamical systems. The context for discrete-time systems naturally arises in disciplines such as ecology and economics, where the state of a system at a given moment t entirely determines its subsequent state after a defined interval, for instance, at $t + 1$.

1.1.3 Evolution Operator

A pivotal element of a dynamical system is an evolutionary principle that governs the state x_t of the system at time t, with the proviso that the initial state x_0 is known. The most comprehensive approach to specify this evolution entails considering, for a given $t \in T$, a mapping ϕ_t defined within the state space X:

$$\phi_t : X \to X$$

which transforms an initial state $x_0 \in X$ into a state $x_t \in X$ at time t:

$$x_t = \phi_t(x_0).$$

Commonly referred to as the evolution operator of the dynamical system, the map ϕ_t is often called the evolution operator of the dynamical system. In most cases, it is defined indirectly and can be computed only approximately. In the continuous-time case, the family $\{\phi^t\}_{t\in T}$ of evolution operators is called a flow.

Note that $\phi^t x$ might not be defined for all pairs $(x, t) \in X \times T$. Dynamical systems where the evolution operator ϕ^t is defined for both $t \geq 0$ and $t < 0$ are known as invertible. In these systems, the initial state x_0 completely determines not only the future states but also the past behavior of the system. However, it's also useful to consider dynamical systems where the future behavior for $t > 0$ is entirely determined by the initial state x_0 at $t = 0$, but the history for $t < 0$ cannot be unambiguously reconstructed. These noninvertible dynamical systems have evolution operators defined only for $t \geq 0$ (i.e., $t \in \mathbb{R}^1_+$ or $\mathbb{Z}_+$). In the continuous-time case, they are referred to as semiflows.

Additionally, $\varphi^t x_0$ might be defined only locally in time, for instance, for $0 \leq t < t_0$, where t_0 depends on $x_0 \in X$. A significant example of such behavior is a "blow-up," where a continuous-time system in $X = \mathbb{R}^n$ approaches infinity within a finite time, i.e.,

$$||\varphi^t x_0|| \to +\infty, \text{ as } t \to t_0.$$

The evolution operators exhibit two key properties that demonstrate the deterministic nature of dynamical systems. First,

$$\varphi^0 = \text{id}, \tag{1.1}$$

where id is the identity map on X ($\text{id}\, x = x$ for all $x \in X$). This indicates the system's state remains unchanged spontaneously.

The second property of the evolution operators reads

$$\varphi^{t+s} = \varphi^t \circ \varphi^s, \tag{1.2}$$

meaning

$$\varphi^{t+s} x = \varphi^t(\varphi^s x),$$

for all $x \in X$ and $t, s \in T$, where both sides of the equation are defined. This property implies that the system's evolution over $t + s$ units of time is equivalent to evolving over s units first, then over t units. This reflects that the system is autonomous, with unchanging governing laws over time.

A discrete-time dynamical system with integer t is fully described by a single map $f = \varphi^1$, known as the "time-one map." Using (1.1), we get

$$\varphi^2 = \varphi^1 \circ \varphi^1 = f \circ f = f^2,$$

where f^2 is the second iterate of f. Similarly,

$$\varphi^k = f^k$$

for all $k > 0$. If the system is invertible, this holds for $k \leq 0$ with $f^0 = \text{id}$.

For many systems, $\varphi^t x$ is a continuous function of $x \in X$, and if $t \in \mathbb{R}^1$, it is continuous in time as well. Continuity is defined with respect to the metric or norm in X. Additionally, many systems on $\mathbb{R}^n$ or smooth manifolds in $\mathbb{R}^n$ are such that $\varphi^t x$ is smooth in both x and t. These are called smooth dynamical systems.

1.1.4 Dynamic System

Now we are able to give a formal definition of a dynamical system.

Definition 1.1 A dynamical system is a triplet $\{T, X, \varphi^t\}$, where T is the time set, X is the state space, and $\varphi^t : X \to X$ is a parameterized family of evolution operators indexed by $t \in \mathbb{R}$, satisfying the properties $\varphi^0(x) = x$ and $\varphi^{t+s}(x) = \varphi^t(\varphi^s(x))$.

Remark 1.1 Within the dynamical system $\{T, X, \varphi^t\}$, when the time set T is continuous, i.e., $T = \mathbb{R}^1$, the system is termed a continuous-time dynamical system; when $T = \mathbb{Z}$, it becomes a discrete-time dynamical system.

Example 1.1 $e^{tA}x$ is a dynamic system.

Proof Assume $\varphi_t(x)$ is a differentiable dynamical system. Let

$$f(x) = \frac{d}{dt}\varphi_t(x)|_{t=0} = \lim_{\Delta t \to 0} \frac{\varphi_{\Delta t}(x) - \varphi_0(x)}{\Delta t},$$

we obtain a differential equation

$$\frac{dx}{dt} = f(x). \tag{1.3}$$

Assuming that $\varphi_t(x)$ is the solution of Eq. (1.3), and substituting $\varphi_t(x)$ into both sides of the Eq. (1.3). Substituting into the left side gives

$$\frac{d\varphi_t(x)}{dt} = \lim_{\Delta t \to 0} \frac{\varphi_{t+\Delta t}(x) - \varphi_t(x)}{\Delta t},$$

and substituting into the right side gives

$$f(\varphi_t(x)) = \lim_{\Delta t \to 0} \frac{\varphi_{\Delta t}(\varphi_t(x)) - \varphi_0(\varphi_t(x))}{\Delta t}.$$

From the Definition 1.1, we have

$$\varphi_{\Delta t}(\varphi_t(x)) = \varphi_{t+\Delta t}(x) \quad \text{and} \quad \varphi_0(\varphi_t(x)) = \varphi_t(x).$$

Additionally, it satisfies the initial condition

$$\varphi_t(x)|_{t=0} = x.$$

The dynamical equations that characterize continuous flows are conventionally formulated as differential equations. In the realm of ordinary differential equations (leaving aside the domain of partial differential equations temporarily), these equations manifest in diverse structures: single-variable equations of higher orders are observed, within which one can distinguish between autonomous (those that do not explicitly incorporate time within the equation) and nonautonomous types (those that explicitly include time as a variable). Moreover, there are multivariate systems of ordinary differential equations. Importantly, it is recognized that all prevalent forms of nonlinear ordinary differential equations can be recast into an autonomous system of first-order ordinary differential equations. This transformation is achievable by considering each derivative of higher-order autonomous equations as an independent variable in its own right.

Definition 1.2 Consider the following differential equations,

$$\frac{dx}{dt} = f(t, x), \quad f \in C(G = I \times D \subseteq \mathbb{R} \times \mathbb{R}^n, \mathbb{R}^n), \tag{1.4}$$

and

$$\frac{dx}{dt} = f(x), \quad f \in C(D \subseteq \mathbb{R}^n, \mathbb{R}^n), \tag{1.5}$$

1. If the vector field defined by $f(t, x)$ relies not only on the initial position x_0 but also on time t, causing the direction at any point within the domain D to be nonunique and subject to temporal variation, then Eq. (1.4) characterizes a nonautonomous system.
2. If the vector field defined by $f(x)$ remains unaffected by time, signifying that at any point x_0 within the domain D, there is a singular direction $f(x_0)$, and then Eq. (1.5) corresponds to an autonomous system.

Consider a second-order ordinary differential equation,

$$\ddot{x} = f(x, \dot{x}). \tag{1.6}$$

Let $y = \dot{x}$; then Eq. (1.6) can be transformed into a system of first-order ordinary differential equations,

$$\begin{cases} \dot{x} = y, \\ \dot{y} = f(x, y). \end{cases}$$

For nonautonomous equations, by treating the explicit time t in the equation as a new variable $z = t$ and introducing a new equation $\dot{z} = 1$. In this way, the original system of n variables of nonautonomous equations becomes a system of autonomous differential equations with $n + 1$ variables.

If the explicit time t in the nonautonomous equations appears in the form of some special function, it can be further simplified based on the characteristics of the function. Taking the forced van der Pol equation for a mechanical oscillator as an example,

$$\ddot{x} + a(x^2 - 1)\dot{x} + \omega^2 x = F\cos(\Omega t), \tag{1.7}$$

where x represents the displacement of the oscillator, a controls the nonlinearity and the strength of the damping (or amplification if negative), ω represents the natural frequency of the undamped oscillator, $F\cos(\Omega t)$ represents an external driving force with amplitude F, and angular frequency Ω.

Since the time factor $\cos\Omega t$ is a solution of the equation,

$$\ddot{x} = -\Omega^2 z,$$

therefore, a new two-variable system of first-order equations can be introduced:

$$\begin{cases} \dot{z} &= u, \\ \dot{u} &= -\Omega^2 z, \end{cases}$$

representing this cosine factor. Thus (1.7) can be transformed into an autonomous system of first-order differential equations

$$\begin{cases} \dot{x} = y, \\ \dot{y} = -\Omega^2 x - a(x^2 - 1)y + Fz, \\ \dot{z} = u, \\ \dot{u} = -\Omega^2 z. \end{cases}$$

For a set of n-dimensional autonomous differential equations, the standard form is

$$\dot{x}_i = f_i(x), \quad i = 1, 2, \ldots, n. \tag{1.8}$$

For nonautonomous differential equations, they can be written as

$$\dot{x} = f(x),$$

where x is a vector in the n-dimensional Euclidean space $\mathbb{R}^n$ ($x \in \mathbb{R}^n$) and the component in its i-th direction is x_i. Then (1.8) can be written as

$$\begin{pmatrix} \dot{x}_1 \\ \dot{x}_2 \\ \vdots \\ \dot{x}_n \end{pmatrix} = \begin{pmatrix} f_1(x_1, x_2, \ldots, x_n) \\ f_2(x_1, x_2, \ldots, x_n) \\ \vdots \\ f_n(x_1, x_2, \ldots, x_n) \end{pmatrix},$$

where $x = (x_1, x_2, \ldots, x_n)^T \in \mathbb{R}^n$ and $f = (f_1, f_2, \ldots, f_n)^T \in \mathbb{R}^n$. The space $\mathbb{R}^n$ spanned by the state variable x_i constitutes the state space.

1.1.5 Steady State

From a geometric perspective, using visual aids to elucidate the properties of a dynamical system facilitates comprehension. The fundamental geometric entities corresponding to the dynamical system $\{T, X, \varphi^t\}$ are the trajectories of the system in the state space and the phase portraits composed of these trajectories.

Definition 1.3 A trajectory originating from x_0 is an ordered subset of the state space X:

$$Or(x_0) = \{x \in X : x = \varphi^t x_0, \text{ for all } t \in T \text{ such that } \varphi^t x_0 \text{ is defined}\}.$$

In a continuous system, a trajectory is a curve parametrized by time t within the state space X, with direction corresponding to the increasing direction of time. In a discrete system, trajectories consist of points listed according to increasing integers in the state space X, with the simplest trajectory being an equilibrium point.

Definition 1.4 A point $x_0 \in X$ is termed an equilibrium point or a fixed point if $\varphi^t x_0 = x_0$ for all $t \in T$.

It can be readily observed that in the phase space, the trajectory at a fixed point does not exhibit a well-defined slope,

$$\frac{dx_i}{dx} = \frac{dx_i}{dt} / \frac{dx}{dt} = 0/0. \tag{1.9}$$

Equation (1.9) indicates either that no trajectory passes through the given point or that more than one (potentially infinitely many) trajectories do pass through it. Such points with undefined slopes are known as singular points and are sometimes also referred to as critical points. In other words, fixed points, stationary points, equilibrium points, steady points, singular points, and critical points are different terms for the same concept. In phase space, all points other than singular points have a well-defined slope; such points are termed ordinary points or analytic points. Fixed points come in various types, and the different types of fixed points and their stability are crucial in determining the behavior of dynamical systems, especially the behavior in the vicinity of these fixed points. These topics will be introduced sequentially in the following sections.

Theorem 1.1 *In the phase space, excluding singular points, all trajectories are nonintersecting.*

Proof Since

$$\frac{dx_i}{dx_k} = \frac{dx_i}{dt} / \frac{dx_k}{dt} = \frac{f_i(x_j)}{f_k(x_j)}, \quad j = 1, 2, \ldots, n.$$

Given that Eq. (1.8) is autonomous, the values of f_i and f_k are uniquely determined at each point in the phase space, thereby ensuring that the slope of the trajectory at any given point is also well-defined. Consequently, this implies that trajectories in the phase space cannot intersect.

However, it is important to note that (1) if Eq. (1.8) is nonautonomous, then f_i and f_k are also functions of time. For a given point in the phase space, the values of f_i and f_k can vary at different times t, which means that trajectories at the same point can have different slopes and, thus, can intersect. (2) If the trajectories of a high-dimensional system ($n > 3$) are projected onto a lower-dimensional space, such as a two-dimensional plane, then it is natural for different spatial trajectories to intersect in their planar projections.

Another relatively simple form of trajectories is a limit cycle.

Definition 1.5 A limit cycle, denoted as L_0, is a nonequilibrium trajectory on which each point $x_0 \in L_0$ satisfies $\varphi^{t+T_0}x_0 = \varphi^t x_0$ for some $T_0 > 0$ and all $t \in T$.

The minimal value of T_0 satisfying this condition is designated as the period of the limit cycle L_0. In the context of continuous-time dynamical systems, a limit cycle that exists without the presence of any other limit cycles in its vicinity is termed a stable limit cycle. Importantly, each individual trajectory itself can be considered an invariant set. Formally, an invariant set within the dynamical system $\{T, X, \varphi^t\}$ is defined as a subset S of X for which, if an initial point x_0 belongs to S, then under the action of the evolution operators φ^t, all subsequent points $\varphi^t x_0$ remain within S for all $t \in T$.

1.2 Bifurcations in Dynamical System

Bifurcation theory constitutes the mathematical exploration of alterations in the qualitative or topological configuration within a given family, encompassing elements like the integral curves of a vector field family and the solutions of a differential equation family. Typically applied to the mathematical scrutiny of dynamic systems, bifurcation emerges when a minor, smooth modification to a system's parameter values (referred to as bifurcation parameters) triggers an abrupt and qualitative or topological transformation in its behavior. Bifurcations manifest in continuous systems (characterized by ordinary differential equations, delay differential equations or partial differential equations) and discrete systems (captured by maps).

1.2.1 Bifurcation in Continuous System

Consider a parameter-dependent continuous system:

$$\dot{x} = f(x, \alpha), x \in \mathbb{R}^n, \alpha \in \mathbb{R}^n, \tag{1.10}$$

where $x \in \mathbb{R}^n$ and $\alpha \in \mathbb{R}^n$ denote the state variable and parameter. Assuming $x = x_0$ is a dynamic system's equilibrium point (fix point), J denotes the respective Jacobian matrix evaluated at x_0.

Definition 1.6 For system (1.10), if all eigenvalues of J have nonzero real parts, the equilibrium point x_0 is termed a hyperbolic equilibrium point.

Let $\lambda = (\lambda_1, \lambda_2, \ldots, \lambda_n)^T \in \mathbb{R}^n$ be the Jacobian matrix of $f_x(x, \alpha_0)$ at an equilibrium point of the dynamical system. Assume the function $f_x(x, \alpha_0)$ is continuously differentiable with respect to x in the neighborhood. The following types of bifurcations of system (1.10) are discussed based on the eigenvalues of the Jacobian matrix.

- Fold bifurcation occurs when $\lambda_1 = 0$ and the remaining eigenvalues have nonzero real parts.
- Hopf bifurcation is characterized by a pair of complex conjugate eigenvalues $\lambda_{1,2} = \pm i\omega_0, \omega_0 > 0$, with all other eigenvalues having nonzero real parts.
- Cusp bifurcation occurs when $\lambda_1 = 0$ and the system exhibit a quadratic fold, which generally indicates the system undergoes a cusp catastrophe.
- Bautin bifurcation is also known as a generalized Hopf bifurcation, where $\lambda_{1,2} = \pm i\omega_0, \omega_0 > 0$ and the first Lyapunov coefficient $l_1 = 0$.
- Bogdanov-Takens (BT) bifurcation occurs when the Jordan block of the Jacobian at the bifurcation point has the form $\begin{bmatrix} 0 & 1 \\ 0 & 0 \end{bmatrix}$ for $\lambda_1 = 0$.
- Zero-Hopf bifurcation is when $\lambda_1 = 0, \lambda_{2,3} = \pm i\omega_0, \omega_0 > 0$.
- Hopf-Hopf bifurcation is characterized by two pairs of complex conjugate eigenvalues $\lambda_{1,2} = \pm i\omega_1, \lambda_{3,4} = \pm i\omega_2, \omega_{1,2} > 0$.

For simplicity, it is assumed that the unspecified eigenvalues also have nonzero real parts. It should be noted that the eigenvalues are ordered to satisfy the conditions corresponding to the seven types of bifurcations. These conditions hold regardless of the specific order of the eigenvalues, as long as the conditions are met.

1.2.2 Bifurcation in Discrete System

Consider a parameter-dependent discrete dynamical system given by the equation

$$x \mapsto f(x, \alpha), \quad x \in \mathbb{R}^n, \quad a \in \mathbb{R}^n, \tag{1.11}$$

where the mapping f is smooth with respect to both x and α. Assume that $x_0 = 0$ is a fixed point of the system when $\alpha = \alpha_0$, and let J denote the Jacobian matrix at x_0.

Definition 1.7 For system (1.11), if the moduli of all eigenvalues of J are not equal to 1, the fixed point x_0 is referred to as a hyperbolic fixed point.

A hyperbolic equilibrium point (or fixed point) is considered stable if all eigenvalues of J have real parts less than zero (moduli less than 1). Conversely, it is unstable if at least one eigenvalue of J has a real part greater than zero (modulus greater than 1).

Let $\mu = (\mu_1, \mu_2, \ldots, \mu_n)^T \in \mathbb{R}^n$ be the eigenvalues (multipliers) of the characteristic equation of the Jacobian matrix $f_x(x, \alpha_0)$ at the fixed point. The following types of bifurcations of system (1.10) are discussed based on the eigenvalues of the Jacobian matrix.

- Fold bifurcation occurs when $\mu_1 = 1$.
- Flip bifurcation is characterized by $\mu_1 = -1$.
- Neimark–Sacker bifurcation occurs when $\mu_{1,2} = e^{i\theta_0}, 0 < \theta_0 < \pi$.
- Supercritical pitchfork bifurcation happens when $\mu_1 = 1$.
- Subcritical pitchfork bifurcation is characterized by $\mu_1 = -1$.
- Period-tripling bifurcation occurs when $\mu_{1,2} = e^{i\theta_0}, 0 < \theta_0 < \pi$.
- 1:1 resonance bifurcation identified by $\mu_{1,2} = 1$.
- 1:2 resonance bifurcation occurs when $\mu_{1,2} = -1$.
- 1:3 resonance bifurcation is characterized by $\mu_{1,2} = e^{i\theta_0}, \theta_0 = \frac{2\pi}{3}$.
- 1:4 resonance bifurcation occurs when $\mu_{1,2} = e^{i\theta_0}, \theta_0 = \frac{\pi}{2}$.
- Fold-Flip bifurcation is identified by $\mu_1 = 1, \mu_2 = -1$.

For simplicity, it is assumed that the unspecified moduli are not equal to one. It should be noted that the moduli are arranged to satisfy the conditions corresponding to the types of bifurcations described above. These conditions remain valid regardless of the specific order of the eigenvalues, provided that the conditions are met.

1.2.3 Bifurcation of Limit Cycle

Consider a continuous system defined by Eq. (1.10) with a periodic orbit L_0. Selecting a point $x_0 \in L_0$, we introduce a section Σ within the ring-shaped trajectory of L_0. This section Σ intersects L_0 at a nonzero angle, forming a $n-1$ dimensional hypersurface. From points x in close proximity to x_0 on Σ, trajectories emerge that eventually return to a point $\tilde{x} \in \Sigma$ near x_0. This process gives rise to a mapping $\mathcal{P} : \Sigma \to \Sigma$, where x maps to $\tilde{x}$ as follows:

$$x \mapsto \tilde{x} = \mathcal{P}(x).$$

This mapping $\mathcal{P}$ is known as the Poincaré map corresponding to the ring L_0. Notably, x_0 remains invariant under the Poincaré map: $\mathcal{P}(x_0) = x_0$. Introducing local coordinates $\xi = (\xi_1, \xi_2, \cdots, \xi_{n-1})$ on the section Σ such that $\xi = 0$ corresponds to x_0, the mapping $\mathcal{P}$ can be represented by a locally defined map $\mathcal{P}: \mathbb{R}^{n-1} \to \mathbb{R}^{n-1}$, which maps the ξ corresponding to x to the $\tilde{\xi}$ corresponding to $\tilde{x}$:

$$\mathcal{P}(\xi) = \tilde{\xi}.$$

Consequently, the stability of the ring L_0 is equivalent to the stability of the fixed point $\xi = 0$ of $\mathcal{P}$. If all eigenvalues (multipliers) $\mu_1, \cdots, \mu_{n-1}$ of the Jacobian matrix of P fall within the unit circle, the ring is stable. If at least one multiplier lies on the unit circle, the ring is nonhyperbolic. If, at the parameter value $\alpha = \alpha_0$, the ring exhibits a simple multiplier $\mu_1 = 1$, with all other multipliers outside the unit circle, a fold bifurcation leading to a limit cycle occurs at $\alpha = \alpha_0$. The stability conditions for other branches of the limit cycle are consistent with the discrete system.

For a periodically perturbed system described by

$$\dot{x} = f(t, x), \ (t, x) \in \mathbb{R}^1 \times \mathbb{R}^n. \tag{1.12}$$

wherein $f(t + T_0, x) = f(t, x)$, and it defines a coordinate-autonomous system on the cylindrical manifold $X = \mathbb{S} \times \mathbb{R}^n$:

$$\begin{cases} \dot{t} = 1, \\ \dot{x} = f(t, x). \end{cases} \tag{1.13}$$

A n-dimensional section $\Sigma = \{(x, t) \in X : t = 0\}$ is selected within X, with coordinates $x^{\mathrm{T}} = (x_1, x_2, \cdots, x_n)$. Assuming the solutions $x(t, x_0)$ of (1.13) exist for $t \in [0, T_0]$, we can introduce the Poincaré map:

$$x_0 \mapsto \mathcal{P}(x_0) = x(T_0, x_0),$$

Discretizing the system (1.13), we obtain the corresponding discrete system $\mathbb{Z}, \mathbb{R}^n, \mathcal{P}^k$. Evidently, the fixed points of $\mathcal{P}$ correspond to T_0-periodic solutions of (1.12), the N_0-torus of $\mathcal{P}$ corresponds to $N_0 T_0$-periodic solutions (subharmonic solutions), and so forth. Through this discrete system, various properties of the original system can be studied. When $f(t, x)$ takes on specific forms of periodic functions, a similar approach can be applied to transform nonautonomous systems into other forms of continuous systems, which can then be discretized for analysis. We briefly introduce three types of bifurcation of periodic orbits.

Fold bifurcation, as depicted in Fig. 1.1. When $a = 0$, the Jacobian matrix of the fixed point of the Poincaré map corresponding to the ring possesses a simple multiplier $\mu_1 = 1$. As the parameter changes from left to right, the system transforms from having two rings to a single ring and then collides and vanishes.

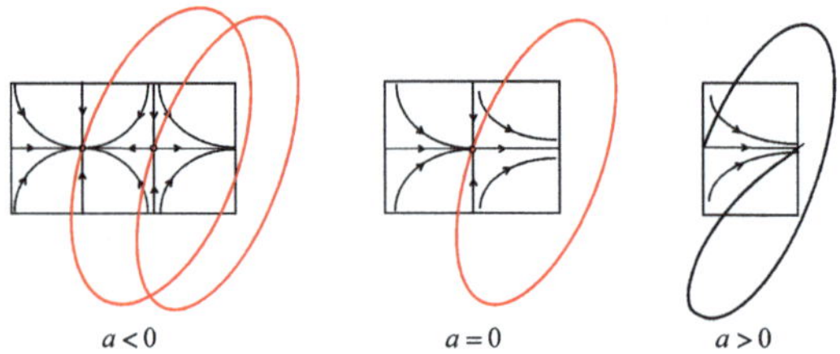

Fig. 1.1 Schematic diagram for fold bifurcation of limit cycle

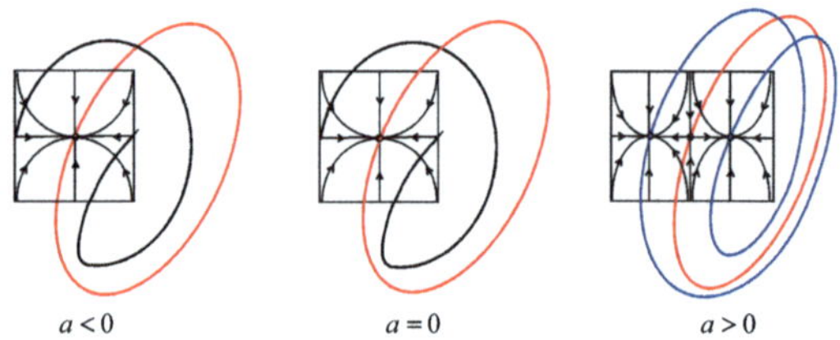

Fig. 1.2 Schematic diagram for flip bifurcation of limit cycle

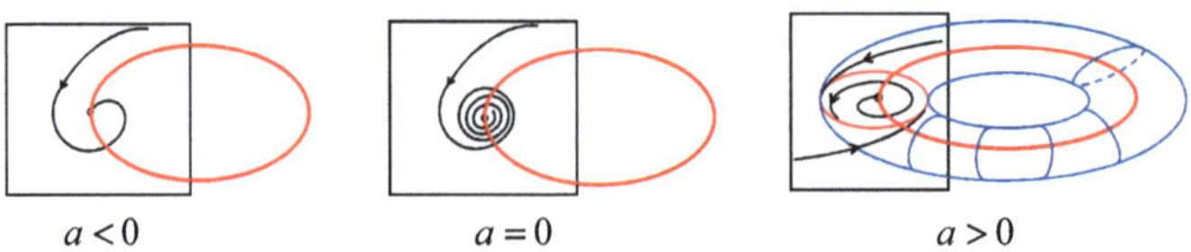

Fig. 1.3 Schematic diagram for Neimark–Sacker bifurcation of limit cycle

Flip bifurcation (period-doubling bifurcation), as shown in Fig. 1.2. At $a = 0$, the fixed point of the Poincaré map corresponding to the ring features a Jacobian matrix with a simple multiplier $\mu_1 = -1$. As the parameter varies from left to right, the ring doesn't vanish but bifurcates into a larger ring at the right end. The period of the larger ring is twice that of the smaller ring, which corresponds to the addition of two two-periodic fixed points introduced by the Poincaré map.

Neimark–Sacker bifurcation (Hopf bifurcation), depicted in Fig. 1.3. At $a = 0$, the fixed point of the Poincaré map corresponding to the ring has a pair of complex multipliers lying on the unit circle, $\mu_{1,2} = e^{\pm i\theta}$. As the parameter progresses from left to right, the ring does not vanish but bifurcates into a quasiperiodic solution, often referred to as a "torus," at the right end. This quasiperiodic solution does not possess a definite period and corresponds to a closed invariant curve that emerges in the Poincaré map. This curve can be observed as the red line on the cross-section in the figure.

1.3 Normal Forms for Bifurcations

Indeed, bifurcations can be categorized into two primary classes for a clearer understanding: global bifurcation and local bifurcation.

- A global bifurcation refers to a significant qualitative change in the behavior of a dynamical system that occurs over a range of parameter values. It involves a transformation of the system's entire state space, impacting its overall dynamics. In a global bifurcation, the stability and nature of multiple equilibria or periodic orbits can change as a parameter varies. The system transitions from one global behavior to another, and these changes are typically captured in bifurcation diagrams that illustrate how the system's characteristics evolve across a parameter range.
- A local bifurcation involves changes in the behavior of a dynamical system that occur near a specific equilibrium point or periodic orbit as a parameter is varied. It is a more localized phenomenon that pertains to the stability and qualitative properties of nearby solutions. Local bifurcations are characterized by the appearance, disappearance, or qualitative transformation of equilibria or periodic orbits as a parameter crosses certain critical values. They provide insights into how small perturbations around an equilibrium can lead to drastic changes in the system's behavior.

We briefly introduce several normal forms for local bifurcations, assuming that $a = 0$ is the bifurcation point.

1.3.1 Center Manifold Theorem

Within generic systems, distinct layers of bifurcation diagrams interact with each other in accordance with specific rules. This underlying organization imparts a semblance of similarity among bifurcation diagrams across various applications. To embark on a discussion of this subject, it is necessary to define the criteria for determining when two dynamical systems possess "qualitatively similar" or equivalent bifurcation diagrams. Let us consider two continuous-time dynamical systems for elucidation:

$$\dot{x} = f(x, \alpha), x \in \mathbb{R}^n, \alpha \in \mathbb{R}^m, \tag{1.14}$$

and

$$\dot{y} = g(y, \beta), y \in \mathbb{R}^n, \beta \in \mathbb{R}^m, \tag{1.15}$$

with smooth right-hand sides and the same number of variables and parameters.

Definition 1.8 Dynamical system (1.14) is called topologically equivalent to a dynamical system (1.15) if

- there exists a homeomorphism of the parameter space $p : \mathbb{R}^m \to \mathbb{R}^m, \beta = p(\alpha)$;
- there is a parameter-dependent homeomorphism of the phase space $h_\alpha : \mathbb{R}^n \to \mathbb{R}^n, y = h_\alpha(x)$ mapping orbits of the system (1.14) at parameter values α onto orbits of the system (1.15) at parameter values $\beta = p(\alpha)$, preserving the direction of time.

Evidently, the homeomorphism p metamorphoses the parametric portrait of system (1.14) into the parametric portrait of system (1.15), and simultaneously, the homeomorphism h_α establishes the correspondence between their respective phase portraits. As per its definition, parameter-dependent systems that are topologically equivalent result in (topologically) identical bifurcation diagrams.

In the context of local bifurcations involving equilibria and fixed points, universal bifurcation diagrams are established through the utilization of topological normal forms. Let us engage in a discourse on this topic within the continuous-time framework, acknowledging that its applicability extends to discrete-time systems as well. At times, it becomes feasible to devise a simplified system (expressed polynomially in terms of ζ_i system

$$\dot{\zeta}_i = g(\zeta, \beta, \delta), \zeta \in \mathbb{R}^n, \beta \in \mathbb{R}^k, \delta \in \mathbb{R}^l, \tag{1.16}$$

which at $\beta = 0$ an equilibrium $\zeta = 0$ satisfying k bifurcation conditions determining a codim k bifurcation of this equilibrium. Here δ is a vector of the coefficients $\delta_i, i = 1, 2, \dots l$ of the polynomials involved in (1.16). Across all scenarios we consider, a finite number of regions emerge within the coefficient space, each corresponding to topologically distinct bifurcation diagrams of Eq. (1.16). In simpler instances, the δ_i adopts only a finite range of integer values. For instance, all coefficients δ_i are equal to 1, except for a single $\delta_{i0} = \pm 1$. In more intricate scenarios, certain components δ_i may assume real values.

The center manifold theorem states that for certain nonlinear differential systems, there exists a special low-dimensional manifold, known as the center manifold. This manifold possesses the property that the dynamic behavior of the system can be effectively approximated as a lower-dimensional dynamical system when observed on this manifold [1].

Consider the differential equation $\dot{x} = f(x)$, where $x \in \mathbb{R}^n$ and the function $f \in C^r$ (with $r \geq 2$) satisfy $f(0) = 0$. Let $x_0 = 0$ be a fixed point, and suppose the Jacobian matrix of f at x_0 has eigenvalues $\lambda_1, \lambda_2, \dots, \lambda_n$, with n_+ eigenvalues having positive real parts, n_0 eigenvalues having zero real parts, and n_- eigenvalues having negative real parts. The center manifold theorem then asserts the existence of a local center manifold $W^c_{\text{loc}}(0)$ associated with the origin $x = 0$ in the following manner.

Theorem 1.2 *There exists a local flow ϕ^t, and this flow, restricted to $W^c_{loc}(0)$, exhibits that $x = 0$ is a fixed point for ϕ^t within this local manifold. Specifically, for any point $x \in U$, where U is a neighborhood around 0, the trajectory of x under ϕ^t remains in U as $t \to \infty$ (and $t \to -\infty$ for trajectories entering U), thus mapping $\phi^t(x) \to W^c_{loc}(0)$.*

Consider the differential system given by

$$\begin{cases} \dot{u} = Bu + g(u, v), \\ \dot{v} = Cv + h(u, v), \end{cases}$$

where $u \in \mathbb{R}^{n_0}$, $v \in \mathbb{R}^{n_+ + n_-}$, B is an $n_0 \times n_0$ matrix and C is an $(n_+ + n_-) \times (n_+ + n_-)$ matrix. In this configuration, the nonlinear functions g and h can be approximated by their Taylor expansions, and both are in C^r class with $r \geq 2$.

The extended center manifold theorem asserts the existence of a mapping W^c, characterized by $W^c = \{(u, v) : v = V(u)\}$ where $V : \mathbb{R}^{n_0} \to \mathbb{R}^{n_+ + n_-}$ and $V(0) = 0$ with V having derivatives of order $O(\|u\|^2)$. For u in a small neighborhood near zero, the dynamics of the system can be described by the reduced system:

$$\begin{cases} \dot{u} = Bu + g(u, V(u)), \\ \dot{v} = Cv. \end{cases}$$

This reduced system provides an approximation where the center manifold W^c can be effectively analyzed, and it serves as a foundational model to explore the stability properties and dynamic behaviors near the equilibrium point.

1.3.2 Normal Forms for Bifurcations in Continuous System

When $\alpha = 0$, the system (1.10) simplifies to

$$\dot{u} = Au + F(u), \quad u \in \mathbb{R}^n,$$

where A is the Jacobian matrix at the equilibrium point and $F(u)$ represents the higher-order nonlinear terms. The Taylor expansion of $F(u)$ around $u = 0$ includes terms up to the fifth order and is given by

$$\begin{aligned} F(u) = {} & \frac{1}{2}F_2(u, u) + \frac{1}{6}F_3(u, u, u) + \frac{1}{24}F_4(u, u, u, u) \\ & + \frac{1}{120}F_5(u, u, u, u, u) + O(\|u\|^6). \end{aligned}$$

The components F_2, F_3, F_4, and F_5 are constructed using the corresponding partial derivatives of F, represented as follows:

$$F_2^i(u, v) = \sum_{j,k=1}^{n} \left. \frac{\partial^2 F_i(x)}{\partial x_j \partial x_k} \right|_{x=0} u_j v_k,$$

$$F_3^i(u, v, w) = \sum_{j,k,l=1}^{n} \left. \frac{\partial^3 F_i(x)}{\partial x_j \partial x_k \partial x_l} \right|_{x=0} u_j v_k w_l,$$

$$F_4^i(u, v, w, y) = \sum_{j,k,l,m=1}^{n} \left.\frac{\partial^4 F_i(x)}{\partial x_j \partial x_k \partial x_l \partial x_m}\right|_{x=0} u_j v_k w_l y_m,$$

$$F_5^i(u, v, w, y, z) = \sum_{j,k,l,m,s=1}^{n} \left.\frac{\partial^5 F_i(x)}{\partial x_j \partial x_k \partial x_l \partial x_m \partial x_s}\right|_{x=0} u_j v_k w_l y_m z_s,$$

where $i = 1, 2, \ldots, n$. These expansions are critical for modeling the nonlinear interactions within the system, providing a detailed basis for analyzing the dynamics near the equilibrium.

Saddle-Node bifurcation: For a system described by the matrix A, if the nonlinearity F exhibits a quadratic term, the dynamics near the bifurcation point can be approximated by

$$\dot{x} = bx^2 + O(\|x\|^3),$$

where the coefficient b is determined by the formula $b = -\frac{1}{2}\langle p, F_2(q, q)\rangle$, with p, q being the left and right eigenvectors of A corresponding to the zero eigenvalue. These vectors satisfy $Aq = 0$, $A^T p = 0$, and $\langle p, q\rangle = 1$.

Hopf bifurcation: In this scenario, the system involves complex conjugate eigenvalues $\lambda_{1,2} = \pm i\omega_0$, $\omega_0 > 0$. The normal form near the bifurcation point for this case is

$$\dot{z} = i\omega_0 z + l_1|z|^2 z + O(\|z\|^4),$$

where l_1 is a coefficient influenced by Lyapunov constants, calculated as follows:

$$\begin{aligned} l_1 =& \frac{1}{2\omega_0}\mathrm{Re}\{\langle p, F_3(q, q, q)\rangle - 2\langle p, F_2(q, A^{-1}F_2(q, q))\rangle \\ &+ \langle p, F_2(q, (2i\omega_0 I_n - A)^{-1}F_2(q, q))\rangle\}, \end{aligned} \tag{1.17}$$

and the vectors p, q meet the conditions $Aq = i\omega_0 q$, $A^T p = -i\omega_0 p$, with $\langle p, q\rangle = 1$, where the inner product in $\mathbb{C}^2$ is defined as $\langle p, q\rangle = p_1 q_1 + p_2 q_2$.

Cusp bifurcation: If the F in the matrix A includes a cubic term, the dynamics near the bifurcation point can be described by

$$\dot{x} = cx^3 + O(\|x\|^4),$$

where the coefficient c is determined as follows:

$$c = \frac{1}{6}\langle p, F_3(q, q, q) - 3F_2(q, A^{-1}F_2(q, q))\rangle.$$

Here, p and q are the left and right eigenvectors of A related to the zero eigenvalue, satisfying $Aq = 0$, $A^T p = 0$, and $\langle p, q\rangle = 1$. The matrix A^{-1} is used to denote the Moore-Penrose inverse, assuming A is not invertible.

Bautin bifurcation: For a system exhibiting complex conjugate eigenvalues $\lambda_{1,2} = \pm i\omega_0$, $\omega_0 > 0$, the normal form near the bifurcation point takes the form

$$\dot{z} = i\omega_0 z + l_2|z|^2 z + O(\|z\|^6),$$

where l_2 is computed for Lyapunov stability analysis and is given by

$$\begin{aligned} l_2 =& \frac{1}{12\omega_0}\text{Re}\Big\{\langle p, F_5(q,q,q,q,q) + F_4(q,q,q,a_1) + 3F_4(q,q,q,a_2) \\ &+ 6F_4(q,q,a_2,a_2) + 3F_3(q,a_3,a_3) + 3F_3(q,q,a_4) + 6F_3(q,a_1,a_1) \\ &+ 6F_3(q,a_2,a_2) + 2F_2(q,a_5) + 3F_2(q,a_6) + 3F_2(a_1,a_3) + 6F_2(a_2,a_4)\rangle\Big\}, \end{aligned}$$

where the terms a_1 through a_6 are defined as follows:

$$\begin{aligned} a_1 &= (2i\omega_0 I_n - A)^{-1} F_2(q,q), \\ a_2 &= -A^{-1} F_2(q,q), \\ a_3 &= (3i\omega_0 I_n - A)^{-1}[F_3(q,q,q) + 3F_2(q,a_1)], \\ a_4 &= (i\omega_0 I_n - A)^{NA}[F_3(q,q,q) + F_2(q,a_1) + 2F_2(q,a_2) - 2c_1 q_1], \\ a_5 &= (2i\omega_0 I_n - A)^{-1}[F_4(q,q,q,q) + 3F_3(q,q,a_1) + 3F_2(a_1,a_2)], \\ a_6 &= -A^{-1}[F_4(q,q,q,q) + 4F_3(q,q,a_2) + 3F_3(q,a_1,a_1) + 2F_2(a_2,a_2)], \end{aligned}$$

and

$$c_1 = \frac{1}{2}\langle p, F_3(q,q,q) + F_2(q,(2i\omega_0 I_n - A)^{-1} F_2(q,q))\rangle.$$

Bogdanov–Takens bifurcation: The system exhibits critical eigenvalues $\lambda_{1,2} = 0$. The normal form near the bifurcation point is

$$\begin{cases} \dot{x}_1 = x_2, \\ \dot{x}_2 = \alpha_1 x_1^2 + \alpha_2 x_1 x_2 + O(\|x\|^3), \end{cases}$$

where the coefficients α_1 and α_2 are calculated using

$$\alpha_1 = \frac{1}{2}\langle p_1, F_2(q_0,q_0)\rangle, \quad \alpha_2 = \langle p_0, F_2(q_0,q_0)\rangle + \langle p_1, F_2(q_0,q_1)\rangle,$$

and the eigenvectors p_0, q_0, p_1, q_1 satisfy the following conditions

$$Aq_0 = 0, \; Aq_1 = q_0, \; A^T p_0 = 0, \; A^T p_1 = p_0, \; \langle p_0, q_0 \rangle = 1, \; \langle p_1, q_1 \rangle = 1.$$

Zero–Hopf bifurcation: Consider a system where the matrix A exhibits mixed critical eigenvalues—$\lambda_1 = 0$ and $\lambda_{2,3} = \pm i\omega_0$, $\omega_0 > 0$. The dynamics near the bifurcation point can be described by

$$\dot{x}_1 = b_1 x_1^2 + b_2 x_2^2 + b_3 x_1^3 + b_4 x_1 x_2^2 + O(\|x_1, x_2\|^4),$$
$$\dot{x}_2 = \omega_0 x_1 + d_1 x_1 x_2 + d_2 x_2^2 + d_3 x_2^3 + O(\|x_1, x_2\|^4),$$

where the coefficients are defined as

$$b_1 = \frac{1}{2}\langle p_0, F_2(q_0, q_0)\rangle,$$
$$b_2 = \langle p_0, F_2(q_1, q_1)\rangle,$$
$$b_3 = \frac{1}{6}\left\langle p_0, F_3(q_0, q_0, q_0) + 3F_2\left(q_0, -A^{-1}F_2(q_0, q_0)\right)\right\rangle,$$
$$b_4 = \left\langle p_0, F_3(q_0, q_1, q_1) + F_2\left(q_1, (-i\omega_0 I_n - A)^{-1}F_2(q_0, q_1)\right)\right\rangle,$$
$$d_1 = \langle p_1, F_2(q_0, q_1)\rangle,$$
$$d_2 = \frac{1}{2}\left\langle p_1, F_3(q_0, q_0, q_1) + 2F_2\left(q_0, (i\omega_0 I_n - A)^{-1}F_2(q_0, q_1)\right)\right\rangle,$$
$$d_3 = \frac{1}{2}\left\langle p_1, F_3(q_1, q_1, q_1) + 2F_2\left(q_1, (-i\omega_0 I_n - A)^{-1}F_2(q_1, q_1)\right)\right\rangle.$$

The vectors p_0, q_0, p_1, q_1 satisfy the eigenvector relations necessary for critical eigenvectors,

$$Aq_0 = 0, \quad Aq_1 = i\omega_0 q_1,$$
$$A^T p_0 = 0, \quad A^T p_1 = -i\omega_0 p_1,$$
$$\langle p_0, q_0 \rangle = 1, \quad \langle p_1, q_1 \rangle = 1.$$

Hopf–Hopf bifurcation: Given a linear matrix A with a pair of purely imaginary eigenvalues $\lambda_{1,2} = \pm i\omega_1$ and $\lambda_{3,4} = \pm i\omega_2$, where $\omega_1 > \omega_2 > 0$, and no other critical eigenvalues, the system undergoes a Hopf-Hopf bifurcation near the origin. The normal form near the bifurcation point is given by

$$\begin{cases} \dot{x}_1 = i\omega_1 x_1 + m_1|x_1|^2 x_1 + m_2|x_2|^2 x_1 + O(\|x_1, \bar{x}_1, x_2, \bar{x}_2\|^4), \\ \dot{x}_2 = i\omega_2 x_2 + n_1|x_1|^2 x_2 + n_2|x_2|^2 x_2 + O(\|x_1, \bar{x}_1, x_2, \bar{x}_2\|^4), \end{cases}$$

where

$$
\begin{aligned}
m_1 &= \frac{1}{2}\langle p_1, F_3(q_1, q_1, \bar{q}_1) + F_2((2i\omega_1 I_n - A)^{-1} F_2(q_1, q_1), \bar{q}_1) \\
&\quad - 2F_2(A^{-1}F_2(q_1, \bar{q}_1), q_1)\rangle, \\
m_2 &= \langle p_1, F_3(q_1, q_2, \bar{q}_2) + F_2([(i(\omega_1 + \omega_2)I_n - A]^{-1} F_2(q_1, q_2), \bar{q}_2) \\
&\quad + F_2([(i(\omega_1 - \omega_2)I_n - A]^{-1} F_2(q_1, \bar{q}_2), q_2) + F_2(-A^{-1}F_2(q_2, \bar{q}_2), q_1)\rangle, \\
n_1 &= \langle p_2, F_3(q_1, \bar{q}_1, q_2) + F_2([(i(\omega_1 + \omega_2)I_n - A]^{-1} F_2(q_1, \bar{q}_1), q_2) \\
&\quad - F_2(A^{-1}F_2(q_1, \bar{q}_1), q_2) + F_2([-(i(\omega_1 - \omega_2)I_n - A]^{-1} F_2(q_1, q_2), \bar{q}_1)\rangle, \\
n_2 &= \frac{1}{2}\langle p_2, F_3(q_2, \bar{q}_2, q_2) + F_2((2i\omega_2 I_n - A)^{-1} F_2(q_2, q_2), \bar{q}_2) \\
&\quad - 2F_2(A^{-1}F_2(q_2, \bar{q}_2), q_2)\rangle,
\end{aligned}
$$

with eigenvectors q_1, q_2, p_1, p_2 satisfying $Aq_1 = i\omega_1 q_1$, $Aq_2 = i\omega_2 q_2$, $A^T p_1 = i\omega_1 p_1$, $A^T p_2 = i\omega_2 p_2$, $\langle p_1, q_1\rangle = \langle p_2, q_2\rangle = 1$, and $\langle p_1, q_2\rangle = \langle p_2, q_1\rangle = 0$.

1.3.3 Normal Forms for Bifurcations in Discrete System

The map can be rewritten in the following form:

$$
x \mapsto A(\alpha)x + F(x, \alpha), \tag{1.18}
$$

where $A(\alpha)$ is the Jacobian matrix of the map and $F = O(\|x\|^2)$ is a smooth vector function. At $\alpha = 0$, its Taylor expansion is

$$
\begin{aligned}
F(x, 0) &= \frac{1}{2}B(x, x) + \frac{1}{6}C(x, x, x) + \frac{1}{24}D(x, x, x, x) + \frac{1}{120}E(x, x, x, x, x) \\
&\quad + O(\|x\|^6),
\end{aligned}
$$

Similar to the continuous-time systems, here $B(x, y)$, $C(x, y, z)$, $D(x, y, z, u)$, and $E(x, y, z, u, v)$ are symmetric multilinear forms satisfying

$$
\begin{aligned}
B^i(x, y) &= \sum_{j,k=1}^{n} \left.\frac{\partial^2 F_i(\xi, 0)}{\partial \xi_j \partial \xi_k}\right|_{\xi=0} x_j y_k, \\
C^i(x, y, u) &= \sum_{j,k,l=1}^{n} \left.\frac{\partial^3 F_i(\xi, 0)}{\partial \xi_j \partial \xi_k \partial \xi_l}\right|_{\xi=0} x_j y_k u_l,
\end{aligned}
$$

$$D^i(x, y, u, v) = \sum_{j,k,l,m=1}^{n} \left. \frac{\partial^4 F_i(\xi, 0)}{\partial \xi_j \partial \xi_k \partial \xi_l \partial \xi_m} \right|_{\xi=0} x_j y_k u_l v_m,$$

$$E^i(x, y, u, v, w) = \sum_{j,k,l,m,s=1}^{n} \left. \frac{\partial^5 F_i(\xi, 0)}{\partial \xi_j \partial \xi_k \partial \xi_l \partial \xi_m \partial \xi_s} \right|_{\xi=0} x_j y_k u_l v_m w_s, \ i = 1, 2, \dots, n.$$

Fold bifurcation: Suppose the linear matrix A has only one eigenvalue equal to 1, and all other eigenvalues do not have a modulus of 1. Then the map (1.18) undergoes a fold bifurcation near the origin, and its normal form near the bifurcation point is

$$u \mapsto u + au^2 + O(\|u\|^3),$$

where the coefficient of the u^2 term in front, a, is given by

$$a = \frac{1}{2}\langle p, B(q, q)\rangle,$$

which determines the subcriticality of the fold bifurcation. Here, p and q are eigenvectors of the matrix A. A is the linearized matrix, and A^T is its transpose. At $\lambda = 1$, $Aq = q$, $A^T p = p$, and $\langle p, q\rangle = 1$. If $a > 0$, the bifurcation is subcritical, and the nonstationary branch (1.12) is not determined and can lead to disappearance or instability. If $a < 0$, the opposite is true. Further, if $a = 0$, the map (1.12) undergoes a cusp bifurcation near the origin, and its normal form near the bifurcation point is

$$u \mapsto u + cu^3 + O(\|u\|^4),$$

where

$$c = \frac{1}{6}\langle p, C(q, q, q) - 3B(q, (A - I_n)^{-1} B(q, q))\rangle.$$

Period-doubling bifurcation: Suppose the matrix A has only one eigenvalue equal to -1, and all other eigenvalues do not have a modulus of 1. Then the map (1.18) undergoes a period-doubling bifurcation near the origin, and its normal form near the bifurcation point is

$$\xi \mapsto -\xi + c\xi^3 + O(\|\xi\|^4),$$

where the coefficient of the ξ^3 term, c, determines the subcriticality of the period-doubling bifurcation, and p and q are eigenvectors of the matrix A. At $\lambda = -1$, $Aq = -q$, $A^T p = p$, and $\langle p, q\rangle = 1$. If $c > 0$, the period-doubling bifurcation leads to a stable period-2 orbit, and the nonstationary branch (1.12) is not determined and can lead to disappearance or instability. If $c < 0$, the opposite is true. Further, if $c = 0$, the map (1.12) undergoes a generalized flip bifurcation near the origin, and its normal form near the bifurcation point is

$$\xi \mapsto -\xi + d_1\xi^5 + O(\|\xi\|^6),$$

where

$$d_1 = \frac{1}{120}\langle p, 5B(q, e_1) + 10B(e_2, e_3) + 10C(q, q, e_3) + 15C(q, e_2) \\ + 10D(q, q, q, q) + E(q, q, q, q, q)\rangle,$$

and e_1, e_2, e_3 satisfy the following equations

$$(A - I_n)e_1 = 4B(q, e_3) + 3B(e_2, e_2) + 6C(q, q, e_3) + D(q, q, q, q),$$

$$-(A + I_n)e_2 = 3B(q, e_2) + 6C(q, q, q) + 3B(q, q),$$

$$-(A + I_n)e_3 = 2B(q, q) + 3C(q, q, q) + 3B(q, e_3).$$

Neimark–Sacker bifurcation: If the linear matrix A has a pair of complex conjugate eigenvalues with modulus 1, i.e., $\lambda_{1,2} = e^{\pm i\theta_0}$, where $0 < \theta_0 < \pi$ and there are no other eigenvalues on the unit circle, the map undergoes a Neimark–Sacker bifurcation. Assume $q \in \mathbb{C}^n$ corresponds to the eigenvalue λ_1, with $Aq = e^{i\theta_0}q$ and $A\bar{q} = e^{-i\theta_0}\bar{q}$. The eigenvector p corresponding to $e^{-i\theta_0}$ satisfies $A^T p = e^{-i\theta_0}p$ and $\langle p, q\rangle = 1$.

Under the non-resonance condition $e^{ik\theta_0} \neq 1, k = 1, 2, 3, 4$, the normal form near the bifurcation point is given by

$$z \mapsto e^{i\theta_0}z(1 + d|z|^2) + O(|z|^4),$$

where

$$d = \frac{1}{2}\mathrm{Re}\{e^{-i\theta_0}[\langle p, C(q, q, \bar{q})\rangle + 2\langle p, B(q, (I_n - A)^{-1}B(q, \bar{q}))\rangle \\ + \langle p, B(\bar{q}, (e^{2i\theta_0}I_n - A)^{-1}B(q, q))\rangle]\}.$$

If $d < 0$, the Neimark–Sacker bifurcation results in the appearance of an unstable invariant circle. If $d > 0$, the opposite occurs, indicating a stable scenario.

1:1 Resonance : If the matrix A has a double eigenvalue $\lambda_1 = \lambda_2 = 1$ at the fixed point x_0, then the map (1.18) undergoes a 1:1 resonance bifurcation. In this case, there exist eigenvectors q_0 and q_1 corresponding to the eigenvalue 1, satisfying the conditions $Aq_0 = q_0$ and $Aq_1 = q_1 + q_0$. The adjoint eigenvectors p_0 and p_1 of the transposed matrix A^T satisfy $A^T p_0 = p_0$ and $A^T p_1 = p_1 + p_0$, with the orthogonality conditions $\langle p_0, q_0\rangle = \langle p_1, q_1\rangle = 1$ and $\langle p_0, q_1\rangle = \langle p_1, q_0\rangle = 0$.

After a series of transformations, the normal form near the 1:1 resonance is

$$\begin{pmatrix} \xi_1 \\ \xi_2 \end{pmatrix} \mapsto \begin{pmatrix} \xi_1 + \xi_2 \\ \xi_2 + a\xi_1^2 + b\xi_1\xi_2 \end{pmatrix} + O(\|\xi\|^3),$$

where

$$a = \frac{1}{2}\langle p_1, B(q_0, q_0)\rangle \quad \text{and} \quad b = \langle p_0, B(q_0, q_0)\rangle + \langle p_1, B(q_0, q_1)\rangle.$$

For the bifurcation to occur at the 1:1 resonance, the nondegeneracy conditions $a \neq 0$ and $b - a \neq 0$ must be satisfied. The sign of $(b-a)a$ determines the bifurcation diagram. Near the 1:1 resonance, there are fold and Neimark–Sacker bifurcations, and the bifurcation curves intersect at the 1:1 resonance.

1:2 Resonance : If the matrix A has a double eigenvalue $\lambda_1 = \lambda_2 = -1$ at the fixed point x_0, then the system (1.12) undergoes a 1:2 resonance bifurcation. In this case, there exist eigenvectors q_0 and q_1 corresponding to the eigenvalue -1, satisfying the conditions $Aq_0 = -q_0$ and $Aq_1 = -q_1 + q_0$. The adjoint eigenvectors p_0 and p_1 of the transposed matrix A^T satisfy $A^T p_0 = -p_0$ and $A^T p_1 = -p_1 - p_0$, with the orthogonality conditions $\langle p_0, q_0\rangle = \langle p_1, q_1\rangle = 1$ and $\langle p_0, q_1\rangle = \langle p_1, q_0\rangle = 0$. After a series of transformations, the normal form near the 1:2 resonance is

$$\begin{pmatrix} \xi_1 \\ \xi_2 \end{pmatrix} \mapsto \begin{pmatrix} -\xi_1 + \xi_2 \\ -\xi_1 + c\xi_1^3 + d\xi_1^2\xi_2 \end{pmatrix} + O(\|\xi\|^4),$$

where the critical coefficients are

$$\begin{aligned} c =& \frac{1}{6}\langle p_1, C(q_0, q_0, q_0) + 3B(q_0, (I_n - A)^{-1}B(q_0, q_0))\rangle, \\ d =& \frac{1}{2}\langle p_1, 2B(q_0, h_{11}) + B(q_1, h_{20}) + C(q_0, q_0, q_1)\rangle \\ &+ \langle p_0, 3B(q_0, h_{20}) + C(q_0, q_0, q_0)\rangle, \end{aligned}$$

and h_{11}, h_{20} are determined by

$$(A - I_n)h_{11} = -2B(q_0, q_0), \quad (A - 2I_n)h_{20} = -B(q_0, q_0).$$

For the bifurcation to occur at the 1:2 resonance, the nondegeneracy conditions $c \neq 0$ and $d+3c \neq 0$ must be satisfied. The sign of $d+3c$ determines the bifurcation diagram. Near the 1:2 resonance, there are period-doubling and Neimark–Sacker bifurcations, and these bifurcation curves intersect at the 1:2 resonance.

1:3 Resonance :If the matrix A has eigenvalues $\lambda_{1,2} = \frac{\pm\sqrt{3}i-1}{2}$ at the fixed point x_0, then the system (1.18) undergoes a 1:3 resonance bifurcation. In this case, there exist eigenvectors q and $\bar{q}$ corresponding to the eigenvalues λ_1 and λ_2, respectively, satisfying the conditions: $Aq = e^{i\theta_0}q$ and $A\bar{q} = e^{-i\theta_0}\bar{q}$. The adjoint eigenvectors p and $\bar{p}$ of the transposed matrix A^T satisfy $A^T p = e^{-i\theta_0}p$ and $A^T\bar{p} = e^{i\theta_0}\bar{p}$, with $\langle p, q\rangle = 1$. After a series of transformations, the normal form near the 1:3 resonance is

$$\xi \mapsto e^{i\theta_0}\xi + b\xi^2 + c|\xi|^2\xi + O(|\xi|^4),$$

where the critical coefficients are

$$\begin{aligned} b =& \frac{1}{2}\langle p, B(\bar{q}, q)\rangle, \\ c =& \frac{1}{2}\langle p, C(q, q, \bar{q}) + 2B(q, (I_n - A)^{-1}B(q, \bar{q})) \\ &- B(\bar{q}, (e^{2i\theta_0}I_n - A)^{-1}(2b\bar{q} - B(q, q)))\rangle. \end{aligned}$$

For the bifurcation to occur at the 1:3 resonance, the nondegeneracy condition $b \neq 0$ must be satisfied. The sign of $\frac{2}{3}(e^{2i\theta_0}c/|b|^2 - 1)$ determines the bifurcation diagram. Near the 1:3 resonance, there is a coexistence of stable and unstable limit cycles, forming a nested structure.

1:4 Resonance: If the matrix A has eigenvalues $\lambda_{1,2} = \pm i$ at the fixed point x_0, then the system (1.18) undergoes a 1:4 resonance bifurcation. In this case, there exist eigenvectors q and $\bar{q}$ corresponding to the eigenvalues λ_1 and λ_2, respectively, satisfying the conditions: $Aq = iq$ and $A\bar{q} = -iq$. The adjoint eigenvectors p and $\bar{p}$ of the transposed matrix A^T satisfy $A^T p = -ip$ and $A^T\bar{p} = ip$, with $\langle p, q\rangle = 1$. After a series of transformations, the normal form near the 1:4 resonance is

$$\xi \mapsto i\xi + c|\xi|^2\xi + d|\xi|^3\xi + O(|\xi|^4),$$

where the critical coefficients are

$$c = \frac{1}{2}\langle p, C(q, q, \bar{q}) + 2B(q, (I_n - A)^{-1}B(q, \bar{q})) - B(\bar{q}, (I_n + A)^{-1}B(q, q))\rangle,$$

$$d = \frac{1}{6}\langle p, C(q, q, q) + 2B(q, (I_n - A)^{-1}B(q, q)) - B(q, (I_n + A)^{-1}B(q, q))\rangle.$$

For the bifurcation to occur at the 1:4 resonance, the nondegeneracy conditions $c \neq 0$ and $d \neq 0$ must be satisfied. The sign of c determines the stability of the bifurcation. Near the 1:4 resonance, there is a coexistence of period-four limit cycles, forming a nested structure.

Fold-flip Bifurcation :If the matrix A has two eigenvalues $\lambda_1 = 1$ and $\lambda_2 = -1$ at the fixed point x_0, then the system (1.18) undergoes a fold-flip bifurcation. In this case, there exist eigenvectors q_1 and q_2 corresponding to the eigenvalues 1 and $-$ 1, respectively, satisfying the conditions: $Aq_1 = q_1$ and $Aq_2 = -q_2$. The adjoint eigenvectors p_1 and p_2 of the transposed matrix A^T satisfy $A^T p_1 = p_1$

and $A^T p_2 = -p_2$, with the orthogonality conditions $\langle p_1, q_1\rangle = \langle p_2, q_2\rangle = 1$ and $\langle p_1, q_2\rangle = \langle p_2, q_1\rangle = 0$. After a series of transformations, the normal form near the fold-flip bifurcation is

$$\begin{pmatrix} x_1 \\ x_2 \end{pmatrix} \mapsto \begin{pmatrix} x_1 + \beta_1 x_1^2 + \beta_2 x_2^2 + \beta_3 x_1^3 + \beta_4 x_1 x_2^2 + \beta_5 x_1 x_2^2 + \beta_6 x_1 x_2 x_2 + \beta_7 x_2^3 \\ -x_2 + \beta_5 x_1 x_2 + \beta_6 x_1 x_2 + \beta_7 x_2^3 \end{pmatrix}$$
$$+ O(\|x\|^4),$$

where the critical coefficients are

$$\beta_1 = \frac{1}{2}\langle p_1, B(q_1, q_1)\rangle, \quad \beta_2 = \frac{1}{2}\langle p_1, B(q_2, q_2)\rangle,$$
$$\beta_3 = \frac{1}{6}\langle p_1, C(q_1, q_1, q_1) + 3B(q_1, \gamma_1)\rangle,$$
$$\beta_4 = \frac{1}{6}\langle p_1, C(q_1, q_2, q_2) + B(q_1, \gamma_2) + 2B(q_2, \gamma_3)\rangle,$$
$$\beta_5 = \langle p_2, B(q_1, q_2)\rangle,$$
$$\beta_6 = \frac{1}{6}\langle p_2, C(q_1, q_1, q_2) + B(q_2, \gamma_1) + 2B(q_1, \gamma_3)\rangle,$$
$$\beta_7 = \frac{1}{6}\langle p_2, C(q_2, q_2, q_2) + 3B(q_2, \gamma_2)\rangle,$$

with γ_1, γ_2, and γ_3 are determined by

$$(A - I_n)\gamma_1 = \langle p_1, B(q_1, q_1)\rangle q_1 - B(q_1, q_1),$$
$$(A - I_n)\gamma_2 = \langle p_1, B(q_2, q_2)\rangle q_1 - B(q_2, q_2),$$
$$(A + I_n)\gamma_3 = \langle p_2, B(q_1, q_2)\rangle q_2 - B(q_1, q_2).$$

1.4 Application of Dynamical Systems

Dynamical systems find a wide spectrum of applications spanning various domains, encompassing realms like physics, mathematics, biology, economics, and more. In the context of ecology, dynamical systems are harnessed to scrutinize the interplays, energy dynamics, and fluctuations in population sizes across diverse biological entities within an ecosystem. Through the formulation of dynamic models, the process of ecosystem evolution can be emulated, facilitating predictions about species population trends, analysis of ecosystem stability, and its resilience toward external disturbances.

The methodological approach to studying dynamical systems encompasses analytical solutions, numerical simulations, and mathematical derivations. Differential equations, discrete maps, among others, are some prevalent dynamical models that capture the dynamic conduct of continuous and discrete systems. By dissecting features such as equilibrium points, periodic orbits, and attractors inherent to a dynamical system, researchers can glean insights into the intricate mechanisms governing system evolution.

The utility of dynamical systems extends extensively to ecology and life sciences. It serves as a compass for researchers to navigate the intricate realms of species interactions within ecosystems, energy transfers, and population variations. Dynamic models empower the prediction of ecosystem responses, recovery dynamics, and stability profiles, thereby furnishing a scientific scaffold for environmental conservation, safeguarding biodiversity, and orchestrating strategies to control disease proliferation. In sum, the deployment of dynamical systems furnishes potent methodologies and instruments that underpin the inquiry into ecology and life sciences, illuminating the underlying laws and phenomena governing the natural world. Here are some examples [2–10] of powertrain applications in these areas:

Population dynamics and stability analysis: Dynamical systems serve as an invaluable tool for dissecting interactions amid distinct species, encompassing food chains, competition dynamics, and symbiotic relationships. By constructing population dynamics models, researchers forecast the temporal trajectory of species quantities and unravel the oscillatory behavior, chaos, and resilience inherent within-population dynamics.

Ecosystem response and biodiversity conservation: The discerning lens of dynamical systems aids in predicting the systemic retorts and recuperation pathways ensuing from external perturbations within ecosystems. For instance, researchers engineer models that scrutinize the interplay between flora and herbivores within an ecosystem while deciphering anthropogenic activities' repercussions upon ecological equilibrium. Also, dynamical systems offer a strategic compass for optimizing conservation tactics to safeguard biodiversity.

Disease transmission and control: In the life sciences realm, the dynamical systems framework is harnessed to scrutinize the dissemination and mitigation of infectious disorders. Employing a dynamic systems portrayal, disease transmission models predict the proliferation of epidemics, thus forging the bedrock for public health directives and strategies about disease mitigation and containment.

Cancer modeling: The dynamical system paradigm is efficaciously wielded to engender models delineating tumor progression and therapeutic interventions. With a predictive prowess, it anticipates the trajectory of tumor cell counts, tailors therapeutic interventions, and unravels the mechanistic underpinnings steering tumor growth.

Chapter 2
Introduction of Machine Learning

2.1 Machine Learning Concept, Category, and Workflow

2.1.1 Concept and Category

Machine learning (ML) is a field of artificial intelligence (AI) concerned with enabling computer systems to improve their performance on specific tasks by learning from data. Traditional programming usually requires all steps and rules to be explicitly specified; however, in ML, the system learns autonomously by analyzing and recognizing patterns in data and continuously optimizes its own performance based on the data. The core idea of ML is to allow computers to automatically discover laws, trends, and patterns from data in order to make predictions, classification, clustering, and other tasks. This requires training the model with large amounts of data, allowing the model to extract general rules from it, and then applies these rules to new, unseen data. ML can be broken down into several subfields, including but not limited to, supervised learning, unsupervised learning, semi-supervised learning, and reinforcement learning (Fig. 2.1).

Supervised learning constitutes a foundational paradigm within ML, whereby algorithms leverage labeled training data to predict outcomes for new instances. This methodology encompasses both classification and regression tasks. Here are some common supervised learning algorithms and their applications:

- Linear regression: An algorithm for predicting continuous values, suitable for regression problems, such as air quality forecasting.
- Logistic regression: Devised for binary classification endeavors, including scenarios like disease recurrence prediction.
- Decision trees: Applicable to both classification and regression tasks, these algorithms make decisions through a tree structure, rendering their outcomes easily interpretable.

J. Ren, Y. Tao, *Modeling in Life Sciences and Ecology*, Springer Asia Pacific Mathematics Series 4, https://doi.org/10.1007/978-981-95-1038-2_2

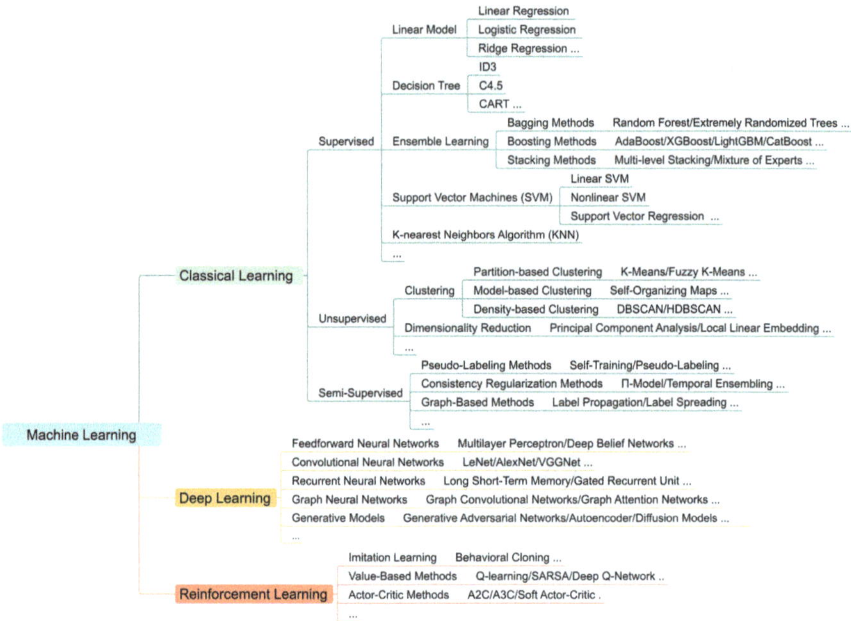

Fig. 2.1 The categories of ML algorithms

- Support vector machines (SVM): Esteemed for classification and regression problems, by finding the best segmentation hyperplane to distinguish different categories.
- K-nearest neighbors algorithm (KNN): Classification or regression based on similarity measures and prediction based on the nearest K neighbors.
- Naive Bayes: Devised for text classification and natural language processing tasks, based on Bayesian theorem for probability inference.

Unsupervised learning addresses the extraction of patterns and structures from unlabeled data, primarily applied to clustering and dimensionality reduction.

- K-means clustering is used to divide data points into K clusters, suitable for clustering problems.
- Hierarchical clustering creates a tree-like cluster structure that automatically determines the number of clusters.
- Principal component analysis (PCA) diminishes data dimensions while retaining essential information, often utilized for feature selection.
- Independent component analysis (ICA) facilitates the recovery of independent signals from mixed signals, finding utility in signal processing.

Semi-supervised learning is a learning paradigm between supervised and unsupervised learning that exploits both labeled (supervised) and unlabeled (unsupervised) data for training. Here are some common semi-supervised learning algorithms:

- Self-training: This is a simple semi-supervised approach in which an initial model is trained using labeled data, and then this model is used to assign pseudo-labels to unlabeled data, and these data are used together with labeled data for retraining model.
- Mixture of Gaussians: Used for clustering problems, combining labeled and unlabeled data by modeling Gaussian mixtures on the data.
- Co-training: Applicable to multi-view problems, the data is divided into multiple views, and each view uses a different feature set. The model is trained on each view, and then the model performance is jointly improved through collaborative updating and label propagation.
- Semi-supervised support vector machines: This is a method for handling unlabeled data by introducing a "noise region" around the classification hyperplane.

Reinforcement learning focuses on how to make the agent take a series of actions in an environment to maximize the cumulative reward. Reinforcement learning can be divided into two categories. One is model-based reinforcement learning, which is a method within the field of reinforcement learning. In this approach, the learning agent attempts to understand the dynamics of the environment by constructing a model of it and utilizing this model to make decisions. This model can take the form of a predictive model for environment state transitions and reward functions, aiding the agent in predicting the outcomes of different actions in various states, along with their potential rewards. The other is model-free reinforcement learning, which learns directly from interactions with the environment. This involves simulating multiple potential scenarios to anticipate the long-term effects of different decisions, thereby making better-informed choices. For complex biological processes, the agent can learn the optimal regulation strategy in continuous attempts to achieve specific biological effects or products.

- Q-Learning: An algorithm based on the value function used to train the agent to learn the optimal strategy through continuous trials.
- Deep reinforcement learning (DRL): A method that combines neural networks and reinforcement learning for complex environments and action spaces.
- Monte Carlo tree search (MCTS): MCTS is an algorithm used for planning and decision-making, particularly suited for model-free scenarios. It constructs a search tree, selecting the most promising actions at each step and then simulates multiple potential future trajectories to evaluate the value of actions.
- Dyna-Q: The Dyna-Q algorithm employs an environment model to facilitate planning and learning. In each interaction, it utilizes real experiences for Q-value updates and trains the model on virtual states to enhance learning.
- World models: This approach combines generative models and reinforcement learning. It employs generative models like variational autoencoders to model the environment's state transitions and reward functions. Subsequently, it learns strategies through planning in simulated environments.

2.1.2 Basic Workflow

The machine learning workflow is illustrated in Fig. 2.2, encompassing pivotal stages: data collection, exploration, preprocessing, modeling, validation, interpretation, and application. Among these, data acquisition holds paramount significance, as the quality of data profoundly influences algorithmic performance. Data exploration involves a comprehensive analysis of the sample data to elucidate its morphological attributes and contextual relevance. Preprocessing activities, such as addressing missing values, attribute coding, data normalization, standardization, feature selection, and principal component analysis, are closely interlinked with data exploration.

Data preprocessing commands a substantial share of the comprehensive procedure. This stage transforms raw data into formats conducive to analytical scrutiny. The choice of the appropriate machine learning approach should be guided by the intrinsic attributes of the underlying data and the specific task at hand. During training, optimal weights/parameters are identified to optimize classification accuracy (for classification tasks) or minimize discrepancies between predictions and actual values (for regression tasks).

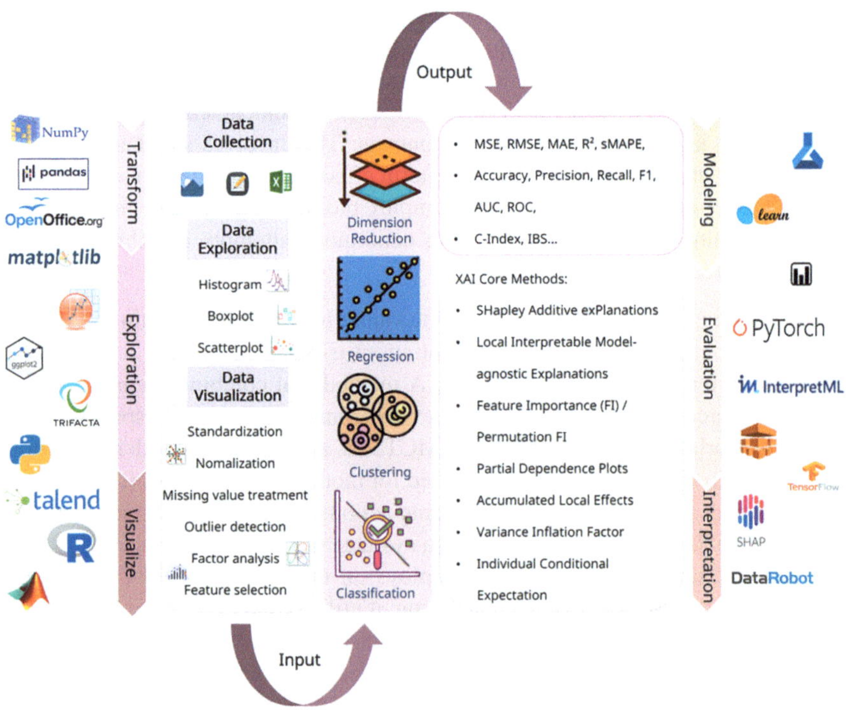

Fig. 2.2 Illustration of the workflow and frequently used tools of ML

The nature of the task will determine suitable model assessment metrics, which gauge the regression model's fitting performance and classification accuracy, respectively. Employing a diverse set of evaluation metrics, tailored to the specific task, is vital for a multidimensional evaluation of model performance. Notably overlooked, model interpretation yields manifold advantages, encompassing the unveiling of latent insights to construct more robust models and enabling accessibility to models for individuals with varying degrees of expertise [11].

2.2 Several Classical Machine Learning Algorithms

2.2.1 Decision Tree

The decision tree (DT) algorithm, represented as a visual tree structure, is a prevalent methodology in data analysis and statistics for both classification and regression tasks [12]. Each internal node in this framework acts as a decision point based on specific attributes, with branches representing decision outcomes and terminal leaf nodes containing final classification results. By analyzing sample data and using tailored algorithms, a decision tree aims to cover comprehensive and mutually exclusive decision paths. In simpler terms, a decision tree is like a tree with branches representing decision conditions. Starting from the main trunk, branches split into finer subdivisions, each combination of conditions forming a unique sequence of decision rules (Fig. 2.3).

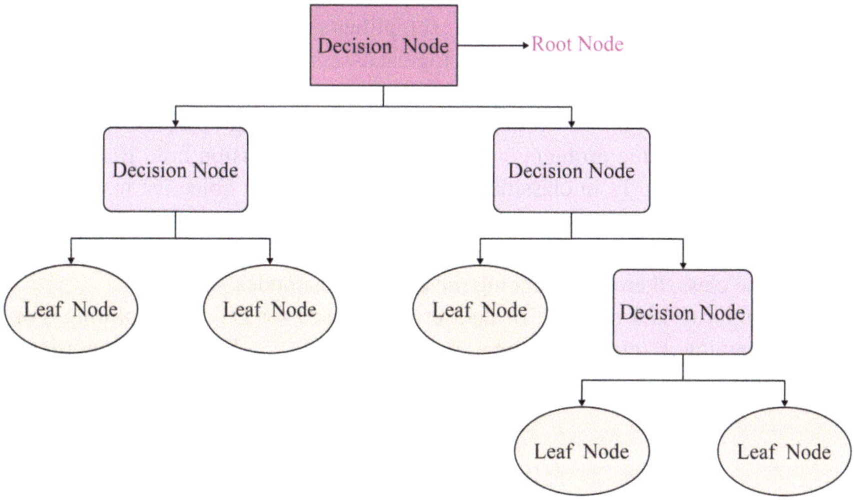

Fig. 2.3 The workflow of a decision tree

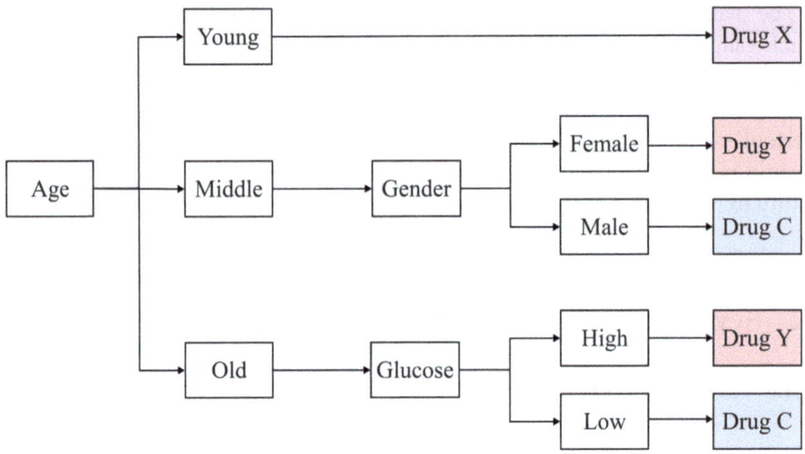

Fig. 2.4 A clinical example to illustrate the use of a decision tree algorithm in treatment selection

Here is a clinical example to illustrate the use of a decision tree algorithm in treatment selection (Fig. 2.4). Suppose a group of patients suffers from a specific disease, and during the treatment process, each patient responds to one of three drugs (Drug X, Drug Y, or Drug C). Medical researchers use a decision tree algorithm to select the appropriate medication through the following steps: For young patients, prescribe Drug X; for middle-aged patients, further consider gender, prescribing Drug C for males and Drug Y for females; for elderly patients, assess blood glucose levels, prescribing Drug Y for high levels and Drug C for low levels. This decision-making process exemplifies the basic concept of a decision tree algorithm, applying a step-by-step logical assessment to tailor treatment to each patient's specific characteristics.

DTs have fundamental formulas related to node splitting and prediction, which depend on whether the problem is a classification or regression task. Below are the basic formulas for DTs in classification task. The primary objective in this task is to select the optimal splitting attribute. Generally, as the splitting process continues, the samples within each branch node of the decision tree should increasingly belong to the same class, thereby enhancing the purity of the nodes.

Gini impurity is a measure of dataset purity used for DT node splitting. For a node containing K classes, assume $p(i)$ $(i = 1, 2, \ldots, K)$ is the proportion of samples belonging to class i in the dataset D, and the Gini impurity is as follows:

$$Gini(D) = 1 - \sum_{i=1}^{K} p(i)^2, \tag{2.1}$$

where $Gini(D)$ is the Gini impurity of the dataset D. $Gini(D)$ ranges between 0 and 1. A smaller value indicates higher purity of the dataset, meaning the sample distribution is skewed toward a particular class. In the context of the DT algorithm, features and thresholds that result in lower Gini impurity are typically chosen for node splitting to achieve better data purification.

Information gain is another commonly used metric in DTs to select the best feature for splitting. The calculation of information gain involves the node's entropy and the reduction in weighted average entropy of the child nodes. Using the same notation as in (2.1), the formula for entropy of the node ($H(X)$) for the dataset D is

$$H(X) = -\sum_{i=1}^{K} p(i)\log_2(p(i)). \tag{2.2}$$

Generally, a higher information gain indicates a greater "purity improvement" obtained by using attribute a for splitting. Therefore, information gain can be used to select the splitting attribute in decision trees, specifically choosing attribute a as follows:

$$a = \arg\max \mathrm{Gain}(D, a).$$

The well-known ID3 decision tree learning algorithm [13] uses information gain as the criterion for selecting splitting attributes.

For regression tasks, the formulas for DTs typically involve node splitting and prediction using mean values. In regression DTs, the splitting criterion can be the reduction in variance, often measured by the mean squared error (MSE). The goal is to find the feature and threshold that minimize the MSE of the child nodes. The MSE for a node t with n_t samples is calculated as

$$MSE(t) = \frac{1}{n_t}\sum_{i=1}^{n_t}(y_i - \bar{y}_t)^2,$$

where y_i is the actual value of the i-th sample in node t, $\bar{y}_t$ is the mean value of the target variable in node t, and n_t is the number of samples in node t. Further, for each leaf node in a regression DT, the prediction value is typically the mean of the target values for all samples in that leaf node. The prediction $\hat{y}_t$ for a leaf node t with n_t samples is given by

$$\hat{y}_t = \frac{1}{n_t}\sum_{i=1}^{n_t} y_i.$$

2.2.2 *Ensemble Learning*

Ensemble learning, also known as multi-classifier systems or committee-based learning, combines multiple learners to accomplish learning tasks. The general structure of ensemble learning involves generating a set of individual learners and then combining them using a specific strategy. These individual learners are typically produced by applying an existing learning algorithm to the training data. When the ensemble consists exclusively of the same type of individual learners—such as a decision tree ensemble composed entirely of decision trees or a neural network ensemble composed entirely of neural networks—it is considered homogeneous. In homogeneous ensembles, the individual learners are referred to as base learners, and the corresponding learning algorithm is called the base learning algorithm.

Conversely, ensembles can also include different types of individual learners, such as a combination of decision trees and neural networks, making the ensemble heterogeneous. In heterogeneous ensembles, individual learners are generated by different learning algorithms, and there is no longer a single base learning algorithm. Instead, these learners are often referred to as component learners or simply as individual learners.

Ensemble learning can involve various strategies, but three have become particularly popular for their ease of implementation and effectiveness in practice: bagging, stacking, and boosting. These methods enhance the performance and robustness of the learning tasks by leveraging the strengths of multiple learners.

2.2.2.1 Boosting

Boosting algorithms are indeed a family of algorithms that can boost weak learners into strong learners. Their mechanism is similar: First, a base learner is trained from the initial training set. Then, based on the performance of the base learner, the distribution of training samples is adjusted, so that the training samples previously misclassified by the base learner receive more attention in the subsequent training. Next, a new base learner is trained based on the adjusted sample distribution. This process is repeated until the number of base learners reaches a predefined value. Finally, these base learners are combined with weights to obtain a strong learner.

AdaBoost is one of the most famous representatives of boosting algorithms, first proposed by Freund and Schapire [14]. A common derivation of AdaBoost is based on an additive model to minimize the exponential loss function. The specific expression for expanding the linear combination of base learners in AdaBoost is

$$H(x) = \sum_{t=1}^{T} \alpha_t h_t(x),$$

where $H(x)$ is the final strong learner, a weighted sum of base learners, T is the total number of base learners, $h_t(x)$ is the t-th base learner, and α_t is the weight assigned to the t-th base learner, determined during the boosting process. The exponential loss function can be represented as

$$L_{\exp}(H|D) = E_{x\sim D}\left[e^{-f(x)\cdot H(x)}\right].$$

where D is the training set, f is the true function, H is the linear combination of base learners, and E represents expectation.

To minimize $L_{\exp}(H|D)$, we take its derivative with respect to $H(x)$,

$$\frac{\partial L_{\exp}(H|D)}{\partial H(x)} = -e^{-f(x)H(x)} f(x).$$

Setting this derivative to zero to find the minimum, we get

$$H(x) = \frac{1}{2} \ln \frac{P(f(x) = 1 \mid x)}{P(f(x) = -1 \mid x)}.$$

Therefore, the final prediction is given by

$$\text{sign}(H(x)) = \text{sign}\left(\frac{1}{2} \ln \frac{P(f(x) = 1 \mid x)}{P(f(x) = -1 \mid x)}\right).$$

This indicates that AdaBoost effectively estimates the log odds of the true class labels and combines the base learners in a way that focuses on the more difficult samples, thus iteratively improving the classification accuracy. By minimizing the exponential loss, AdaBoost ensures that the combined classifier is a strong learner, significantly reducing the classification error compared to individual weak learners.

In the AdaBoost algorithm, the goal is to combine weak classifiers into a strong classifier through iterative rounds. Each iteration selects a base classifier and assigns it a weight to minimize the overall classification error. In the t-th iteration, the strong classifier from the previous $t - 1$ rounds is given by

$$H_{t-1}(x) = \sum_{j=1}^{t-1} \alpha_j h_j(x).$$

The aim is to minimize the loss function by adding a new base classifier $h_t(x)$ with an associated weight α_t, resulting in

$$H_t(x) = H_{t-1}(x) + \alpha_t h_t(x).$$

The loss function at this point is

$$L_{\exp}(H_t) = \sum_{i=1}^{N} \exp(-y_i(H_{t-1}(x_i) + \alpha_t h_t(x_i))).$$

Expanding and reorganizing the loss function yield

$$L_{\exp}(H_t) = \sum_{i=1}^{N} \exp(-y_i H_{t-1}(x_i)) \exp(-\alpha_t y_i h_t(x_i)).$$

Letting $D_i^{(t)} = \exp(-y_i H_{t-1}(x_i))$, the loss function can be rewritten as

$$L_{\exp}(H_t) = \sum_{i=1}^{N} D_i^{(t)} \exp(-\alpha_t y_i h_t(x_i)).$$

To minimize this loss function, the selection of α_t is critical. Taking the derivative with respect to α_t and setting it to zero give

$$\frac{\partial L_{\exp}(H_t)}{\partial \alpha_t} = \sum_{i=1}^{N} D_i^{(t)}(-y_i h_t(x_i)) \exp(-\alpha_t y_i h_t(x_i)) = 0.$$

This equation is generally difficult to solve directly. To simplify, the classification error ϵ_t is introduced

$$\epsilon_t = \sum_{i=1}^{N} w_i^{(t)} I(y_i \neq h_t(x_i)),$$

where $w_i^{(t)} = \frac{D_i^{(t)}}{\sum_{j=1}^{N} D_j^{(t)}}$ is the weight of sample i.

Through logarithmic transformation and algebraic manipulation, the classifier weight update formula is derived as

$$\alpha_t = \frac{1}{2} \ln\left(\frac{1-\epsilon_t}{\epsilon_t}\right). \tag{2.3}$$

Equation (2.3) indicates that when the classification error ϵ_t is small, the classifier weight α_t is large, signifying that the classifier is given more importance in the final combination; conversely, when ϵ_t is large, the classifier weight α_t is small.

Assume that $t-1$ rounds of iterations have been completed, and the algorithm is entering the t-th round. At this point, the strong learner $H_{t-1}(x)$ already includes the weighted combination of the first $t-1$ base learners

$$H_{t-1}(x) = \sum_{j=1}^{t-1} \alpha_j h_j(x).$$

In the t-th iteration, the goal is to minimize the exponential loss function by adding a new base learner $h_t(x)$ and its weight α_t. Specifically, α_t and $h_t(x)$ are chosen to minimize

$$L_{\exp}(H_t|D) = \sum_{i=1}^{N} e^{-y_i(H_{t-1}(x_i)+\alpha_t h_t(x_i))}.$$

Expanding and rewriting the loss function yield

$$L_{\exp}(H_t|D) = \sum_{i=1}^{N} D_i^{(t)} e^{-\alpha_t y_i h_t(x_i)}.$$

where $D_i^{(t)} = e^{-y_i H_{t-1}(x_i)}$ can be considered as a constant resulting from the first $t-1$ iterations. The weights are updated using the following formula:

$$w_i^{(t+1)} = \frac{w_i^{(t)} e^{-\alpha_t y_i h_t(x_i)}}{\sum_{j=1}^{N} w_j^{(t)} e^{-\alpha_t y_j h_t(x_j)}}. \tag{2.4}$$

This normalization ensures that the new weights sum up to 1, maintaining a valid probability distribution. Based on (2.3) and (2.4), the AdaBoost algorithm is as follows:

Algorithm 2.1 Pseudocode for AdaBoost algorithm

1: **Input**: Training dataset $D = \{(x_1, y_1), (x_2, y_2), \ldots, (x_N, y_N)\}$, where x_i is the feature vector of the i-th sample, y_i is the true label of the i-th sample ($y_i \in \{-1, +1\}$); Base learner algorithm; Number of iterations T.
2: **Output**: Final strong classifier $H(x)$.
3: Initialize weights: $w_i^{(1)} = \frac{1}{N}$, for all $i = 1, 2, \ldots, N$.
4: **for** $t = 1$ **to** T **do**
5: Train base learner $h_t(x)$ using weights $\{w_i^{(t)}\}$.
6: Compute error rate: $\epsilon_t = \sum_{i=1}^{N} w_i^{(t)} I(y_i \neq h_t(x_i))$.
7: Compute classifier weight: $\alpha_t = \frac{1}{2} \ln\left(\frac{1-\epsilon_t}{\epsilon_t}\right)$.
8: Update weights: $w_i^{(t+1)} = \frac{w_i^{(t)} \exp(-\alpha_t y_i h_t(x_i))}{\sum_{j=1}^{N} w_j^{(t)} \exp(-\alpha_t y_j h_t(x_j))}$.
9: **end for**
10: Final classifier: $H(x) = \text{sign}\left(\sum_{t=1}^{T} \alpha_t h_t(x)\right)$.
11: **Return** $H(x)$.

2.2.2.2 Bagging

This strategy involves training models using random samples of the dataset, with replacement. Each model is trained independently, and their predictions are combined through averaging or voting. Random forest (RF), proposed by Breiman in 2001, is an extended variant of bagging [15]. RF enhances bagging by incorporating random attribute selection during the training of decision trees. In traditional decision trees, the optimal attribute for splitting is selected from the entire set of attributes at each node. In contrast, RF modifies this process by randomly selecting a subset of attributes from the current node's attribute set before choosing the optimal splitting attribute from this subset.

The parameter k controls the degree of randomness introduced. If $k = d$ (where d is the total number of attributes), the construction of the base decision tree is identical to that of a traditional decision tree. If $k = 1$, a single attribute is randomly selected for splitting.

Unlike bagging, which achieves diversity solely through sample perturbation, RF enhances diversity by also introducing attribute perturbation. During tree construction, it randomly selects a subset of attributes at each node before choosing the best split. This dual source of randomness—sample and attribute perturbation—improves generalization performance by increasing the heterogeneity among the individual trees, leading to better predictive accuracy and robustness.

Given an ensemble of base learners $\{h_1, h_2, \ldots, h_N\}$ where h_i outputs $h_i(x)$, for instance, x, below introduces several common strategies for combining these h_i. For numeric outputs ($h_i(x) \in R$), a popular strategy is averaging.

- **Simple averaging**

$$H(x) = \frac{1}{N}\sum_{i=1}^{N} h_i(x).$$

- **Weighted averaging**

$$H(x) = \sum_{i=1}^{N} w_i h_i(x).$$

where $H(x)$ represents the model's final prediction for sample x and w_i is the weight of base learner h_i, typically requiring $w_i \geq 0$ and $\sum_{i=1}^{N} w_i = 1$. Simple averaging is easy to implement and computationally efficient, but it cannot differentiate the contributions of individual base learners, which may be influenced by poorly performing learners. Weighted averaging, on the other hand, allows assigning different weights based on the performance of base learners, offering more flexibility. However, it requires manual or algorithmic determination of weights, which may be sensitive to improper weight settings and require weight tuning. When all base learners perform equally well ($w_i = 1/r$), the two methods yield the same results.

For classification tasks, the most common strategy for combining the predictions of base learners is using the voting method. Let h_i predict a label from the set of class labels $\{c_1, c_2, \ldots, c_N\}$. We represent the prediction output of h_i on sample z as an N-dimensional vector $h_i^1(x), h_i^2(x), \ldots, h_i^N(x)$, where $h_i^j(x)$ is the output of h_i for class label c_j.

- **Majority voting**

$$H(x) = \begin{cases} c_j, & \text{if } \frac{\sum_{i=1}^{N} h_i^j(x)}{N} > 0.5, \\ \text{reject}, & \text{otherwise}. \end{cases}$$

This method predicts the class label c_j if it receives more than half of the votes; otherwise, it rejects the prediction.

- **Plurality voting**

$$H(x) = \text{argmax}_c \sum_{i=1}^{N} h_i^j(x).$$

This method predicts the class label that receives the most votes among the base learners' predictions.

- **Weighted voting**

$$H(x) = \text{argmax}_c \sum_{i=1}^{N} w_i h_i^j(x),$$

where w_i is the weight assigned to base learner h_i.

Each of these three voting methods has its own advantages and disadvantages. The majority voting method is simple, intuitive, and easy to implement, but it may be sensitive to noisy data. Plurality voting, similarly simple to implement, may be sensitive to class imbalance in the data. Weighted voting, while more flexible and able to assign different weights to base learners based on their performance, requires the determination of these weights, which can be a manual or algorithmic process and may require additional tuning. Therefore, when choosing a voting method, it is important to consider the specific characteristics of the problem and the data distribution to balance the trade-offs and select the most suitable approach.

2.2.3 *Extremely Randomized Trees*

Next, we introduce extremely randomized trees (ExtraTrees), which are developed based on the random forest (RF) methodology [16]. ExtraTrees deviates from RF in two pivotal aspects. Firstly, ExtraTrees typically avoids the bootstrapping technique commonly used by RF for generating each decision tree's training set. Instead, it incorporates all samples from the original training set while still preserving the element of random feature selection. Secondly, during the process of feature partitioning, ExtraTrees randomly selects a feature value for splitting, differing from the predetermined rules used in the RF approach. Although ExtraTrees does not follow the traditional bagging method, it still leverages ensemble learning to enhance model stability and predictive performance. The structure of ExtraTrees is summarized in Fig. 2.5. Below are the brief steps of the ExtraTrees.

First, given a training dataset $D = \{(x_1, y_1), (x_2, y_2), \ldots, (x_n, y_n)\}$, where x_i is the input feature vector and y_i is the corresponding output label. For each node, a subset $K \subset \{1, 2, \ldots, d\}$ of the d features is randomly selected, typically with $|K| = \sqrt{d}$ or $|K| = \log_2(d)$.

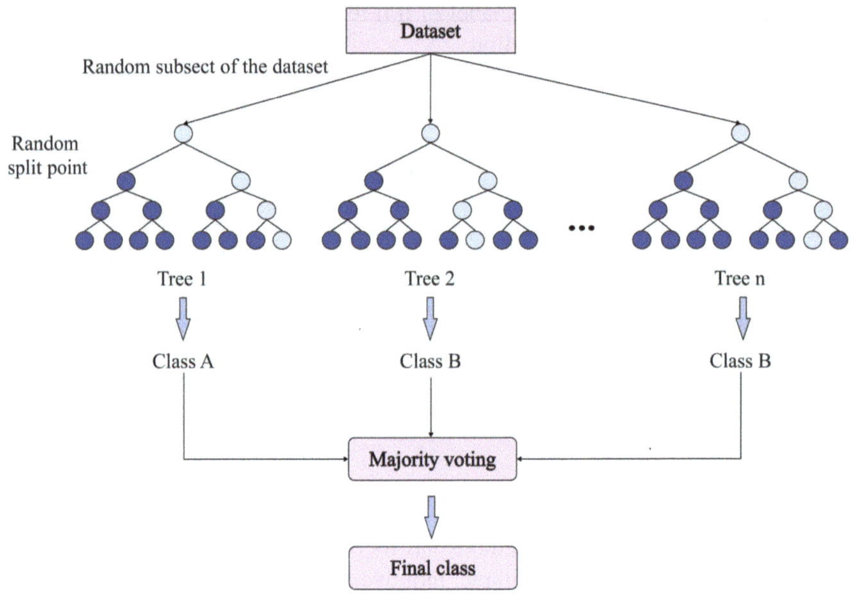

Fig. 2.5 Structure of ExtraTrees

Further, for each selected feature $k \in K$, a split threshold θ_k is randomly chosen, specifically

$$\theta_k \sim \mathcal{U}(a_k, b_k),$$

where a_k and b_k are the minimum and maximum values of feature k.

Next, the feature and threshold pair (k, θ_k) that optimizes the objective function (such as information gain or Gini index) is chosen for node splitting. Assuming feature k and threshold θ_k are selected, the left and right child nodes are $D_{\text{left}} = \{(x_i, y_i) \mid x_i^k \leq \theta_k\}$ and $D_{\text{right}} = \{(x_i, y_i) \mid x_i^k > \theta_k\}$, respectively. The information gain $IG(D, k, \theta_k)$ is calculated as

$$IG(D, k, \theta_k) = H(D) - H(D, k, \theta_k),$$

with

$$H(D) = -\sum_{c=1}^{C} p_c \log_2(p_c),$$

and

$$H(D, k, \theta_k) = \frac{|D_{\text{left}}|}{|D|} H(D_{\text{left}}) + \frac{|D_{\text{right}}|}{|D|} H(D_{\text{right}}),$$

where $H(D)$ is the entropy before the split and $H(D, k, \theta_k)$ is the entropy after the split.

The Gini index $\Delta G(D, k, \theta_k)$ is calculated as

$$\Delta G(D, k, \theta_k) = G(D) - G(D, k, \theta_k),$$

with

$$G(D) = 1 - \sum_{c=1}^{C} p_c^2,$$

and

$$G(D, k, \theta_k) = \frac{|D_{\text{left}}|}{|D|} G(D_{\text{left}}) + \frac{|D_{\text{right}}|}{|D|} G(D_{\text{right}}).$$

Here $G(D)$ is the Gini impurity before the split and $G(D, k, \theta_k)$ is the Gini impurity after the split. This process is recursively repeated for each child node until a stopping criterion (such as maximum tree depth or minimum sample size) is met. After constructing multiple trees, each tree t produces a prediction $h_t(x)$.

For classification tasks, the final prediction is determined by majority voting,

$$\hat{y} = \text{mode}\{h_1(x), h_2(x), \ldots, h_T(x)\}.$$

For regression tasks, the final prediction is obtained by averaging

$$\hat{y} = \frac{1}{T} \sum_{t=1}^{T} h_t(x).$$

To summarize, the ExtraTrees algorithm constructs decision trees by randomly selecting features and split thresholds, leveraging ensemble learning to enhance model stability and predictive performance. Its main advantage lies in introducing more randomness, which reduces the risk of overfitting and increases efficiency in the construction process compared to the RF algorithm.

2.2.4 Support Vector Machine

Support vector machines (SVM) as a supervised learning model find application in data analysis and pattern recognition, particularly excelling in addressing binary classification challenges. In the classification process of SVM, the primary objective is to segregate two distinct classes of data by means of a separation hyperplane,

$$w \cdot x + b = 0, \tag{2.5}$$

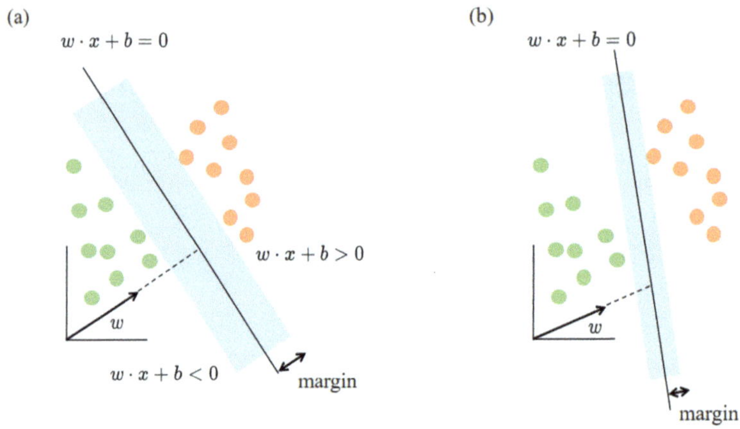

Fig. 2.6 Two potential hyperplanes splitting a set of data

where w represents the weight vector, x is the input feature vector, and b is the bias term. Figure 2.6 shows two potential hyperplanes splitting a set of data. Each has a different value of w and constant b.

The SVM optimization problem aims to find a decision boundary (hyperplane) that maximizes the distance between data points and that boundary. These distances are referred to as support vectors, while also ensuring the maximization of correctly classified data points. This can be mathematically formulated as a convex quadratic programming problem, where the objective is to minimize a loss function, often defined as the sum of distances between support vectors and the decision boundary. This objective is subject to constraints that ensure both correct classification and the maximization of margin between classes. Given the hyperplane (2.5), a new data point x_j can be classified by

$$f(x) = sign(w \cdot x + b).$$

Specifically, for classification labels $y_j \in \{\pm 1\}$, the data to the left or right of the hyperplane in Fig. 2.6 is given by

$$y_j(w \cdot x + b) = sign(w \cdot x + b) = \begin{cases} +1 & \text{orange cycle,} \\ -1 & \text{green cycle.} \end{cases}$$

Thus the classifier y_j is directly influenced by the position of x_j. The optimization goal includes minimizing the number of data points that are incorrectly categorized and increasing the margin that can be achieved. We provide a specified loss function to create the optimization objective function

$$L(y_j, \hat{y}_j) = L(y_j, sign(w \cdot x + b)) = \begin{cases} +1 \text{ if } y_j \neq \text{sign(w} \cdot \text{x+b),} \\ 0 \text{ if } y_j = \text{sign(w} \cdot \text{x+b).} \end{cases}$$

More simply,

$$L(y_j, \hat{y}_j) = \begin{cases} +1 & \text{if data is incorrectly labeled,} \\ 0 & \text{if data is correctly labeled.} \end{cases}$$

Therefore, every incorrectly identified data point results in a loss of one. The sum of the loss functions $L(y_j, \hat{y}_j)$ equals the training error over m data points. In addition to minimizing the loss function, another objective of SVM is to maximize the separation margin between classes. We can then give the optimization problem as

$$\begin{cases} \text{argmin}_{w,b} C \sum_{j=1}^{m} max(0, 1 - y_j, \hat{y}_j) + \frac{1}{2} \parallel w \parallel^2, \\ \min_j |x_j \cdot w| = 1, \end{cases}$$

where $\frac{1}{2}|w|^2$ represents the regularization term and C is a hyperparameter that controls the weight of the loss term. The second term involves the summation of hinge losses over all samples, reflecting the penalty for misclassification, while correctly classified samples do not affect the optimization.

2.2.5 K-Nearest Neighbor

The K-nearest neighbors algorithm (KNN) is a widely used machine learning algorithm for both classification and regression tasks. The core principle of this algorithm involves calculating the distances between a given new sample and the training samples. By identifying the K closest training samples, the algorithm then utilizes the class labels of these K samples (for classification problems) or the values of the target variable (for regression problems) to predict or estimate the new sample's outcome [17].

A notable characteristic of KNN, compared to other learning methods, is its apparent lack of an explicit training process. In fact, KNN is a well-known representative of lazy learning. Lazy learning techniques involve merely storing the samples during the training phase, resulting in zero training time overhead. The actual processing is deferred until a test sample is received. In contrast, methods that involve learning and processing samples during the training phase are referred to as eager learning. Below is the pseudocode for KNN.

Algorithm 2.2 Pseudocode for KNN algorithm

Load the dataset.
Initialize k (number of neighbors).
for each instance x_i in the dataset **do**
    Calculate the distance between the query instance and x_i.
    Store the distance and the index of x_i.
end for
Sort the distances in ascending order.
Select the first k distances.
Retrieve the labels of the selected k instances.
if classification **then**
    Return the mode of the k labels.
else
    Return the mean of the k labels.
end if

2.2.6 Gaussian Naive Bayes

Gaussian Naive Bayes (GNB) is a specialized form of the Naive Bayes classifier that assumes Gaussian distributions for the features. While the general Naive Bayes classifier can handle various distributions, Gaussian Naive Bayes specifically models the likelihoods using the normal distribution. This makes it particularly effective when the features follow a Gaussian distribution, leveraging the independence assumption to simplify computations and achieve robust classification performance.

To provide a comprehensive introduction to Gaussian Naive Bayes, we will first review Bayesian decision theory and the concept of the Bayes optimal classifier. Bayesian decision theory is a foundational statistical framework for decision-making that relies on probabilities. In the context of classification tasks, it evaluates all pertinent probabilities and associated losses to determine the optimal class label. The essential principle involves leveraging these probabilities and the respective misclassification losses to identify the most suitable class label. In the following, we will focus on multi-class classification tasks.

Assume there are N possible class labels, denoted as $\mathcal{Y} = \{c_1, c_2, \ldots, c_N\}$. Let λ_{ij} be the loss incurred by classifying a sample with true class c_j as c_i. Based on the posterior probability $P(c_j \mid x)$, the expected loss for classifying a sample x as c_i is given by

$$R(c_i \mid x) = \sum_{j=1}^{N} \lambda_{ij} P(c_j \mid x).$$

Our task is to find a decision rule $h : \mathcal{X} \to \mathcal{Y}$ that minimizes the overall risk,

$$R(h) = \mathbb{E}_x[R(h(x) \mid x)].$$

Clearly, for each sample x, if the decision rule $h(x)$ minimizes the conditional risk $R(h(x) \mid x)$, then the overall risk $R(h)$ will also be minimized. This leads to the Bayes decision rule: to minimize the conditional risk, choose the class label that minimizes $R(c \mid x)$ for each sample x, that is,

$$h^*(x) = \arg\min_{c \in \mathcal{Y}} R(c \mid x).$$

Here, h^* is called the Bayes optimal classifier, and the corresponding overall risk $R(h^*)$ is called the Bayes risk. $1 - R(h^*)$ represents the best possible accuracy achievable by any classifier, serving as the theoretical upper bound of model accuracy. Specifically, the goal is to minimize the classification error rate; hence, the conditional risk becomes

$$R(c \mid x) = 1 - P(c \mid x).$$

Therefore, the Bayes optimal classifier becomes

$$h^*(x) = \arg\max_{c} P(c \mid x).$$

We now introduce the derivation process of the Gaussian Naive Bayes classifier. The Bayes optimal classifier chooses the class with the highest posterior probability $P(c \mid x)$. According to Bayes' Theorem, the posterior probability is given by

$$P(c \mid x) = \frac{P(x \mid c)P(c)}{P(x)}.$$

Since $P(x)$ is the same for all classes, it can be omitted in the decision rule

$$h^*(x) = \arg\max_{c} P(x \mid c)P(c).$$

For Gaussian Naive Bayes, we assume that the likelihood $P(x \mid c)$ follows a Gaussian distribution for each feature x_i,

$$P(x_i \mid c) = \frac{1}{\sqrt{2\pi\sigma_{c,i}^2}} \exp\left(-\frac{(x_i - \mu_{c,i})^2}{2\sigma_{c,i}^2}\right),$$

where x_i is the i-th feature, $\mu_{c,i}$ is the mean of the i-th feature for class c, and $\sigma_{c,i}$ is the standard deviation of the i-th feature for class c.

Assuming feature independence, d is the number of features. The combined likelihood for all features $x_1, x_2, \ldots, x_d$ given class c is

$$P(x \mid c) = \prod_{i=1}^{d} P(x_i \mid c).$$

Substituting the Gaussian distribution for each feature, we get

$$P(x \mid c) = \prod_{i=1}^{d} \frac{1}{\sqrt{2\pi\sigma_{c,i}^2}} \exp\left(-\frac{(x_i - \mu_{c,i})^2}{2\sigma_{c,i}^2}\right).$$

Therefore, the classifier for GNB becomes

$$h^*(x) = \arg\max_c \left(P(c) \prod_{i=1}^{d} \frac{1}{\sqrt{2\pi\sigma_{c,i}^2}} \exp\left(-\frac{(x_i - \mu_{c,i})^2}{2\sigma_{c,i}^2}\right)\right).$$

Taking the logarithm (which preserves the arg max), we have

$$h^*(x) = \arg\max_c \left(\log P(c) + \sum_{i=1}^{d} \log\left(\frac{1}{\sqrt{2\pi\sigma_{c,i}^2}}\right) - \sum_{i=1}^{d} \frac{(x_i - \mu_{c,i})^2}{2\sigma_{c,i}^2}\right).$$

This simplifies to

$$h^*(x) = \arg\max_c \left(\log P(c) - \sum_{i=1}^{d} \frac{\log(2\pi\sigma_{c,i}^2)}{2} - \sum_{i=1}^{d} \frac{(x_i - \mu_{c,i})^2}{2\sigma_{c,i}^2}\right).$$

2.2.7 *Multilayer Perceptron*

A multilayer perceptron (MLP) is a type of artificial neural network that consists of multiple layers of neurons: an input layer, one or more hidden layers, and an output layer [18]. Each neuron in one layer is connected to every neuron in the subsequent layer, making the network fully connected. MLPs are powerful models used for various tasks such as classification and regression due to their ability to learn complex mappings from inputs to outputs.

The input layer is the first layer in an MLP and receives the raw input data. Each node in this layer represents one feature of the input data. For example, if the input data has n features, the input layer will consist of n nodes. This layer simply passes the input data to the next layer without performing any computations.

The hidden layers are the core computational units of the MLP. Each hidden layer consists of neurons that apply a transformation to the data received from the previous layer. This transformation involves computing a weighted sum of the inputs, adding a bias term, and then applying a nonlinear activation function. For a single neuron j in layer l, the weighted sum of inputs to neuron j in layer l is

$$z_j^{(l)} = \sum_{i=1}^{n^{(l-1)}} w_{ji}^{(l)} a_i^{(l-1)} + b_j^{(l)},$$

with

$$a_j^{(l)} = \phi(z_j^{(l)}),$$

where $w_{ji}^{(l)}$ is the weight between neuron i in layer $l-1$ and neuron j in layer l, $a_i^{(l-1)}$ is the activation of neuron i in layer $l-1$, and $b_j^{(l)}$ is the bias term for neuron j in layer l. ϕ is the activation function, which could be the sigmoid function,

$$\sigma(z) = \frac{1}{1+e^{-z}},$$

hyperbolic tangent (tanh) function,

$$\tanh(z) = \frac{e^z - e^{-z}}{e^z + e^{-z}},$$

or rectified linear unit (ReLU),

$$\text{ReLU}(z) = \max(0, z).$$

The output layer is the final layer in the MLP and produces the network's prediction. In a classification task, the output layer typically uses a softmax function to convert the raw scores from the final hidden layer into probabilities for each class. The softmax function for class k is given by

$$\text{softmax}_k(\mathbf{z}) = \frac{e^{z_k}}{\sum_{j=1}^{K} e^{z_j}},$$

where $\mathbf{z}$ is the input vector to the output layer and K is the number of classes. In a regression task, the output might be a single neuron representing the predicted value, typically without applying a nonlinear activation function to allow for continuous output values [19], where $\mathbf{z}$ is the input vector to the output layer and K is the number of classes. In a regression task, the output might be a single neuron representing the predicted value, typically without applying a nonlinear activation function to allow for continuous output values.

Training an MLP involves adjusting the weights and biases of the network to minimize a loss function, which measures the difference between the predicted outputs and the actual targets. This training process is carried out using an algorithm called backpropagation, which consists of two main phases: forward propagation and backward propagation [19]. During forward propagation, the input data is passed through the network to compute the output. The loss function is then evaluated to quantify the prediction error. For instance, the mean squared error loss for a regression task is given by

$$L(\mathbf{y}, \hat{\mathbf{y}}) = \frac{1}{n}\sum_{i=1}^{n}(y_i - \hat{y}_i)^2,$$

where $\mathbf{y}$ is the vector of actual values, $\hat{\mathbf{y}}$ is the vector of predicted values, and n is the number of samples.

Backward propagation, or backprop, is used to compute the gradients of the loss function with respect to each weight and bias in the network. These gradients are calculated using the chain rule of calculus and are then used to update the weights and biases in a way that reduces the loss. This process is typically repeated for many iterations over the training data, gradually improving the network's performance. The weight update rule using gradient descent is given by

$$w_{ji}^{(l)} \leftarrow w_{ji}^{(l)} - \eta \frac{\partial L}{\partial w_{ji}^{(l)}},$$

where η is the learning rate.

2.3 Model Evaluation

The assessment of ML models is a critical facet of model development, ensuring their reliability, generalization, and effectiveness across diverse datasets. The effectiveness of ML models is objectively assessed by thorough assessment procedures, enabling well-informed judgments on their use in practical applications. In the following subsection, we introduce several evaluation methods for regression and classification model. These methods mainly serve three purposes:

1. Performance assessment:
   - How is the model performing?
2. Model optimization:
   - Is the model fitting well and can it be improved?
   - Which model best approximates our data points?
3. Statistical decision-making:
   - Is the model good enough for us to use?

- Has this model passed our stringent hypothesis testing standards?

2.3.1 Evaluation Methods for Regression

The effectiveness and precision of a regression model's predictions are evaluated using evaluation methods. When predicting a numerical result and the target variable is continuous, regression models are employed. Here are a few typical techniques for testing regression models. Assume n is the number of data points, y_i is the actual value of the target variable for the ith data point, and $\hat{y}_i$ is the predicted value of the target variable for the ith data point.

Mean squared error (MSE) is one of the most widely used metrics for regression evaluation. The core concept of MSE is to penalize larger errors more heavily, thereby encouraging the model to focus on reducing significant prediction errors. By squaring the errors, larger deviations contribute more substantially to the total error. It can be calculated as

$$\mathrm{MSE} = \frac{1}{n}\sum_{i=1}^{n}(y_i - \hat{y}_i)^2.$$

MSE offers several advantages: It is highly intuitive as it amplifies the impact of large errors, thus improving overall prediction accuracy by making the model more attentive to substantial discrepancies. Additionally, MSE is easy to differentiate, providing significant benefits in mathematical processing, particularly in optimization algorithms like gradient descent due to its simple derivative with respect to predicted values. However, MSE has notable limitations. It is highly sensitive to outliers, as the squaring of errors means that a single outlier can significantly skew the MSE value, potentially leading the model to deviate from accurate predictions for the bulk of the data. Furthermore, the unit of MSE is the square of the unit of the original data, making it less intuitive to interpret. For example, if the target variable is in dollars, the MSE will be in squared dollars, which is less straightforward compared to metrics like root mean squared error (RMSE).

RMSE is the square root of MSE. It is a common metric that provides the error in the same units as the target variable and can be expressed as follows:

$$\mathrm{RMSE} = \sqrt{\frac{1}{n}\sum_{i=1}^{n}(y_i - \hat{y}_i)^2}.$$

Compared to MSE, RMSE has the advantage that the error's unit is the same as the unit of the original data, making its results more intuitive and easier to interpret. Additionally, due to the squaring effect, RMSE assigns higher weight to larger errors, effectively reflecting the impact of substantial prediction errors during model evaluation. However, RMSE has notable drawbacks. Like MSE, it is highly sensitive to outliers, as the presence of outliers can significantly skew the RMSE value due to the squared terms. Furthermore, RMSE might overly penalize the model by amplifying larger prediction errors, which can lead the model to focus excessively on reducing these large errors, potentially at the expense of optimizing for smaller errors.

MAE measures the average of the absolute discrepancies between the actual values and the predicted values. The formula is

$$\mathrm{MAE} = \frac{1}{n}\sum_{i=1}^{n} |y_i - \hat{y}_i|.$$

MAE has several advantages, including robustness to outliers and interpretability. Because MAE treats all errors equally without squaring them, it is less sensitive to extreme values, making it a more robust metric in the presence of outliers. Additionally, MAE is easy to understand and interpret as it provides a straightforward average of the absolute errors, measured in the same units as the original data. However, MAE also has some disadvantages. It can be more difficult to optimize than MSE, because its derivative is not continuous at zero, complicating the use of gradient-based optimization methods. Furthermore, MAE is less sensitive to large errors, which, while beneficial in the presence of outliers, can be a drawback if the goal is to penalize large errors more heavily.

Coefficient of determination (R^2) is a crucial metric in the evaluation of regression models. It is widely used to assess the model's ability to explain the variability in the data. The formula is

$$\mathrm{R}^2 = 1 - \frac{\sum_{i=1}^{n}(y_i - \hat{y}_i)^2}{\sum_{i=1}^{n}(y_i - \bar{y})^2}.$$

R^2 is a dimensionless metric that provides a standardized way to compare the performance of different models or the same model across different datasets, offering a clear and intuitive measure of how well the model explains the variability of the response data. However, R^2 has several limitations: It was originally designed for linear regression models and may not be effective for nonlinear models; it is sensitive to outliers as the squared terms in its calculation can significantly skew its value; and it can be artificially inflated by adding more explanatory variables, even if they do not significantly improve the model's predictive power.

Symmetric mean absolute percentage error (sMAPE) estimates the percentage difference between projected and actual values. In contrast to conventional MAPE, sMAPE takes taking into account both overestimation and underestimation errors.

$$\mathrm{sMAPE} = \frac{100}{n}\sum_{i=1}^{n}\frac{|(y_i - \hat{y}_i)|}{(|y_i| + |\hat{y}_i|)/2}.$$

sMAPE has several advantages, including scale independence and balanced error measurement. As a percentage error metric, sMAPE is scale-independent, making it useful for comparing performance across different datasets and models with varying scales. It considers both overestimations and underestimations equally, ensuring that the errors are not biased toward one direction. Additionally, the results are expressed as percentages, making them easy to understand and communicate. However, sMAPE also has notable disadvantages. It can be problematic when the actual values are zero, leading to division by zero and making the metric undefined. It is also sensitive to small denominators, which can produce excessively high percentage errors and distort the overall evaluation. Moreover, the symmetric nature of sMAPE makes the optimization process more complex compared to simpler metrics like MAE or MSE, as it involves both the predicted and actual values in the denominator.

It's often a good idea to use a combination of metrics to comprehensively understand your model's performance. This combined perspective makes judgments regarding model refinement, feature selection, and probable outlier treatment possible.

2.3.2 *Evaluation Methods for Classification*

The following subsection introduces evaluation metrics used in classification tasks, which are fundamentally based on the confusion matrix. These metrics provide a comprehensive assessment of a model's performance, including its ability to correctly classify instances, balance precision and recall, and measure overall effectiveness.

Typically, one category of interest is referred to as the positive category, while the other is the negative category. When positive and negative samples are analyzed separately, four key indicators are generated. These indicators are comprehensive and nonoverlapping, fully encompassing both the predicted and actual labels of the dataset. Consequently, they can be systematically organized into a matrix representation, as shown in Fig. 2.7. Below are the detailed explanations of the four indicators.

- True positive (TP): This refers to cases where the true label of the model is positive, the prediction is positive, and the prediction is correct;
- True negative (TN): This indicator represents cases where the true label of the model is negative, the prediction is negative, and the prediction is correct;
- False positive (FP): In this case, the true label of the model is negative, but the prediction is positive, resulting in an incorrect prediction;
- False negative (FN): This indicator occurs when the true label of the model is positive, but the prediction is negative, resulting in an incorrect prediction.

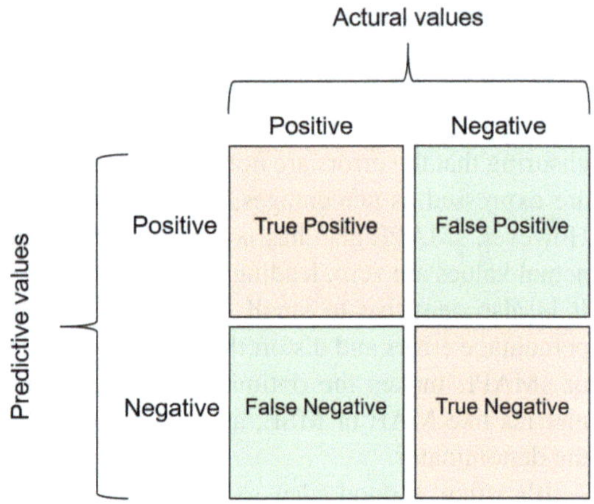

Fig. 2.7 Confusion matrix

Accuracy represents the proportion of correctly classified samples out of the total number of samples. It is calculated as

$$\text{Accuracy} = \frac{\text{TP} + \text{TN}}{\text{TP} + \text{FP} + \text{TN} + \text{FN}}.$$

Accuracy is a very general metric that is suitable for situations where the sample classes are relatively balanced, effectively reflecting the predictive capability of the model in such scenarios. However, while accuracy is intuitive, it may not be a good performance metric in certain situations. For instance, in datasets with highly imbalanced positive and negative samples, accuracy can be misleading. Additionally, accuracy does not distinguish between the impacts of false positives and false negatives, which can be problematic in applications sensitive to error types, such as disease diagnosis, where solely relying on accuracy might mask the model's true performance.

Precision, also known as the positive predictive value, represents the proportion of correctly predicted positive samples out of all samples predicted as positive. It is calculated as follows:

$$\text{Precision} = \frac{\text{TP}}{\text{TP} + \text{FP}}.$$

Precision is particularly important in scenarios where the cost or consequences of false positives (incorrectly predicting negative instances as positive) are severe. For example, in spam detection, misclassifying an important email as spam (a false positive) can result in the loss of crucial information. Therefore, improving precision is critical in such cases. However, precision has its limitations. One major drawback is that it ignores FN, focusing solely on the samples predicted as positive and neglecting the positive samples incorrectly classified as negative. This can lead to a model that avoids predicting positives to enhance precision. Additionally, precision is unsuitable for imbalanced data. In datasets with highly imbalanced positive and negative samples, precision can provide misleading results. For instance, if positive samples are extremely rare, a high precision might not accurately reflect the model's overall performance, as it could be achieved by predicting only a very few positive samples.

Recall, also known as sensitivity or true positive rate, refers to the proportion of correctly identified positive samples out of all actual positive samples. It is calculated as follows:

$$\text{Recall} = \frac{\text{TP}}{\text{TP} + \text{FN}}.$$

Recall is particularly important in scenarios where the consequences of missing positive instances are severe. For example, in medical diagnosis, failing to diagnose a patient with a disease can lead to worsening of the condition or even endanger the patient's life. Similarly, in legal surveillance systems, missing a genuine violation can render legal enforcement ineffective. However, recall has its limitations. One major drawback is that it ignores FP, focusing solely on how well the model captures positive instances while neglecting the negative samples incorrectly classified as positive. This can lead to a model that predicts more samples as positive to increase recall, resulting in a higher false positive rate. Additionally, recall can be misleading in balanced datasets; if a model simply predicts all samples as positive, it would achieve a recall of 100%, but such a model would have no practical predictive value.

The F1 score is defined as the harmonic mean of precision and recall, combining them into a single metric. It is calculated as follows:

$$\text{F1 Score} = \frac{2 \times \text{Precision} \times \text{Recall}}{\text{Precision} + \text{Recall}}.$$

The F1 score is particularly useful in scenarios where the dataset is highly imbalanced between positive and negative samples. For example, in medical diagnosis, where actual cases of disease are far fewer than healthy cases, using precision or recall alone might not fully reflect the model's performance. Similarly, in fraud detection, genuine fraudulent activities are typically rare compared to normal activities, and the F1 score can better evaluate the model's ability to identify the minority class. However, while the F1 score is a useful metric, it has some limitations. One major drawback is that it ignores TN and other considerations related to FP. The F1 score primarily focuses on the performance of predicting the positive class, which might not be comprehensive enough for scenarios requiring attention to both positive and negative predictions. Additionally, the F1 score does not differentiate the importance of precision and recall. In some applications, such as certain commercial settings, avoiding false-positive predictions might be more critical than missing true positive cases. In such situations, the F1 score might need to be adjusted, or other more suitable metrics might need to be considered.

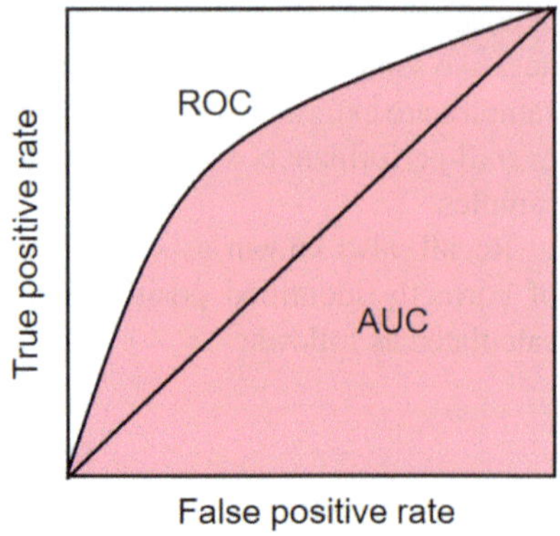

Fig. 2.8 ROC-AUC classification evaluation metric

The receiver operating characteristic (ROC) curve is a graphical representation of a classifier's performance [20]. The area under the curve (AUC) is the area under the ROC curve (Fig. 2.8), which is calculated as follows:

$$\mathrm{AUC} = \int_{x=0}^{1} \mathrm{ROC}(x)dx. \tag{2.6}$$

The AUC value provides a single numeric measure that quantifies a model's ability to distinguish between two classes (positive and negative). The AUC ranges from 0 to 1: An AUC of 1 indicates that the model perfectly discriminates between all positive and negative samples; an AUC of 0.5 suggests that the model's predictions are no better than random guessing; and an AUC below 0.5 indicates that the model performs worse than random guessing, typically signaling serious issues with the model.

ROC and AUC are particularly suitable for evaluating and comparing model performance in imbalanced datasets. Because they do not depend on a specific classification threshold, they are especially useful in applications such as medical diagnostic tests and credit risk assessments, where precise control of false-positive rates and true positive rates is crucial. However, despite their power as evaluation tools, ROC curves and AUC have limitations. In extremely imbalanced datasets, the ROC curve can be overly optimistic, potentially showing a high true positive rate even with a significant number of false positives. Additionally, ROC curves and AUC do not account for the differing costs of false positives and false negatives. In practical applications, the consequences of these errors can vary significantly, and ROC and AUC do not directly incorporate these cost differences, which may be critical for an accurate assessment of model performance.

2.3.3 Example Evaluation Methods for Survival Analysis Model

Next, we introduce two metrics for evaluating survival analysis models: the concordance index (C-index), Brier score (BS), and integrated Brier score (IBS). These metrics are used to assess the duration until the event occurrence and the likelihood of the event occurrence, respectively [21, 22].

The C-index quantifies the agreement between the predicted outcomes and the actual observed values across all valid observations in a survival analysis study. It is calculated by organizing the subjects into random pairs within groups of n. For any given pair, if the patient with a higher predicted survival probability indeed outlives the one with a lower predicted probability, the predicted values are considered concordant with the actual outcomes, resulting in a consistency index $c_i = 1$ for that pair. Conversely, if the predicted values are inconsistent with the actual outcomes, the consistency index $c_i = 0$. The C-index ranges from 0 to 1, with higher values indicating superior predictive performance of the model. The mathematical expression for calculating the C-index is as follows:

$$\text{C-index} = \frac{\sum c_i}{\text{number of valid pairs}},$$

where c_i denotes the consistency index for each valid pair.

The BS is defined as the mean squared difference between the predicted and actual values. Smaller values of this score suggest greater accuracy of the prediction model, making it a frequently employed metric for assessing model calibration. The formula to calculate the BS is as follows:

$$\text{BS} = \frac{1}{m}\sum_{i=1}^{m}\sum_{j=1}^{l}(y_{ij} - \bar{y_{ij}})^2,$$

where m denotes the sample size, l denotes the number of categories, y_{ij} denotes the model prediction, and $\bar{y_{ij}}$ denotes the true observation.

The IBS is an extension of the Brier Score that integrates the BS over time, providing a single measure of model performance over the entire follow-up period. It is particularly useful for evaluating models in the context of time-to-event data. The formula for calculating the IBS is

$$\text{IBS} = \int_0^T \text{BS}(t)\,dt,$$

where $\text{BS}(t)$ is the Brier Score at time t. This integration over time allows for a comprehensive assessment of the model's predictive accuracy throughout the entire study period rather than at a single time point.

2.4 Feature Selection

Feature selection is a crucial component of feature engineering [23]. While ML algorithms have made significant progress and research in predicting drug-target relationships, the quality of classification results still heavily relies on the effectiveness of feature selection. Choosing representative features can yield satisfactory results even with general classification algorithms. Therefore, the accurate and efficient acquisition of representative and deep-level features is a key focus in ML research. On the other hand, high-dimensional data often leads to imprecise predictions, commonly referred to as the curse of dimensionality. This is primarily due to the existence of noise and redundant information within high-dimensional data, which can adversely affect the accuracy of classification results. Hence, reducing the dimensionality of feature data to avoid the curse of dimensionality and mitigate the negative impact of noise on model classification, while obtaining effective features that meet practical requirements and enhancing algorithm prediction accuracy, becomes a critical issue in the research process.

The choice of feature selection method depends on factors like the dataset's characteristics, the nature of the problem, and the preferred algorithm. While feature selection can significantly improve model performance, it's important to validate the selected features' impact through cross-validation and to consider potential trade-offs between reduced dimensionality and the potential loss of information. In essence, effective feature selection is instrumental in optimizing model efficiency, interpretability, and predictive accuracy. Feature selection methods can be broadly categorized into three types: filter methods, wrapper methods, and embedded methods.

2.4.1 Filter Method

Filter methods for feature selection rely on measures of information, distance, or correlation, frequently utilizing statistical tests such as the Chi-square test, ANOVA test, or correlation coefficient. These tests assist in identifying features that exhibit a high correlation with the target variable. One significant advantage of filter methods is their independence from the biases of specific models, as they concentrate solely on the intrinsic properties of the data. This characteristic enables the same subset of features to be applied across various algorithms. Furthermore, filter methods are computationally efficient, especially when employing information or distance-based measures, rendering them suitable for the rapid generation of feature subsets. They are also adept at managing large datasets.

The variance filtering method is a commonly used feature selection technique. It primarily focuses on the variability of features. If a feature has a small variance, it indicates limited variation and contains relatively less information. Therefore, features with low variance are considered to have limited predictive ability and are often eliminated.

Correlation analysis is commonly employed to examine the degree and direction of interdependence between two or more random variables. The coefficient values can range from -1 to 1. A correlation coefficient of -1 describes a perfect negative, or inverse, correlation, with values in one series rising as those in the other decline, and vice versa. A coefficient of 1 shows a perfect positive correlation, or a direct relationship. A correlation coefficient of 0 means there is no linear relationship. When the absolute value of the correlation coefficient between two variables exceeds 0.7, they are usually considered to have strong multicollinearity. The inclusion of highly collinear variables in model training can lead to negative effects such as instability and degradation of accuracy. The Pearson correlation coefficient, Spearman rank correlation coefficient, and Kendall rank correlation coefficient are commonly used to measure this interdependence. Below, we introduce two classic correlation methods

Different types of correlation coefficients are used to assess correlation based on the properties of the compared data. By far, the most common is the Pearson coefficient r_{xy}, which is the covariance of the two variables divided by the product of their standard deviations.

Definition 2.1 Given paired data $\{(x_1, y_1), \ldots, (x_n, y_n)\}$ consisting of n pairs, r_{xy} is defined as

$$r_{xy} = \frac{\sum_{i=1}^{n}(x_i - \bar{x})(y_i - \bar{y})}{\sqrt{\sum_{i=1}^{n}(x_i - \bar{x})^2}\sqrt{\sum_{i=1}^{n}(y_i - \bar{y})^2}},$$

where n is sample size, x_i, y_i are the individual sample points indexed with i, $\bar{x} = \frac{1}{n}\sum_{i=1}^{n} x_i$ (the sample mean), and analogously for $\bar{y}$.

From the formula above, we can deduce that the Pearson correlation coefficient is the ratio of covariance to standard deviation, indicating certain requirements for the data. Firstly, experimental data is typically assumed to be paired samples originating from populations with a normal distribution. Why is such an assumption common? Because after calculating the Pearson correlation coefficient, methods like t-tests are often employed for its significance testing. However, t-tests rely on the assumption that the data follows a normal distribution. Secondly, substantial differences between experimental data should be avoided, as the Pearson correlation coefficient is sensitive to the influence of outliers. Consider the example of heart rate and running. If an individual's cardiac health is compromised, heart issues may arise during running, leading to abnormal heart rate measurements (excessively fast, slow, or even zero). Including such outliers in the correlation analysis can significantly distort the calculated results. In summary, when applying the Pearson correlation coefficient, attention must be paid to factors like data distribution and outliers to ensure the reliability and accuracy of the computed outcomes.

The Spearman correlation coefficient, often referred to as the Spearman rank correlation coefficient, does not impose any assumptions on the distribution of the original variables, making it a nonparametric statistic. It can also be calculated for data that follows the Pearson correlation coefficient, although its statistical power might be slightly lower. The term "rank" implies an order or ranking, and Spearman's coefficient is determined based on the ranked positions of the original data. This representation eliminates the constraints present when computing the Pearson correlation coefficient. Let's take a look at its calculation formula.

Definition 2.2 Assume d_i represents the difference between the ranks of paired observations, n is the number of observations, and the Spearman correlation coefficient ρ_S is defined as

$$\rho_S = 1 - \frac{6\sum_{i=1}^{n} d_i^2}{n(n^2-1)}.$$

The Spearman correlation coefficient is valuable for assessing the strength and direction of monotonic relationships between variables. This makes it especially advantageous when the relationship isn't linear or when data doesn't adhere to the normality assumptions required by the Pearson correlation coefficient. For example, in the analysis of biological experimental data, especially when studying datasets from diverse omics fields, the Spearman correlation coefficient finds frequent application.

Kendall rank correlation [24] is a nonparametric test that assesses the degree of association between two variables, relying on ordinal data and pair scores. Unlike the Pearson correlation coefficient, Kendall's tau doesn't assume linearity or require the data to follow a specific distribution. Instead, it focuses on the concordant and discordant pairs of observations' ranks, making it suitable for nonparametric data and cases where the exact numerical difference between values is less important than their ordinal relationships.

Definition 2.3 Kendall correlation (τ) is calculated as

$$\tau = \frac{2(N_c - N_d)}{n(n-1)}, \tag{2.7}$$

where N_c is the number of concordant, N_d is the number of discordant, and n is the number of observations.

In various domains, including biology, economics, and social sciences, Kendall's tau is employed to analyze data with a natural ranking order or when linear relationships might not hold. This is particularly beneficial for uncovering patterns and dependencies within such datasets, providing insights into how variables relate to each other without the constraints of parametric assumptions.

The partial correlation coefficient is a statistical measure used to assess the linear relationship between two variables while controlling for the influence of other correlated variables. It can be employed to analyze the relationship between two variables and eliminate the potential impact of other interfering variables. The calculation of the partial correlation coefficient is based on a multivariate linear regression model, and it quantifies the strength of the linear relationship between two variables given the conditions of other correlated variables.

Definition 2.4 For given random variables X,Y and control random variable $Z = \{Z_1, Z_2, \ldots, Z_n\}$, partial correlation coefficient $\rho_{XY \cdot Z}$ is defined as the linear correlation between two residuals, respectively, resulting from linear regression of X with Z and linear regression of Y with Z. Since the n-order partial correlation can be easily computed from $(n-1)$-order partial correlations, $\rho_{XY \cdot Z}$ can be computed with a dynamic programming method with a complexity of $O(n_3)$. Specially, when $n = 1$,

$$\rho_{XY \cdot Z} = \frac{\rho_{XY} - \rho_{XZ} \cdot \rho_{YZ}}{\sqrt{1 - \rho_{XZ}^2 \sqrt{1 - \rho_{YZ}^2}}}, \tag{2.8}$$

where ρ_{XY} means linear correlation coefficient.

In essence, the partial correlation coefficient reveals the correlation between two variables when the effects of other variables are held constant. This is especially useful when studying complex relationships in multivariate data, as it allows researchers to assess the direct association between two variables while accounting for potential confounding factors.

2.4.2 *Wrapper Method*

The wrapper method generates a new model for each feature subset and records its performance. The subset with the best performance is used for training and testing the final algorithm. Unlike filter methods, which use distance or information-based measures for feature selection, wrapper methods rely on the model's performance to determine the most significant attributes. The flowchart depicted in Fig. 2.9 illustrates this process. Common techniques in the wrapper approach include forward selection, backward elimination, and recursive feature elimination.

Forward selection is an iterative and greedy process that starts with an empty set of features. At each iteration, the most significant feature, determined by its correlation with the target variable, is added to the model. This process continues until no further improvement in model performance is observed. Below is a detailed explanation of the forward selection method.

1. Start with an empty feature set: $S = \emptyset$.
2. At each iteration, evaluate the performance of the model by adding one feature from the remaining set of features.

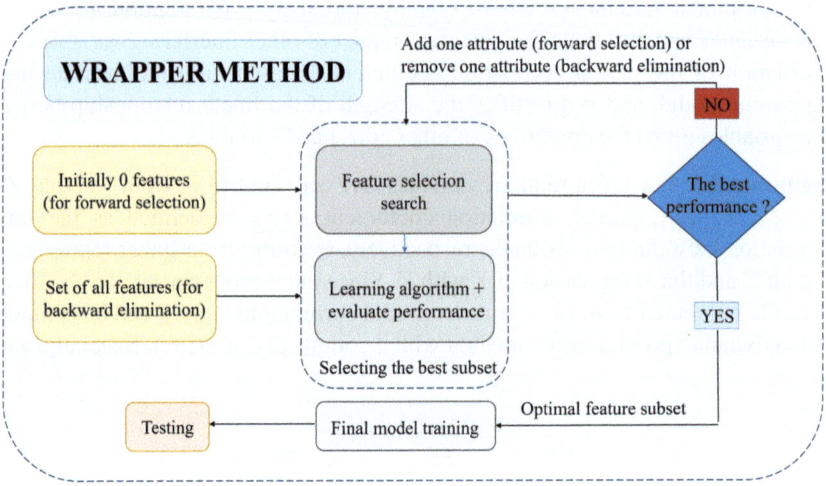

Fig. 2.9 Flowchart of the wrapper method

3. For each unselected feature $f \in F \setminus S$, compute the performance measure $\text{Perf}(M(S \cup \{f\}))$ of the model M with the feature set $S \cup \{f\}$.
4. Select the feature f^* that maximizes the performance measure and update the feature subset, where

$$f^* = \arg\max_{f \in F \setminus S} \text{Perf}(M(S \cup \{f\})),$$

$$S \leftarrow S \cup \{f^*\}.$$

5. If $\text{Perf}(M(S \cup \{f^*\})) \leq \text{Perf}(M(S))$, stop the process. Repeat the process until no further improvement in model performance is observed.

As the name suggests, backward elimination starts with all features present in the dataset. In each iteration, the least significant variable is removed. This process continues until further removal of features no longer improves the model's performance. Unlike filter methods that select features based on their individual correlation with the target variable, backward elimination removes features in order of increasing correlation. Here are the step-by-step processes.

1. Given a dataset X with features $x_1, x_2, \ldots, x_n$ and target variable y, and an initialized model M.
2. Train the model M on the dataset X, and calculate the initial performance metric $P = \text{Performance}(M(X, y))$.
3. Set a threshold δ for the minimum improvement in performance.

4. Repeat the following steps until further removal of features does not improve performance:

   a. For each feature x_i in X, train the model M on the dataset X excluding feature x_i (denoted as X_{-i}).
   b. Calculate the performance metric $P_{-i} = \text{Performance}(M(X_{-i}, y))$.
   c. Identify the feature x_{min} with the least impact on performance:

   $$x_{\text{min}} = \arg\max_{x_i}(P_{-i} - P).$$

   d. If $\max(P_{-i} - P) < \delta$, stop the iteration.
   e. Remove the feature x_{min} from the dataset X:

   $$X = X \setminus \{x_{\text{min}}\}.$$

   f. Update the performance metric $P = P_{-\text{min}}$.

5. Return the remaining features in X.

Recursive feature elimination (RFE): RFE is a feature selection technique that recursively removes the least important features from a model to improve its performance. The goal is to select a subset of features that contribute the most to predicting the target variable. Here is the step-by-step process for RFE:

1. Initialize the dataset X with features $x_1, x_2, \ldots, x_n$ and target variable y, and initialize the model M.
2. Train the model M on the dataset X, and calculate the importance score $\text{IMP}(x_i)$ for each feature x_i.
3. Identify the feature with the smallest importance score, remove it from the dataset X, and update X to $X \setminus \{x_{\text{min}}\}$, where $x_{\text{min}} = \arg\min_{x_i} \text{IMP}(x_i)$.
4. Repeat steps 2 and 3 until the desired number of features k is reached or another stopping criterion is met.
5. The final model includes the remaining features in X.

2.4.3 Embedded Method

The embedded method is also model-based; however, it conducts feature selection simultaneously with model training rather than iteratively constructing features and training the model. This method leverages machine learning models such as support vector machines and random forests to guide feature selection, based on specific regression or classification functions that define the selection criteria. The embedded method evaluates the contribution of features to the model during the selection process. The flowchart depicted in Fig. 2.10 illustrates this process. Consequently,

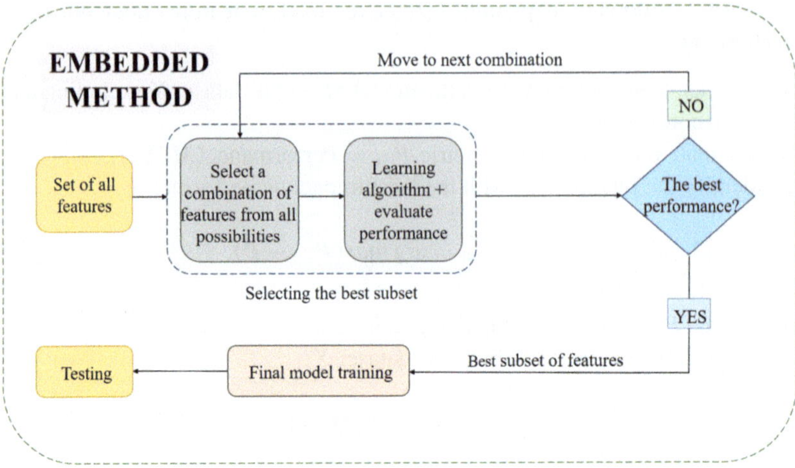

Fig. 2.10 Flowchart of the embedded method

it can enhance the model's performance more effectively compared to the filter method.

In feature engineering, the embedded method can be directly employed to handle features, circumventing the need for the filter method. By integrating feature selection within the model training process, the embedded method can reduce computation time and improve model performance, thus simplifying feature selection. Unlike the filter method, which requires analyzing multiple statistical measures, the embedded method offers distinct advantages. However, it can be time-consuming when the algorithm itself is complex. Therefore, when dealing with large datasets, the filter method should be prioritized.

Some well-known algorithms that utilize embedded feature selection include the following:

1. Least absolute shrinkage and selection operator (Lasso): This penalty promotes autonomous feature selection by encouraging tiny coefficients and driving specific coefficients to zero. Zero coefficient features are basically left out of the model.
2. Random forest: An ensemble learning technique builds several decision trees. Only some characteristics are considered at each split while building each tree. According to how much they help reduce impurity, the algorithm rates the value of features, enabling less significant information to be organically deemphasized.
3. Gradient boosting machines create a group of ineffective learners. GBM incrementally adds decision trees to enhance model performance. It allocates weights to features at each iteration depending on how well they helped minimize the loss function, implicitly favoring more significant features.
4. Elastic net: It is a linear regression method that combines the regularization terms L1 (Lasso) and L2 (Ridge). This combination strikes a compromise between the

feature sparsity of Lasso and the stability of Ridge by allowing for both feature selection and coefficient shrinkage.

Take $L1$ regularization problems using proximal gradient descent (PGD) as an example [25]. Let ∇ denote the differential operator. For the optimization objective,

$$\min_x f(x) + \lambda \|x\|_1,$$

if $f(x)$ is differentiable and ∇f satisfies the Lipschitz condition; i.e., there exists a constant $L > 0$ such that

$$\|\nabla f(x) - \nabla f(y)\| \leq L\|x - y\|,$$

for all x, y, then in the vicinity of x_k, $f(x)$ can be approximated using the second-order Taylor expansion as

$$\hat{f}(x) \approx f(x_k) + \langle \nabla f(x_k), x - x_k \rangle + \frac{L}{2}\|x - x_k\|^2 + \text{const},$$

where const is a constant independent of x and $\langle \cdot, \cdot \rangle$ denotes the inner product.

Clearly, the minimum value of the expression above is

$$x_{k+1} = x_k - \frac{1}{L}\nabla f(x_k).$$

Thus, minimizing $f(x)$ using gradient descent is equivalent to minimizing the quadratic function $\hat{f}(x)$ at each step. Extending this idea to the objective $\min_x f(x) + \lambda\|x\|_1$, each iteration should be

$$x_{k+1} = \arg\min_x \left\{ \frac{L}{2} \left\| x - \left(x_k - \frac{1}{L}\nabla f(x_k) \right) \right\|^2 + \lambda \|x\|_1 \right\}.$$

For each component x_i, this can be broken down into

$$x_{i,k+1} = \begin{cases} z^i - \lambda/L, & z^i > \lambda/L, \\ 0, & |z^i| \leq \lambda/L, \\ z^i + \lambda/L, & z^i < -\lambda/L, \end{cases}$$

where $z = x_k - \frac{1}{L}\nabla f(x_k)$ and $x_{i,k+1}$ and z^i represent the i-th components of x_{k+1} and z, respectively.

Using these steps, PGD method can effectively solve LASSO and other optimization problems based on $L1$ norm regularization.

2.5 The Issue of Imbalanced Data and Its Solutions

Data, algorithms, and computing power are often referred to as the three pillars of deep learning. The importance of data is self-evident, and the problem of imbalanced data samples is a major challenge in data processing. This section addresses the issue of imbalanced samples, providing solutions from both the data and algorithm perspectives.

In classical machine learning assumptions, it is often assumed that the number of samples in each class of the training set is balanced. However, in practical scenarios, training sample data is often imbalanced. For example, in a binary image classification problem, an extreme case could be a training set with 295 positive samples and only 5 negative samples. In such cases of imbalanced class data, failure to address the imbalance may result in the model exhibiting "under-learning" phenomena on the classes with fewer samples, potentially leading to a complete loss of predictive ability for the negative samples on the test set. In addition to common classification and regression tasks, imbalanced sample problems also exist in tasks such as image semantic segmentation and depth estimation, which involve pixel-level tasks.

There are generally two methods to address the problem of imbalanced samples: firstly, from the data perspective, including data sampling methods and class-balanced sampling methods, and secondly, from the algorithm perspective, involving cost-sensitive methods.

2.5.1 *Data-Level Approaches*

When dealing with imbalanced data, two main strategies are commonly employed at the data level: data augmentation and sampling methods. Data augmentation directly modifies the quantity and diversity of data samples, while sampling methods aim to balance the training set sent to the model, ensuring a more consistent number of samples for each class. Sampling methods at the data level primarily consist of two strategies:

1. Data resampling method: These occur during data preprocessing and alter the overall number and distribution of the training set.
2. Class-balanced sampling method: These occur during data loading (referring to loading into the model, not from disk) and use sampling strategies to ensure that the number of samples from different classes sent to the model for training is approximately balanced.

Data augmentation is an effective approach that balances the dataset by increasing sample quantity or changing sample features. These methods effectively increase the diversity of the dataset, improving the performance of the model on imbalanced datasets. Common data augmentation techniques include the following:

- Random flipping: Flip images horizontally or vertically to increase dataset diversity.
- Random rotation: Randomly rotate images by a certain angle to give the model rotational invariance.
- Random cropping: Randomly crop different regions from images to add spatial diversity to samples.
- Random scaling: Randomly scale the size of images to add scale variation to the data.
- Brightness, contrast, and saturation adjustment: Randomly adjust the brightness, contrast, and saturation of images to add color diversity to samples.
- Adding noise: Add random noise to images to make the model robust to noise.
- Color space transformation: Convert images from RGB space to other color spaces (e.g., HSV, LAB) to add color information diversity.
- Blurring: Apply blur to images to make the model more robust to changes in image details.

Besides, data-driven approaches are employed to balance class distribution in the input dataset by adjusting the class ratio. There are several strategies for data resampling (Fig. 2.11).

1. Over-sampling (up-sampling): This method involves increasing the number of samples in the minority class to match the majority class. It can be done through techniques like random over-sampling, SMOTE (synthetic minority over-sampling technique), or ADASYN (adaptive synthetic sampling).
2. Under-sampling (down-sampling): This method involves reducing the number of samples in the majority class to match the minority class. One approach is random under-sampling, where samples from the majority class are randomly removed. Another approach is to use more sophisticated methods like cluster-based over-sampling, ENN (edited nearest neighbors), or Tomek links to intelligently select which samples to remove.

For deep learning and computer vision tasks, under-sampling can be implemented by controlling the number of randomly selected images per batch during training. This ensures that each batch has a balanced number of positive and negative samples, even if the overall dataset is imbalanced. For example, in a binary classification task where the original data distribution has a batch-wise average ratio of positive to negative samples of 9:1, under-sampling could involve randomly selecting one positive sample for every nine positive samples in each batch. These sampling methods can be used individually or combined to achieve better balance in the dataset and improve the performance of machine learning models on imbalanced data [26].

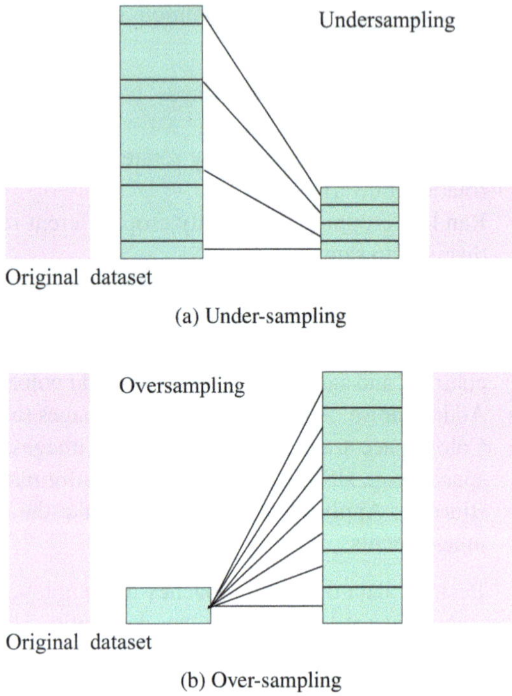

Fig. 2.11 Illustration of under-sampling and over-sampling

2.5.2 *Algorithmic (Loss Function) Level Approaches*

The essence of the class imbalance problem lies in the phenomenon of "under-learning" in machine learning, where classes with fewer samples are inadequately learned. This is often manifested by the relatively low weight assigned to the losses of the smaller class samples. A natural solution is to increase the penalty for misclassifying samples from the smaller class, directly incorporating this cost into the objective function (loss function), which is the essence of "cost-sensitive" methods. These methods essentially adjust the model's attention on the smaller classes.

One such method is the focal loss, which was introduced based on the binary cross-entropy (CE) loss function to address the severe imbalance between positive and negative samples in one-stage object detection. Focal loss reduces the weight of a large number of easy negative samples during training, effectively performing hard sample mining. It has been found effective in one-stage object detection, though its performance in multi-class classification is not guaranteed.

In focal loss, a modulating factor is introduced on top of the cross-entropy loss function, reducing the loss of high-confidence (easily classified) samples. The focal loss function is defined as follows:

$$\mathrm{FL}(p_t) = -\alpha_t(1 - p_t)^{\gamma} \log(p_t).$$

Here, p_t is the model's predicted probability for sample x_i, α_t is a sample weight adjustment factor, and γ is a tuning parameter, typically a positive value to adjust the weight of misclassified samples.

Focal loss provides two key properties. It reduces the weight of easy samples in the loss function, increasing the weight of incorrectly classified samples. Besides, the parameter γ smoothly adjusts the weight reduction rate of easy samples. When γ is small, focal loss behaves more like CE loss. As γ increases, the effect of the modulating factor also increases, with the best model performance often achieved at an optimal value of γ. In summary, the modulating factor in focal loss reduces the contribution of easy samples to the loss, focusing the model's attention on harder-to-learn samples.

In practice, we often use focal loss with a balanced form,

$$\mathrm{FL}(p_t) = -\alpha_t(1 - p_t)^{\gamma} \log(p_t).$$

Lin et al. [27] found in experiments that this form slightly improves accuracy compared to the non-balanced form. The experiments show that the best performance is achieved when γ is set to 2 and α is set to 0.25.

Here is a representative cost-sensitive learning algorithm, AdaCost [28]. This algorithm modifies the AdaBoost algorithm to increase the focus on minority classes, thereby improving their classification accuracy. The main steps are as follows. Assume N is the total number of samples.

1. Initialization: Initialize the weights of all samples.

$$w_i^{(1)} = \frac{1}{N}, \quad \text{for } i = 1, 2, \ldots, N.$$

2. Iteration: For each boosting round $t = 1, 2, \ldots, T$.

- Train a weak classifier $h_t(x)$ using the weighted training samples.
- Calculate the error rate ϵ_t,

$$\epsilon_t = \sum_{i=1}^{N} w_i^{(t)} \cdot I(y_i \neq h_t(x_i)),$$

where I is the indicator function that is 1 if $y_i \neq h_t(x_i)$ and 0 otherwise.

- Compute the cost-sensitive factor β_t,

$$\beta_t = \frac{\epsilon_t}{1 - \epsilon_t}.$$

- Update the weights for the next round,

$$w_i^{(t+1)} = w_i^{(t)} \cdot \beta_t^{(1-I(y_i=h_t(x_i)))\cdot C(y_i)},$$

where $C(y_i)$ is the cost associated with the true class y_i. Typically, higher costs are assigned to minority class samples to give them more weight.

3. Normalization: Normalize the weights, so that they sum to 1,

$$w_i^{(t+1)} = \frac{w_i^{(t+1)}}{\sum_{j=1}^{N} w_j^{(t+1)}}.$$

4. Final model: The final boosted classifier is a weighted majority vote of the weak classifiers

$$H(x) = \text{sign}\left(\sum_{t=1}^{T} \alpha_t h_t(x)\right)$$

where α_t is the weight assigned to each weak classifier, typically computed as $\alpha_t = \log\left(\frac{1}{\beta_t}\right)$.

2.6 Model Interpretation

Interpretation pertains to the analytical elucidation or presentation conveyed in simplified language. In the context of a model, interpretability denotes the model's capacity to be expressed in a comprehensible manner, accessible to human understanding. Specifically, it involves the capability to transform the predictive process of the model into rules defined by logical relationships. Interpretability is often subjective, varying among individuals, making it challenging to quantify uniformly. Our objective is to aspire to ML models that can articulate like a human and reason like a human. When the model's interpretation aligns with our cognitive processes and thought patterns, and it can distinctly expound the prediction journey from input to output, the model's interpretability is deemed satisfactory.

Depending on distinct usage scenarios and users, model interpretability can broadly categorized as intrinsically explainable and post hoc explainable. Intrinsic interpretability pertains to the inherent simplicity of the model's structure. Users can readily discern the internal framework of the model, and the model's outcomes carry an inherent interpretive effect. The model possesses interpretability as a core feature since its design. Post hoc interpretability involves the implementation of specific techniques to enhance the model's interpretability subsequent to its training. This method aims to extract insights acquired by the model. Certain models possess intricate structures that obscure the rationale behind their outcomes, making it challenging for users to deduce the internal reasoning process. The model's outputs lack explanatory language and typically only provide predicted values. In such instances, the model is not inherently interpretable. Post hoc interpretability involves refining the model's interpretability through diverse post-training analytical methods.

2.6.1 Feature Importance

The relevance or contribution of each feature or predictor variable in producing predictions is referred to as feature importance in ML. It aids in discovering the most important information for decision-making by helping us understand which attributes have the most effect on the model's outputs. Depending on the type of model and the nature of the data, there are several approaches for determining feature significance. Here are some prevalent approaches.

2.6.1.1 Feature Importance from Model Coefficients

The coefficients associated with each feature in linear models, such as linear regression or logistic regression, clearly represent their relevance: the higher the coefficient value, the greater the effect on the expected result.

The coefficients in a linear regression model play a critical role in assessing feature relevance. Each coefficient $beta_i$ is connected with a specific characteristic X_i and indicates the change in the target variable (dependent variable) caused by a one-unit change in the related feature, assuming all other features remain constant. For instance, if we have a linear regression model with features $X_1, X_2, X_3, \ldots, X_n$, and their corresponding coefficients are $\beta_1, \beta_2, \beta_3, \ldots, \beta_n$, the feature significance of X_i may be determined as follows:

$$\mathrm{FI}(X_i) = |\beta_i|.$$

By calculating the absolute value of the coefficient, we guarantee that the magnitude of the significance is examined regardless of whether it is positive or negative. A higher absolute value of β_i suggests that the related attribute X_i has a higher significance.

2.6.1.2 Feature Importance from Tree-Based Models

Tree models, such as DTs, RFs, and GBDTs, construct models by splitting features at various nodes. These models employ different methods to calculate feature importance, including Gini impurity reduction, information gain, and split gain.

Gini importance (reduction in Gini impurity) calculates the total reduction in Gini impurity for a feature used in splitting nodes in a decision tree. The Gini impurity measures the likelihood of an incorrect classification of a random sample if it were labeled according to the distribution of labels in the dataset. The formula for the

Gini impurity is

$$\text{Gini}(D) = 1 - \sum_{i=1}^{C} (p_i)^2.$$

where p_i is the proportion of samples belonging to class i in dataset D and C is the number of classes.

Similar to Gini importance, information gain measures the reduction in entropy when a dataset is split on a feature. Entropy is a measure of the disorder or uncertainty in the dataset. The formula for entropy is

$$H(D) = -\sum_{i=1}^{C} p_i \log_2(p_i)$$

where p_i is the proportion of samples belonging to class i in dataset D.

In gradient boosting trees (such as XGBoost, CatBoost, and LightGBM), feature importance is typically calculated by the split gain for each tree. The split gain is the improvement in the loss function (such as mean squared error for regression or log loss for classification) due to a split on a feature. The overall feature importance is the sum of split gains across all trees and splits where the feature is used.

Take CatBoost as an example, it calculates feature importance by summing the gains from all splits where a feature is used across all trees in the model. For a feature f_i, its importance score in CatBoost can be computed using the following formula:

$$\text{importance}(f_i) = \sum_{t=1}^{T} \sum_{s \in S_t(f_i)} \text{gain}(s),$$

where T is the total number of trees in the model, $S_t(f_i)$ is the set of all nodes in the t-th tree where feature f_i is used for splitting, and $\text{gain}(s)$ is the improvement in the loss function due to the split at node s using feature f_i.

The following pseudocode describes the steps to calculate feature importance in CatBoost.

2.6.1.3 Feature Importance from LASSO Regression Model

Least absolute shrinkage and selection operator (LASSO) is a regression technique that introduces penalties for large coefficient values, effectively selecting the most relevant features and setting others to zero. In LASSO regression, feature importance is determined through the regularization process that adds a penalty term to the linear regression objective function. The LASSO penalty encourages sparsity

Algorithm 2.3 Pseudocode for CatBoost model's feature importance

Load the dataset.
Initialize the CatBoost model with desired parameters.
Split the dataset into training and testing sets.
Fit the CatBoost model on the training set.
Initialize an array to store feature importance scores.
for each feature f_i in the dataset **do**
    Compute the importance score for the feature:

$$\text{importance}(f_i) = \sum_{t=1}^{T} \sum_{s \in S_t(f_i)} \text{gain}(s)$$

    where gain(s) is the gain of feature f_i at node s in tree t.
end for
Sort the features by their importance scores in descending order.
Return the sorted list of features and their importance scores.

in the model by pushing some of the coefficient values to exactly zero, effectively selecting only the most important features and setting others to zero.

The LASSO regression objective function is given by

$$\min \frac{1}{2n} \sum_{i=1}^{n} (y_i - \beta_0 - \sum_{j=1}^{p} x_{ij}\beta_j)^2 + \lambda \sum_{j=1}^{p} |\beta_j|,$$

where n is the number of samples in the dataset, p is the number of features (predictors) in the model, y_i is the target variable (response) for the i-th sample, x_{ij} is the value of the j-th feature for the i-th sample, β_0 is the intercept term of the linear regression model, β_j is the coefficient associated with the j-th feature, and λ is the regularization parameter that controls the strength of the penalty.

In LASSO regression, the regularization term $\lambda \sum_{j=1}^{p} |\beta_j|$ is in charge of feature selection. More coefficients are shrunk to zero as the value of λ grows, thereby removing the related features from the model. The LASSO model considers characteristics with nonzero coefficients after regularization to be the most relevant.

2.6.2 *Variance Inflation Factor*

The variance inflation factor (VIF) serves as a crucial metric for identifying multicollinearity within a multiple regression model. Multicollinearity arises when two or more independent variables exhibit high correlation, leading to challenges in interpreting the coefficients and potentially compromising the model's predictive accuracy.

The VIF quantifies the impact of collinearity on the variance of an estimated regression coefficient. Specifically, it assesses how much the variance "inflates" for each independent variable compared to what it would be if that variable were uncorrelated with the other predictors. A substantial VIF value for a specific independent variable indicates a strong collinear relationship with other variables, indicating redundant information shared with the rest of the predictors.

To calculate the VIF for a particular independent variable, say X_i, a linear regression model is fitted with X_i as the response variable, utilizing all other independent variables as predictors. The VIF for X_i is then computed as the reciprocal of the R^2 value obtained from this regression model. Mathematically, the VIF can be expressed as

$$\mathrm{VIF}(X_i) = \frac{1}{1 - R^2}.$$

In a multivariate regression model, the VIF is a useful statistic for detecting multicollinearity among independent variables. A VIF score of 1 denotes that there is no collinearity, since it suggests that correlations with other predictors are not inflating the variance of the calculated regression coefficient. While larger values show stronger correlations between the variables, VIF values greater than 1 suggest varying degrees of collinearity. A general rule of thumb is that a VIF greater than 5 or 10 for a particular variable should raise red flags, because it denotes a serious multicollinearity problem that could affect the accuracy of regression coefficient estimations and the performance of the model as a whole.

Next, we utilize the Boston Housing dataset provided by the scikit-learn library in Python to showcase the code that employs VIF for analyzing the mechanism of predicting median house prices. This dataset encompasses housing information from various neighborhoods in the Boston area, with each neighborhood characterized by 14 features that encompass a variety of factors influencing housing prices. These factors include crime rate, average number of rooms per dwelling, proportion of owner-occupied units, weighted distances to the five Boston employment centers, and more. The target variable is the median value of owner-occupied homes in thousands of dollars for the corresponding neighborhood. The simple Python code and the results are shown below.

The VIF results indicate that most features exhibit low to moderate multicollinearity, with VIF values generally below 5, such as per capita crime rate (CRIM) at 1.79, proportion of residential land zoned for lots over 25,000 sq. ft. (ZN) at 2.30, proportion of non-retail business acres per town (INDUS) at 3.99, and average number of rooms per dwelling (RM) at 1.93. However, some features show higher multicollinearity. Specifically, index of accessibility to radial highways (RAD) and full-value property tax rate per 10,000 (TAX) have notably high VIF values of 7.48 and 9.01, respectively, suggesting significant multicollinearity issues. The intercept (const) has an extremely high VIF of 585.27, which is typical and not concerning. Overall, while some predictors like nitric oxide concentration (NOX) at 4.39, proportion of owner-occupied units built before 1940 (AGE) at

Algorithm 2.4 Calculate VIF for each feature in the Boston dataset

```
import pandas as pd
import statsmodels.api as sm
from statsmodels.stats.outliers_influence import variance_inflation_factor
% Load a sample dataset
from sklearn.datasets import load_boston
boston = load_boston()
data = pd.DataFrame(boston.data, columns=boston.feature_names)
% Add the target variable to the dataset
data['PRICE'] = boston.target
% Calculate VIF for each feature
X = data.drop('PRICE', axis=1)
X = sm.add_constant(X)
% Add a constant term for the intercept
vif_data = pd.DataFrame()
vif_data["feature"] = X.columns
vif_data["VIF"] = [variance_inflation_factor(X.values, i) for i in range(X.shape[1])]
print(vif_data)
```

Table 2.1 VIF for each feature

Feature	VIF
const	585.265238
CRIM	1.792192
ZN	2.298758
INDUS	3.991596
CHAS	1.073995
NOX	4.393720
RM	1.933744
AGE	3.100826
DIS	3.955945
RAD	7.484496
TAX	9.008554
PTRATIO	1.799084
B	1.348521
LSTAT	2.941491

3.10, and weighted distances to five Boston employment centers (DIS) at 3.96 have moderate multicollinearity, RAD and TAX require closer attention due to their high multicollinearity, which could potentially impact the model's stability and interpretability (Table 2.1).

2.6.3 Partial Dependence Plot

A partial dependence plot (PDP) is a visualization technique used in machine learning to understand the relationship between a specific feature (variable) and the predicted outcome of a model. It shows how the predicted outcome changes as the value of the selected feature varies while keeping other features constant [29]. The partial dependence function for regression can be defined as

$$\hat{f}_S(x_S) = E_{X_c}[\hat{f}_S(x_S, X_c)] = \int \hat{f}(x_S, X_c) dP(X_c), \tag{2.9}$$

where x_S denotes variables that the partial dependence function needs to visualize and X_c represents the variables used by the ML model $\hat{f}$. Generally, x_S includes one or two variables in the set S.

The partial function $\hat{f}_S(x_S)$ is estimated by calculating averages in the training data, also known as Monte Carlo method:

$$\hat{f}_S(x_S) = \frac{1}{n} \sum_{i=1}^{n} \hat{f}(x_S, x_C^{(i)}).$$

The partial dependence function provides valuable insights into the average marginal effect on the prediction for specific values of features S. To calculate this function, we use $x_C^{(i)}$, which represents the actual feature values from the dataset for the features not of interest, and n, which is the total number of instances in the dataset. It is essential to note that the PDP relies on the crucial assumption that the features in C are independent of the features in S. Any violation of this assumption could lead to partial dependence plot averages that incorporate improbable or even impossible data points (see drawbacks).

In the context of classification tasks, where the ML model produces probabilities, the partial dependence plot allows us to observe the probability of a particular class given different values for feature(s) in S. Handling multiple classes can be achieved by drawing one line or plot for each class. The partial dependence plot is widely regarded as a global method, since it analyzes all instances in the dataset. By doing so, it provides comprehensive insights into the overall relationship between a feature and the predicted outcome of the ML model.

However, there are indeed certain limitations, especially when dealing with high-dimensional feature data.

- Limited feature representation: Partial dependence plots are indeed constrained in visualization, because the human visual system struggles to effectively process images beyond three dimensions. This complexity arises when attempting to simultaneously represent the relationships between multiple features and predictions in the plot. As the number of features in a high-dimensional dataset increases, the exploration of relationships between all features and predictions becomes time-consuming and impractical.
- Unaccounted feature correlations: When constructing partial dependence plots, it is often assumed that the target feature and other features are independent. This implies that the calculation of the partial dependence function does not consider correlations between features. However, real-world data often exhibits various

inter-feature relationships. When correlations exist between nontarget and target features, partial dependence plots might inaccurately reflect these relationships, leading to biases in model interpretation.

Interpreting a PDP involves analyzing the shape of the curve. Here are a few scenarios.

- Linear relationship: If the curve is relatively straight, it indicates a linear relationship between the feature and the prediction. Changes in the feature value lead to proportional changes in the prediction.
- Nonlinear relationship: If the curve is not straight, it suggests a nonlinear relationship. The shape of the curve helps understand how the prediction changes with varying feature values.
- Interaction effects: When including more than one feature in the PDP, interactions between features can be observed, for example, if the effect of one feature changes depending on the value of another feature.

A 2D PDP is an extension of the concept of PDP that can visualize the interaction between two features and the predicted outcome of a machine learning model. It provides a deeper understanding of how the predicted outcome changes when both selected features vary simultaneously while keeping other features constant. Interpreting a 2D PDP involves analyzing the shape and contours of the surface plot:

- Flat surface: If the surface is relatively flat, it suggests that the interaction between the two features has minimal impact on the prediction. The outcome is primarily influenced by the individual features.
- Curved surface: A curved or nonlinear surface indicates that the interaction between the two features significantly influences the prediction. The shape of the curve can provide insights into the nature of the interaction.
- Interaction patterns: Patterns in the contours or gradients of the surface plot reveal how changes in one feature impact the prediction differently based on the value of the other feature.

2.6.4 Accumulated Local Effect

Similar to partial dependence plots, the accumulated local effects (ALE) plot is also a model interpretation method, but it focuses on specific instances rather than overall trends. This graphical representation illustrates how the predicted value of the model changes with the variation of one or multiple features, given a particular sample. Specifically, the ALE achieves this by progressively altering the values of one or more features on the given sample and observing the corresponding changes in the predicted values, thus showcasing the localized impact of these features on the prediction outcome.

In the previous section, we introduce the mathematical foundation of PDPs, which is essentially the calculation of the marginal expectation of the target feature. Subsequently, we derive the equation for ALE.

The marginal plot involves fixing the target feature at a specific value and then integrating over the nontarget features (in the case of continuous variables) to calculate the conditional expectation of the target feature's impact on the prediction values. Therefore, Eq. (2.9) can be transformed as follows:

$$\begin{aligned}\hat{f}(x_S) &= E_{X_C|X_S}[\hat{f}(X_S, X_C)|X_S = x_S)],\\ &= \int_{x_c} \hat{f}(x_S, x_C) P(x_C|x_S) dx_c.\end{aligned} \tag{2.10}$$

To isolate the impact of the target feature, we perform a differentiation operation on the target feature. Suppose the target feature represents a continuous numerical variable, then integrate the Eq. (2.10), resulting in the following expression:

$$\begin{aligned}\hat{f}(x_S) &= \int_{x_S-\delta}^{x_S+\delta} E_{X_C|X_S}[\hat{f}(X_S, X_C)|X_S = x_S)]dx_S,\\ &= \int_{x_S-\delta}^{x_S+\delta} \int_{x_c} \hat{f}(x_S, x_C) P(x_C|x_S) dx_c dx_S.\end{aligned} \tag{2.11}$$

Centering assists in focusing attention on the relative impact of features on the predicted outcomes, without being affected by absolute values. Performing centering operations to make the average effect of the data 0, the ALE equation is as follows:

$$\begin{aligned}\hat{f}(x_S) &= \int_{x_S-\delta}^{x_S+\delta} E_{X_C|X_S}[\hat{f}(X_S, X_C)|X_S = x_S)]dx_S - \bar{f}(x_S),\\ &= \int_{x_S-\delta}^{x_S+\delta} \int_{x_c} \hat{f}(x_S, x_C) P(x_C|x_S) dx_c dx_S - \bar{f}(x_S),\end{aligned} \tag{2.12}$$

where $\bar{f}(x_S)$ is the mean of $\hat{f}(x_S)$.

Taking into account the calculations presented for ALE, we can summarize three advantages of ALE plots as follows. When estimating ALE values, the estimations remain unbiased, resulting in equally unbiased visual representations. Furthermore, due to ALE's capability to handle inter-feature relationships, the visuals generated remain unaffected by joint effects. Further, ALE computations outpace PDP in speed, necessitating fewer calculations than PDP. Besides, similar to PDP, ALE also intuitively illustrates how the target feature influences the model's predictions. Due to its isolation from the effects of correlated variables, ALE offers more precise interpretation. The ALE plot's curve is centered, with its values representing

Algorithm 2.5 ALE plot example with random forest

```
import pandas as pd
from sklearn.datasets import load_boston
from sklearn.model_selection import train_test_split
from sklearn.ensemble import RandomForestRegressor
import matplotlib.pyplot as plt
from alepython import ale_plot

% Load the dataset
boston = load_boston()
data = pd.DataFrame(boston.data, columns=boston.feature_names)
data['PRICE'] = boston.target

% Split the dataset
X = data.drop('PRICE', axis=1)
y = data['PRICE']
X_train, X_test, y_train, y_test = train_test_split(X, y, test_size=0.2, random_state=42)

% Train the model
model = RandomForestRegressor(n_estimators=100, random_state=42)
model.fit(X_train, y_train)

% ALE plot for a feature
ale_plot(model, X_train, 'RM')
plt.show()
```

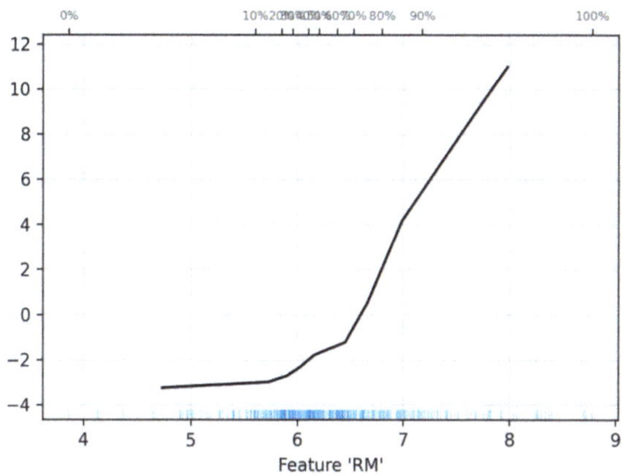

Fig. 2.12 The ALE plot for the "RM" feature (average number of rooms per dwelling) in the Boston Housing dataset

variations in average predicted values, contributing to a clearer and more concise explanation.

Next, we will continue to use the Boston Housing dataset to provide an example of ALE with the corresponding Python code and results (Fig. 2.12). The ALE plot for the "RM" feature (average number of rooms per dwelling) in the Boston Housing dataset indicates that the number of rooms has a significant positive impact on the predicted house prices. Initially, the effect of increasing the number of rooms is minimal, but after a certain point (around 6 rooms), the predicted house prices increase substantially with additional rooms. This suggests that houses with more rooms are valued higher. The density plot at the bottom shows that most data points are concentrated around five to seven rooms per dwelling, indicating that this is the common range in the dataset.

However, the ALE approach does have its limitations. For instance, determining intervals and deciding on an appropriate number of intervals present challenges that require further research and exploration. These challenges can complicate the practical application of ALE. ALE plots are unable to provide local explanations for individual samples. If users seek to understand whether changes in feature values for different samples yield similar changes in predictions, ALE cannot provide an answer.

2.6.5 Local Interpretable Model-Agnostic Explanation

Local interpretable model-agnostic explanation (LIME) is a method used to explain predictions made by machine learning models. Its objective is to provide local explanations for model predictions, specifically within the vicinity of a particular input. In comparison to global explanations, LIME focuses more on elucidating the behavior of the model for individual instances or within localized regions, resulting in explanations that are more easily comprehensible and interpretable. The LIME algorithm requires input of a sample to be explained and a pretrained complex model. Below is the complete set of algorithm steps.

1. Sampling around prediction. For continuous features, LIME generates a specific number of new values around the continuous feature values of the input sample using a standard normal distribution $N(0, 1)$. For categorical features, sampling is performed based on the distribution observed in the training set. For instance, if the input sample is $x^* = (x_1, x_2, \ldots, x_p)$ with p features, where the i-th feature is continuous and has a standard deviation δ_i across the training set, and its value in the sample is x_i, a new sample $(z_1, z_2, \ldots, z_p)$ is generated, where $z_i = a_i \times \delta_i + x_i$ around the sample using is a random number generated from a standard normal distribution $N(0, 1)$. This process is repeated N times to generate N new samples $z_k^*(k = 1, 2, \ldots, N)$.
2. Labeling new samples. The N new samples $z_k^*(k = 1, 2, \ldots, N)$ are passed through the pretrained complex model f to obtain N predicted outcomes. Then the new generated samples is $f(z_1^*, z_2^*, \ldots z_N^*)$.
3. Computing weights based on distances. Weights are assigned to the new generated samples based on their distances to the input sample. Samples that are closer to the input sample are assigned higher weights, as they better represent the input sample. An exponential kernel function is often used to define the weights, typically as

$$\prod_{z_k^*} = \exp \frac{-D(x^*, z_k^*)^2}{\delta^2},$$

where $D(x^*, z_k^*)^2$ represents the distance between the input sample X^* and the generated sample z_k^* and δ is a parameter controlling the spread of the weights. From the formula of this kernel function, we can observe that the closer the distance, the larger the value of $\prod_{z_k^*}$.

4. Feature selection followed by linear regression fitting. To obtain a simpler model for result interpretation, feature selection methods like Forward Selection and Lasso Path are often used to first select features, followed by specifying the exact number of features to retain. Assuming that only $p'(p' \leq p)$ ultimately features are used for interpretation, the selected features can be represented as $z' = (z_{(1)}, z_{(2)}, z_{(p')})$. Here, $z_{(1)}$ may not be equal to z_1. The second step involves using the selected p' features to fit a linear regression model. The assumed linear regression model used for interpretation is

$$g(z') = \omega_g,$$

where $\omega_g = (\omega_0, \omega_1, \ldots, \omega_p')$ represents the coefficients of the linear model. To calculate the coefficients of the linear model, LIME employs a weighted squared loss function, often written as

$$L(f, g, \prod_{x^*}) = \sum_{k=1}^{N} \prod_{x^*}(z_k^*)(f(z_k^*) - g(z_k'))^2,$$

where $f(x_i)$ is the complex model's prediction for the original input x_i, $g(z_i')$ is the prediction of the linear model using the selected features, and ω_i are the previously computed weights based on distances.

Next, we will continue to use the Boston Housing dataset to provide an example of LIME with the corresponding Python code and results (Fig. 2.13). The provided LIME explanation for a specific prediction in the Boston Housing dataset shows that the model predicts a house price of 14.80 (in thousands of dollars) for the given instance. The most significant negative contributors to the prediction are a high percentage of lower status population (LSTAT $>$ 16.37), which decreases the predicted value by 5.38, and fewer rooms per dwelling (RM $<=$ 5.89), which decreases it by 4.44. Higher crime rates (CRIM $>$ 3.20), a lower proportion of Black residents (B $<=$ 375.47), and higher pupil-teacher ratios (18.70 $<$

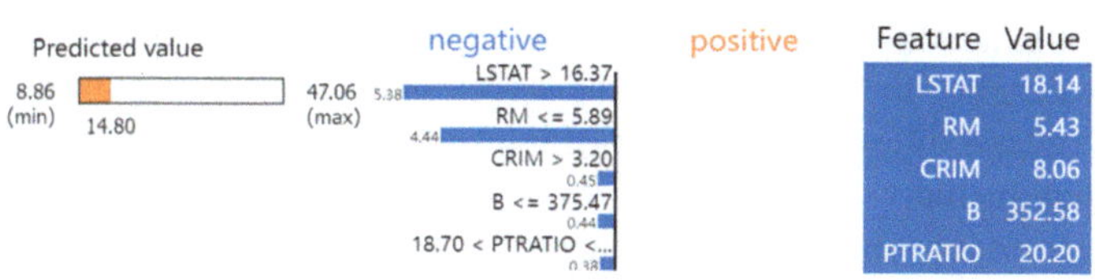

Fig. 2.13 LIME explanation for a specific prediction in the Boston Housing dataset

Algorithm 2.6 Using LIME to explain a prediction in the Boston Housing dataset

```
import lime
import lime.lime_tabular
import pandas as pd
from sklearn.datasets import load_boston
from sklearn.model_selection import train_test_split
from sklearn.ensemble import RandomForestRegressor

% Load the dataset
boston = load_boston()
data = pd.DataFrame(boston.data, columns=boston.feature_names)
data['PRICE'] = boston.target

% Split the dataset
X = data.drop('PRICE', axis=1)
y = data['PRICE']
X_train, X_test, y_train, y_test = train_test_split(X, y, test_size=0.2, random_state=42)

% Train the model
model = RandomForestRegressor(n_estimators=100, random_state=42)
model.fit(X_train, y_train)

% Initialize LIME explainer
explainer = lime.lime_tabular.LimeTabularExplainer(
    training_data=X_train.values,
    feature_names=X_train.columns,
    mode='regression'
)

% Explain a single prediction
i = 25 % Example instance index
exp = explainer.explain_instance(X_test.iloc[i].values, model.predict, num_features=5)
exp.show_in_notebook(show_all=False)
```

PTRATIO < 20.20) also negatively impact the prediction by smaller amounts. The actual feature values for this instance are LSTAT: 18.14, RM: 5.43, CRIM: 8.06, B: 352.58, and PTRATIO: 20.20.

A notable advantage of LIME lies in its model-agnostic nature, rendering it applicable to any type of model, including black box models. Moreover, it offers intuitive and easy-to-understand explanations that assist individuals in grasping the decision-making process of the model. However, LIME has certain limitations. Firstly, it exhibits explanation instability, which arises from the randomness introduced during the sampling of the target sample for explanation. When the model is refitted following resampling, slight variations in feature importance may emerge within individual samples. Secondly, the exclusive reliance on normal sampling overlooks the interdependencies between features, potentially leading to the generation of unrealistic samples. Lastly, the determination of appropriate kernel functions (weights) requires multiple iterations to ensure the reasonableness of the obtained explanations.

2.6.6 Shapley Additive Explanations

SHapley Additive exPlanations (SHAP) by Lundberg and Lee [30] is a tool for measuring variables' importance and understanding the global model structure based on combining local explanations of each prediction. It is inspired by the Shapley values from cooperative game theories [31]. In game theory, a single cooperation of multiple participants can lead to a single output. For a participant (Evelyn), Evelyn's marginal contribution is equal to the difference between the output of her participation in this cooperation and the output of her nonparticipation in this cooperation, the larger Evelyn's marginal contribution, the larger her impact on this cooperation. We can calculate the marginal contribution of each participant in such a way to measure the size of each participant's impact on the output of this cooperation.

The above game theory idea is introduced into SHAP here: The cooperation is treated as modeling, and the participants in the cooperation are treated as characteristics in the modeling. SHAP evaluates feature importance by coalitions of one or more features to predict a quantity of interest. It finds the difference in predictions between coalitions that do and do not include a given feature and then use such differences to quantify the magnitudes and directions of each feature's global and local contributions. For a specific sample, assume that it is characterized by three features (z_1, z_2, z_3), and the marginal contribution of z_1 is equal to the model prediction with z_1—the model prediction without z_1. Since there are many cases of feature combinations corresponding to whether z1 is included in the model or not (e.g., $f(z_1, z_2) - f(z_2)$, $f(z_1, z_3) - f(z_3)$, $f(z_1, z_2, z_3) - f(z_2, z_3)$), the marginal contribution results under each feature combination need to be considered. Therefore, we can first calculate the marginal contribution of z_1 under each feature combination case and the probability of occurrence of that combination, then calculate the expected value of marginal contribution under different feature combinations, and finally measure the importance of z1 comprehensively.

SHAP starts from an additive explanation model such as

$$g(z') = \phi_0 + \sum_{i=1}^{m} \phi_i z'_i,$$

where g is the explanation model, $z \in (0, 1)^m$, indicating whether the respective feature can be observed (denoted as 1) or not (denoted as 0), m is the number of predictor variables, z'_i is a coalition vector, $\phi_i \in \mathbb{R}$ is the importance value of the ith predictor, ϕ_0 is the prediction outcome without any predictor (i.e., the baseline outcome).

SHAP describes the following three desirable properties.

1. **Local accuracy**: The explanation model should match the original model. That is,

$$\hat{f}(z) = g(z') = \phi_0 + \sum_{i=1}^{m} \phi_i z'_i.$$

2. **Missingness**: If the simplified inputs represent feature presence, then missingness requires features missing in the original input to have no impact. That is,

$$z'_i = 0 \Rightarrow \phi_i = 0.$$

3. **Consistency**: Let $\hat{f}_x(z') = \hat{f}(h_x(z'))$ and z'_{-j} indicate that $z'_j = 0$. For any two models f and f' that satisfy

$$\hat{f}'_x(z') - \hat{f}'_x(z'_{-j}) \geq \hat{f}_x(z') - \hat{f}_x(z'_{-j}),$$

for all inputs $z' \in \{0, 1\}^M$, then

$$\phi_j(\hat{f}', x) \geq \phi_j(\hat{f}, x).$$

If a model changes so that the marginal contribution of a feature value increases or stays the same (regardless of other features), the Shapley value also increases or stays the same.

To calculate SHAP values, we define $f_x(S) = E[f(x)|x_S]$, where S is the set of possible subsets of input features (referred to as coalitions in Shapley values) and $E[f(x)|x_S]$ represents the conditional expectation of the subset S of input features. The following diagram explains how we derive. For any sample prediction, the Shapley value ϕ_i is

$$\phi_i = \sum_{S \subseteq m \setminus i} \frac{|S|!(M - |S| - 1)!}{M!} [f(S \cup i) - f(S)], \tag{2.13}$$

where $f(S \cup i)$ and $f(S)$ are the model outcomes with and without the ith predictor, and S is a subset of features from the model. Values of Φ_i that are greater (less) than zero refer to the positive (negative) effect of the variable i, which increases (decreases) the predicted value above (below) the base value.

Let's initiate the interpretation from a single sample and understand how Shapley Values influence the predictions of a complex model. Assume there are four feature variables in the dataset, denoted as $z = (z_1, z_2, z_3, z_4)$. After making predictions using the original model, we obtain the model's prediction value $f(z)$. Subsequently, we compute the average of all model prediction values within the

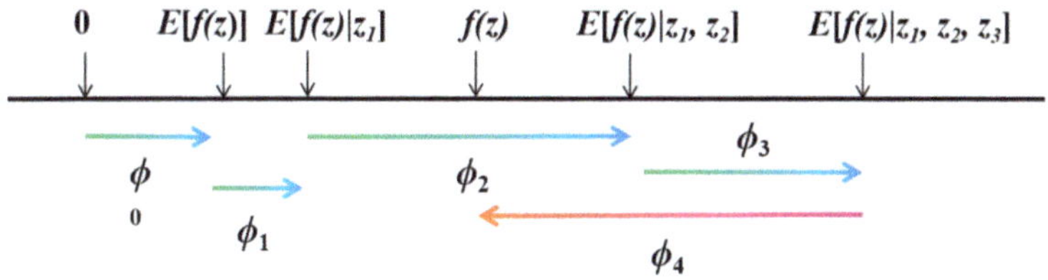

Fig. 2.14 SHAP breakdown plot

dataset, denoted as the expectation $E[f(z)]$, representing the baseline, i.e., ϕ_0. After calculating the corresponding Shapley Values p_1, p_2, p_3 and p_4, we can begin the explanation.

For the entire dataset, ϕ_0 remains fixed, representing the average of model predictions. It can be positive, zero, or negative, serving as a baseline value. Figure 2.14 employs Shapley values to explain the differences between the sample prediction and the average prediction. The four rightward arrows indicate positive Shapley values, implying that those features positively influence the original model's prediction. The leftward arrow represents a negative Shapley value, indicating a negative impact on the original model's prediction. In Fig. 2.14, since $\phi_2 > \phi_3 > \phi_1 > 0 > \phi_4$, features z_1, z_2, and z_3 have a positive impact on $f(x)$, with z_2 having the strongest influence. The presence of z_4 has a negative impact on $f(x)$. Ultimately, due to the combined effects of these four features, the original model's prediction for the sample x deviates from the average prediction and aligns with $f(x)$.

In addition to enhancing machine learning interpretability and explainability, SHAP values have versatile applications. Imagine we have a machine learning model that predicts algae blooms based on features such as pH, dissolved oxygen, water temperature, and nutrient levels.

1. Model debugging: Analyzing SHAP values helps pinpoint biases or outliers within the data that may lead to erroneous predictions. For example, if certain temperature or nutrient concentration readings consistently lead to incorrect predictions, SHAP values can highlight these issues, allowing for data correction or model adjustment.
2. Feature importance: Recognizing and eliminating low-impact features can lead to a more refined and efficient model. By using SHAP values, we can determine which environmental variables (e.g., pH, dissolved oxygen) have minimal impact on predicting algae blooms. Removing these features can streamline the model, reducing complexity and improving performance.
3. Anchoring explanations: SHAP values elucidate individual predictions by highlighting the pivotal features that influenced the outcome. For instance, if the model predicts an algae bloom, SHAP values can show that high nutrient levels and warm temperatures were the main factors, helping users, such as environmental scientists, understand and trust the model's decision-making process.

4. Model summaries: SHAP values generate a comprehensive overview of the model via a SHAP value summary plot, showcasing the most crucial features across the entire dataset. In the context of algae bloom prediction, this might reveal that nutrient levels, water temperature, and sunlight exposure are the top features influencing the model's predictions.
5. Bias detection: SHAP value analysis assists in identifying whether certain features disproportionately impact specific groups, enabling the identification and reduction of bias within the model. For example, if the model's predictions are biased toward certain geographic regions due to uneven data distribution, SHAP values can detect this, prompting a review of the data collection process.
6. Fairness audit: SHAP values serve as a tool to assess a model's fairness and ethical implications. Ensuring the model does not unfairly favor predictions for certain water bodies over others can be crucial for regulatory compliance and equitable resource management.
7. Regulatory approval: SHAP values facilitate regulatory approval by providing explanations for the model's decisions. In the context of algae bloom prediction, regulatory bodies may require transparent models that can explain why certain areas are flagged for potential blooms. SHAP values provide these explanations, supporting the model's credibility and transparency.

2.7 Application of Machine Learning

Before entering the realm of machine learning, the conventional approach to handling multivariate functions was numerical integration. In the realm of statistical physics, we often consider the ability to compute function integrals involving millions of variables as a given, yet we frequently overlook the incredible nature of this capability. This achievement owes itself to the development of Monte Carlo algorithms and variance reduction techniques over the course of many years [32]. Approximating high-dimensional functions poses an even more formidable task. The success of machine learning is far from trivial, and while neural network models were discovered long ago, it's only recently that we've come to fully appreciate their enormous potential in approximating multivariate functions. However, in a short span of time, we've witnessed significant advancements in long-standing issues with the assistance of machine learning.

Within the domain of environmental monitoring and climate change research, ML distinguishes itself through its adeptness in efficiently processing voluminous environmental sensor data. Furthermore, the progressively amplified deployment of ML within climate model analysis and weather forecasting further augments the precision of apprehending trends and reverberations linked to climate change. It adopts a data-driven, bottom-up approach to characterize the complex nonlinear relationships between variables in environmental systems.

In the realm of genomics and bioinformatics research, the widespread integration of ML assumes an instrumental role. It serves to fortify the identification of gene functionalities, anticipation of protein structures, and even the discovery of novel drug targets. Moreover, ML's proficiency is conspicuous in its scrutiny of extensive biological datasets encompassing gene expression profiles and protein-protein interaction networks. This profound analysis unveils concealed patterns and correlations that lie dormant within these intricate datasets.

In the arena of medical diagnosis and personalized therapy, ML assumes a pivotal mantle, demonstrating its utility across a spectrum that spans image analysis, clinical data mining, and genomic research, among other facets. This panorama introduces innovative pathways for heightening the precision of disease diagnosis, forecasting risks, and tailoring therapeutic protocols to individual needs. The ramifications of such progressions reverberate widely, bestowing a transformative impact upon patient outcomes and catalyzing innovation within the medical realm.

These paradigmatic applications distinctly underscore the diverse scope and potential of ML within the domains of ecology and life sciences. Through the judicious application of ML methodologies, researchers and professionals stand poised to glean a deeper appreciation of the nuanced intricacies underlying ecology and biology. In this pursuit, they extend invaluable contributions to environmental conservation, the anticipation of maladies, and the progression of therapeutic methodologies.

Chapter 3
Ecological Modeling with Nonlocal Delay

In nature, the survival and reproduction of populations are influenced by various factors such as habitat, food supply, and mating opportunities. When survival conditions become limited, animals actively disperse to broader habitats in search of additional resources and mating opportunities. Furthermore, the time-lagged activities during specific stages, such as hatching, maturation, and gestation, cause changes in population density to unfold gradually over time rather than occurring instantaneously. Therefore, incorporating environmental heterogeneity, individual mobility, and time delays into ecosystem dynamics models is both necessary and meaningful. This approach significantly enhances the accuracy and realism of the models.

Gourley and Britton proposed a nonlocal term grounded on the notion that intraspecific competition at a specific location depends not only on the population density at that point but also on the average population density in its vicinity [33]. For $t > 0$ and $x \in (0, l\pi)$, the spatiotemporal delay terms are defined by

$$(M_i * u)(x,t) = \int_{-\infty}^{t} \int_{0}^{l\pi} G(x, y, t-s) m_i(t-s) u(y, s) \mathrm{d}y \mathrm{d}s, \; i \in \mathbb{N}^+. \tag{3.1}$$

This type of delay term calculates the probability of an individual having been at point y at the earlier time s, given that it is at point x at the current time t. The function $m_i : [0, \infty) \to \mathbb{R}^+$ is the delay kernel, which weights the population t time units in the past and satisfies $m_i(t) \geq 0$ for all $t \geq 0$. The function $G : (0, l\pi) \times (0, l\pi) \times (0, \infty) \to \mathbb{R}^+$ is a probability density function, which is chosen from the appropriately normalized solution of the heat equation

$$\begin{cases} G_t = dG_{yy}, & y \in (0, l\pi), \\ G_y = 0, & y = 0, \; l\pi, \\ G(x, y, 0) = \delta(x-y), & x, y \in [0, l\pi]. \end{cases} \tag{3.2}$$

J. Ren, Y. Tao, *Modeling in Life Sciences and Ecology*, Springer Asia Pacific Mathematics Series 4, https://doi.org/10.1007/978-981-95-1038-2_3

These functions satisfy the normalization conditions

$$\int_0^{\infty} m_i(t)\mathrm{d}t = 1 \text{ and } \int_0^{l\pi} G(x, y, t)\mathrm{d}x = 1, \quad y \in (0, l\pi), \quad t \geq 0. \tag{3.3}$$

This chapter primarily introduces two population ecological dynamics systems that incorporate nonlocal time delays involving spatial factors into reaction-diffusion equations. This approach not only simulates the dynamic changes of populations in different spatial locations but also captures the time-delay effects caused by environmental changes or internal population behaviors. Consequently, it provides more reliable scientific evidence for ecological conservation and resource management.

3.1 Dynamics of a Diffusive Plankton Model with Nonlocal Delays

A primary aim within pelagic ecology is to comprehend the interactions between available nutrients (i.e., the base of essential resources), phytoplankton, and zooplankton. To facilitate understanding of these complex interactions, nutrient-phytoplankton-zooplankton (NPZ) models have been introduced. They constitute a subset of the more encompassing ecosystem models and function as the most rudimentary mathematical portrayal of a pelagic ecosystem.

It is known that nutrient recycling and matter decomposition take time. These delays result in a scenario where the development of the marine ecosystem cannot be entirely determined by its current state but also relates to the conditions over a preceding period. Concurrently, individual survival rules, which include predation behavior, active resource-seeking, and passive risk avoidance, inevitably lead to diffusion and migration activities that often alter the spatial structure of the marine ecosystem. For these reasons, the combined effects of spatial diffusion and time delay are essential in revealing the driving mechanisms and long-term evolution of marine pelagic ecosystems. Thus, this section will introduce an NPZ model with nonlocal delays mainly from [34].

3.1.1 NPZ Model with Nonlocal Delays

The compartments in the NPZ model correspond to the dissolved nutrient, phytoplankton abundance, and zooplankton abundance. The quantity of each compartment is determined by its concentration of the limiting biomass, typically nitrogen. Following the lead of Franks [35], Kloosterman et al. assumed that five basic processes govern the ecosystem [36]:

1. Phytoplankton nutrient uptake: Phytoplankton absorb dissolved nutrients from their environment to grow and reproduce. This uptake is critical for their survival and directly impacts their biomass concentration.
2. Zooplankton grazing on phytoplankton: Zooplankton feed on phytoplankton, transferring energy and nutrients up the food chain. This grazing pressure controls phytoplankton populations and influences nutrient cycling within the ecosystem.
3. Death of phytoplankton: Phytoplankton experience mortality due to various factors, including natural senescence, predation, and adverse environmental conditions. Their death contributes organic matter back to the nutrient pool.
4. Death of zooplankton: Zooplankton also undergo mortality, contributing to the detrital pool and subsequently to nutrient recycling. Their death is a key factor in the nutrient dynamics of the ecosystem.
5. Nutrient recycling: Nutrient recycling involves the conversion of organic matter from dead phytoplankton and zooplankton back into dissolved nutrients available for uptake by phytoplankton. This process is essential for maintaining nutrient availability in the ecosystem.

The recycling of nutrients does not happen immediately but occurs after an inevitable delay, as dictated by the distribution η. This distribution is nonnegative and normalized such that

$$\int_0^\infty \eta(u)\,du = 1.$$

This delay represents the time required for organic matter to decompose and nutrients to become available again in the dissolved form. The inclusion of this delay is crucial for accurately modeling the temporal dynamics of nutrient availability and ecosystem productivity.

The model in [36] was written as

$$\begin{cases} \dfrac{dN}{dt} = -\mu P f(N) + \delta \displaystyle\int_0^\infty Z(t-u)\eta(u)du + \lambda \int_0^\infty P(t-u)\eta(u)du \\ \qquad + (1-r)g \int_0^\infty Z(t-u)h(P(t-u))\eta(u)du, \\ \dfrac{dP}{dt} = \mu P f(N) - gZh(P) - \lambda P, \\ \dfrac{dZ}{dt} = rgZh(P) - \delta Z. \end{cases} \tag{3.4}$$

Here N, P, and Z denote the concentrations of nutrient, phytoplankton, and zooplankton. μ means the maximum birth rate of phytoplankton, λ denotes the death rate of phytoplankton, g determines the maximum grazing rate of zooplankton, r represents the assimilation efficiency of zooplankton, $r \in (0, 1]$ δ is the death rate of zooplankton, and τ is the mean delay. The function $f(N)$ represents the nutrient uptake function of phytoplankton, while $h(P)$ represents the grazing function of zooplankton. These functions are linked to the growth of phytoplankton and zooplankton, and thus they should be zero at zero and positive for other values.

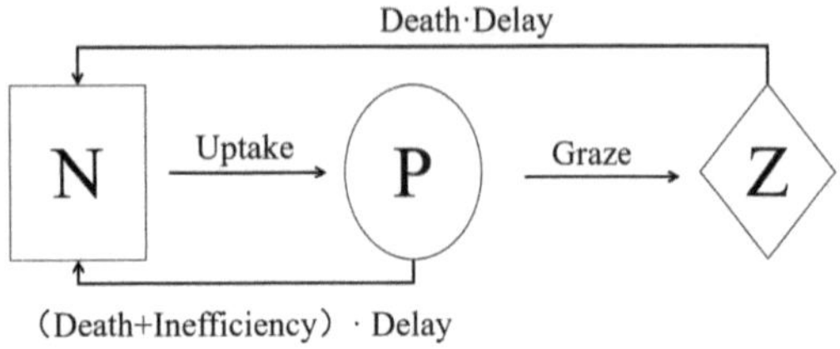

Fig. 3.1 Simplified schematic of model

Various modeling studies have suggested that these functions exhibit an increasing trend, implying that more nutrients lead to more significant growth. Additionally, the functions are bounded above, indicating a limit to the growth of organisms, known as saturation. Furthermore, the phytoplankton uptake function is typically assumed to be concave downward. This means that the nutrient uptake rate decreases as the concentration increases, reflecting a diminishing return of growth with higher nutrient levels. Considering these properties, they assumed that the functions f and h are twice continuously differentiable ($\mathbb{C}^2$) and satisfy

$$f(0) = 0, \quad f'(N) > 0, \quad f''(N) < 0, \quad \lim_{N\to\infty} f(N) = 1,$$
$$h(0) = 0, \quad h'(P) > 0, \quad \lim_{P\to\infty} h(P) = 1.$$

Taking into account the spatiotemporal delay in nutrient recycling, a model based on (3.4) was conducted, which incorporates both spatiotemporal delays described by (3.1) and Laplacian diffusion with a rate of d in a one-dimensional water column [34]. The simplified schematic of model is shown in Fig. 3.1. The column spans from $x = 0$ at the top to $x = l\pi$ at the bottom. Additionally, the model adheres to no flux (Neumann) boundary conditions, reflecting the closed nature of the domain. In order to maintain biological plausibility, it is crucial that the system exhibits positivity, with all state variables taking nonnegative values. To ensure this, careful consideration was given to selecting initial conditions that guarantee nonnegativity for each component. Thus, the resulting model can be expressed as follows:

$$\begin{cases} N_t(x,t) = dN_{xx} - \mu P(x,t) f(N(x,t)) + \delta(M_2 * Z)(x,t) + \lambda(M_1 * P)(x,t) \\ \qquad + (1-r)g(M_3 * Zh(P))(x,t), & t > 0, x \in (0, l\pi), \\ P_t(x,t) = dP_{xx} + \mu P(x,t) f(N(x,t)) - gZ(x,t)h(P(x,t)) \\ \qquad - \lambda P(x,t), & t > 0, x \in (0, l\pi), \\ Z_t(x,t) = dZ_{xx} + rgZ(x,t)h(P(x,t)) - \delta Z(x,t), & t > 0, x \in (0, l\pi), \\ \dfrac{\partial N}{\partial x} = \dfrac{\partial P}{\partial x} = \dfrac{\partial Z}{\partial x} = 0, & t > 0, x = 0, l\pi, \\ N(x,t) = N_0(x,t) \geq 0, & t \in (-\infty, 0], x \in (0, l\pi), \\ P(x,t) = P_0(x,t) \geq 0, & t \in (-\infty, 0], x \in (0, l\pi), \\ Z(x,t) = Z_0(x,t) \geq 0, & t \in (-\infty, 0], x \in (0, l\pi). \end{cases} \tag{3.5}$$

The spatiotemporal delay terms are defined by

$$\begin{aligned}(M_{1,2,3} * u)(x,t) &= \int_{-\infty}^{t}\int_{0}^{l\pi} M_{1,2,3}(x,y,t-s)u(y,s)\mathrm{d}y\mathrm{d}s \\ &= \int_{-\infty}^{t}\int_{0}^{l\pi} G(x,y,t-s)m_{1,2,3}(t-s)u(y,s)\mathrm{d}y\mathrm{d}s,\end{aligned} \tag{3.6}$$

where the functions $m_{1,2,3}(t)$ and $G(x,y,t)$ satisfy the normalization conditions (3.3).

Given that model (3.5) represents a closed system, there is neither biomass loss nor gain within the ecosystem. As a result, a conservation law can be derived by summing the equations in (3.5), integrating over the interval $(0, l\pi)$ with respect to x, and applying the Neumann boundary conditions. This yields the following expression

$$\begin{aligned}&\int_0^{l\pi} (P_t(x,t) + N_t(x,t) + Z_t(x,t))\mathrm{d}x \\ &= \frac{\mathrm{d}}{\mathrm{d}t}\Bigg[\lambda \int_0^{l\pi}\int_0^{\infty}\int_t^{t-\eta} M_1(x,y,\eta)P(x,\nu)\mathrm{d}\nu\mathrm{d}\eta\mathrm{d}x \\ &+ \delta \int_0^{l\pi}\int_0^{\infty}\int_t^{t-\eta} M_2(x,y,\eta)Z(x,\nu)\mathrm{d}\nu\mathrm{d}\eta\mathrm{d}x \\ &+ (1-r)g \int_0^{l\pi}\int_0^{\infty}\int_t^{t-\eta} M_3(x,y,\eta)Z(x,\nu)h(P(x,\nu))\mathrm{d}\nu\mathrm{d}\eta\mathrm{d}x\Bigg].\end{aligned} \tag{3.7}$$

The following expression for the total biomass in this ecosystem can be written as

$$\begin{aligned}\tilde{N}_T &= \int_0^{l\pi} (P(x,t) + N(x,t) + Z(x,t))dx \\ &+ \lambda \int_0^{l\pi}\int_0^{\infty}\int_{t-\eta}^{t} m_1(\eta)P(y,\upsilon)\mathrm{d}\upsilon\mathrm{d}\eta\mathrm{d}x \\ &+ \delta \int_0^{l\pi}\int_0^{\infty}\int_{t-\eta}^{t} m_2(\eta)Z(x,\upsilon)\mathrm{d}\upsilon\mathrm{d}\eta\mathrm{d}x \\ &+ (1-r)g \int_0^{l\pi}\int_0^{\infty}\int_{t-\eta}^{t} m_3(\eta)Z(x,\upsilon)h(P(x,\upsilon))\mathrm{d}\upsilon\mathrm{d}\eta\mathrm{d}x,\end{aligned} \tag{3.8}$$

which remains constant in time. Furthermore, if the model allows for a spatially uniform equilibrium solution (N^*, P^*, Z^*), the conservation law (3.8) can be simplified to the following expression

$$N_T = \lambda P^* T_1 + \delta Z^* T_2 + (1-r)gZ^*h(P^*)T_3 + P^* + N^* + Z^*, \tag{3.9}$$

where T_i is the mean delay, defined by

$$T_i = \int_0^\infty \eta m_i(\eta) d\eta, \quad i = 1, 2, 3.$$

There are three types of spatially uniform equilibrium solutions. The first type is the trivial equilibrium, denoted as

$$E_1 = (N_1, P_1, Z_1) = (N_T, 0, 0),$$

where $N_T \geq 0$. This equilibrium represents a scenario where there is no life in the ecosystem, and all the biomass is present in the nutrient.

The second type of equilibrium, denoted as

$$E_2 = (N_2, P_2, 0)$$

with $N_2 > 0$ and $P_2 > 0$, requires the condition $\mu > \lambda$. This condition ensures that f is an increasing function with $0 \leq f(N) < 1$. Biologically, it implies that the maximum growth rate of phytoplankton must surpass its death rate. When this condition is met,

$$N_2 = f^{-1}\left(\frac{\lambda}{\mu}\right),$$

and from the conservation law,

$$P_2 = \frac{N_T - N_2}{1 + \lambda T_1}.$$

The equilibrium E_2 exists if and only if $N_T > f^{-1}\left(\frac{\lambda}{\mu}\right)$.

The third type of equilibrium, represented by (N_3, P_3, Z_3) with $N_3 > 0$, $P_3 > 0$, and $Z_3 > 0$, requires the condition $rg > \delta$. This condition indicates that the product of the maximum growth rate of zooplankton and the assimilation efficiency must exceed its death rate. When this condition holds true,

$$P_3 = h^{-1}\left(\frac{\delta}{rg}\right), \ Z_3 = \frac{(\mu f(N_3) - \lambda) P_3 r}{\delta}.$$

The equilibrium E_3 satisfies an implicit expression for N_3 given by

$$N_3 = N_T - (\lambda T_1 + 1)P_3 + (\delta T_2 + 1)Z_3 + (1 - r)Z_3 \frac{\delta}{r} T_3. \tag{3.10}$$

It is important to note that this equilibrium point exists only if $Z_3 > 0$, which is determined by the condition $\mu f(N_3) - \lambda > 0$.

In summary, the spatially homogeneous equilibria are as follows:

$$E_1 = (N_T, 0, 0),$$

$$E_2 = \left(f^{-1}\left(\frac{\lambda}{\mu}\right), \frac{N_T - f^{-1}(\frac{\lambda}{\mu})}{1+\lambda T_1}, 0\right),$$

$$E_3 = \left(N_3, h^{-1}\left(\frac{\delta}{rg}\right), \frac{(\mu f(N_3) - \lambda) P_3 r}{\delta}\right),$$

where N_3 satisfies (3.10). Notably, due to the monotonicity of f and h, there exists a unique equilibrium of each type for any value of N_T that satisfies the conditions for equilibrium existence.

3.1.2 *Local Stability of a Generalized Reaction-Diffusion Problem*

In the forthcoming sections, the local asymptotic stability of a generalized reaction-diffusion model with spatiotemporal delays in a bounded domain will be analyzed under Neumann boundary conditions. The model's representation is given by the following equations:

$$\begin{cases} \dfrac{\partial v_1}{\partial t} = d\dfrac{\partial^2 v_1}{\partial x^2} + F_1(x, t, v_{1,2,3}, q, M * \bar{\mathbf{f}}(v_2, v_3)), & t > 0, x \in (0, l\pi), \\ \dfrac{\partial v_{2,3}}{\partial t} = d\dfrac{\partial^2 v_{2,3}}{\partial x^2} + F_{2,3}(x, t, v_{1,2,3}, q), & t > 0, x \in (0, l\pi), \\ \dfrac{\partial v_{1,2,3}}{\partial x} = 0, & t > 0, x = 0, l\pi, \\ v_{1,2,3}(x, t) = v_{1,2,3}(x) \geq 0, & t \in (-\infty, 0], x \in [0, l\pi], \end{cases} \tag{3.11}$$

where q represents a set of parameters. The functions F_1, F_2, and F_3 are continuously differentiable and $F(\theta) = 0$ for some θ. These functions may be nonlinear and depend on $v_{1,2,3}$ and $M * \bar{\mathbf{f}}(v_2, v_3)$. The term $M * \bar{\mathbf{f}}(v_2, v_3)$ represents spatiotemporal delays, as defined in (3.1). The model assumes the existence of a nonnegative spatially homogeneous equilibrium (v_1^*, v_2^*, v_3^*) that satisfies

$$F_1(v_1^*, v_2^*, v_3^*, q, M * \bar{\mathbf{f}}(v_2^*, v_3^*)) = 0,$$

$$F_2(v_1^*, v_2^*, v_3^*, q) = 0,$$

$$F_3(v_1^*, v_2^*, v_3^*, q) = 0.$$

By linearizing the system about the equilibrium (v_1^*, v_2^*, v_3^*), we obtain the following equations:

$$\begin{cases} \dfrac{\partial v_1}{\partial t} = d\dfrac{\partial^2 v_1}{\partial x^2} + \mathcal{A}_{11}(q)v_1 + \mathcal{A}_{12}(q, \mathbf{M}) * v_2 + \mathcal{A}_{13}(q, \mathbf{M}) * v_3, \\ \dfrac{\partial v_2}{\partial t} = d\dfrac{\partial^2 v_2}{\partial x^2} + \mathcal{A}_{21}(q)v_1 + \mathcal{A}_{22}(q)v_2 + \mathcal{A}_{23}(q)v_3, \\ \dfrac{\partial v_3}{\partial t} = d\dfrac{\partial^2 v_3}{\partial x^2} + \mathcal{A}_{31}(q)v_1 + \mathcal{A}_{32}(q)v_2 + \mathcal{A}_{33}(q)v_3, \end{cases} \tag{3.12}$$

where $\mathbf{M} = (M_1, \ldots, M_k)$ and $\mathcal{A}_{12}$ are a linear function of the M_i.

Solving the model (3.2), we obtain the expression for $G(x, y, t)$ as

$$G(x, y, t) = \frac{1}{l\pi} + \frac{2}{l\pi}\sum_{n=1}^{\infty} e^{-d\frac{n^2}{l^2}t} \cos\frac{n}{l}x \cos\frac{n}{l}y.$$

Under the homogeneous Neumann boundary conditions, the appropriate trial solution is given by

$$(v_1, v_2, v_3) = (c_1, c_2, c_3)e^{\phi t} \cos\frac{n}{l}x, \; n = 0, 1, 2, \ldots \tag{3.13}$$

Using this trial solution, we can express the term (3.1) as

$$(M_i * v)(x, t) = \int_{-\infty}^{t}\int_0^{l\pi} G(x, y, t-s)m_i(t-s)e^{\phi s} \cos\frac{n}{l}y\mathrm{d}y\mathrm{d}s.$$

After performing some algebraic manipulations, we find that

$$(M_i * v)(x, t) = \mathcal{M}_i(\phi + \frac{dn^2}{l^2})e^{\phi t} \cos nx.$$

Here, $\mathcal{M}_i$ represents the Laplace transform of m_i. It is worth noting that for any kernel m_i, we have

$$\mathcal{M}_i(iR + \frac{dn^2}{l^2}) = \int_0^{\infty} m_i(t)e^{-(iR+\frac{dn^2}{l^2})t}\mathrm{d}t = \mathcal{C}_i(R, n) - i\mathcal{S}_i(R, n), \tag{3.14}$$

where

$$\mathcal{C}_i(R, n) = \int_0^{\infty} m_i(t)e^{-\frac{dn^2}{l^2}t} \cos Rt\mathrm{d}t,$$

$$\mathcal{S}_i(R, n) = \int_0^{\infty} m_i(t)e^{-\frac{dn^2}{l^2}t} \sin Rt\mathrm{d}t.$$

For $i \in \mathbb{N}$ and $n \geq 0$, we can conclude from (3.14) that

$$|C_i(R,n)| \leq 1, \quad |S_i(R,n)| \leq 1,$$

with

$$C_i(0,0) = 1, \quad S_i(0,0) = 0,$$

Moreover, we have

$$\begin{aligned} |Si(R,n)| &\leq \int_0^{\infty} m_i(t) e^{-\frac{dn^2}{l^2}t} |\sin Rt| \mathrm{d}t, \\ &\leq \int_0^{\infty} m_i(t) e^{-\frac{dn^2}{l^2}t} Rt \mathrm{d}t, \\ &\leq R \int_0^{\infty} m_i(t) t \mathrm{d}t = RTi, \end{aligned} \tag{3.15}$$

and

$$\left. \frac{\mathrm{d}S_i(R,n)}{\mathrm{d}R} \right|_{R=0} = \int_0^{\infty} m_i(t) t e^{-\frac{dn^2}{l^2}t} \mathrm{d}t \leq T_i. \tag{3.16}$$

By substituting this trial solution into the linearized equations, we obtain a characteristic equation of the form $\Phi(\phi, n^2, q) = 0$, where ϕ represents the eigenvalue. By substituting (3.13) into (3.12), we can derive the characteristic equation for the linearized system. The characteristic equation is given by

$$\Phi(\phi, n^2, q) = -\phi^3 + \mathcal{B}_2(n^2, q)\phi^2 + \mathcal{B}_1(\phi, n^2, q)\phi + \mathcal{B}_0(\phi, n^2, q) = 0, \tag{3.17}$$

where the coefficients are defined as follows:

$$\mathcal{B}_2(n^2, q) = \mathcal{A}_{11} + \mathcal{A}_{22} + \mathcal{A}_{33} - 3\frac{dn^2}{l^2},$$

$$\begin{aligned} \mathcal{B}_1(\phi, n^2, q) = {} & \mathcal{A}_{12}(\mathcal{M})\mathcal{A}_{21} + \mathcal{A}_{13}(\mathcal{M})\mathcal{A}_{31} + \mathcal{A}_{23}\mathcal{A}_{32} + 2\frac{dn^2}{l^2}(\mathcal{A}_{11} + \mathcal{A}_{22} \\ & + \mathcal{A}_{33}) - \mathcal{A}_{22}\mathcal{A}_{33} - \mathcal{A}_{11}(\mathcal{A}_{22} + \mathcal{A}_{33}) - 3\frac{d^2n^4}{l^4}, \end{aligned}$$

$$\mathcal{B}_0(\phi, n^2, q) = \mathcal{A}_{12}(\mathcal{M})(\mathcal{A}_{23}\mathcal{A}_{31} - \mathcal{A}_{21}\mathcal{A}_{33} + \mathcal{A}_{21}\frac{dn^2}{l^2}) + \mathcal{A}_{13}(\mathcal{M})(\mathcal{A}_{21}\mathcal{A}_{32}$$

$$- \mathcal{A}_{22}\mathcal{A}_{31}) + (\mathcal{A}_{11} - \frac{dn^2}{l^2})((\mathcal{A}_{22} - \frac{dn^2}{l^2})(\mathcal{A}_{33} - \frac{dn^2}{l^2})$$
$$- \mathcal{A}_{32}\mathcal{A}_{23}).$$

Here, $\mathcal{B}_{0,1}$ depends on ϕ through $\mathcal{M}$.

The stability analysis of the equilibrium relies on the roots of the characteristic equation (3.17). A locally stable equilibrium is characterized by having all roots of the characteristic equation in the left half of the complex plane. To determine the number of roots in the right half of the complex plane, we can use Lemma 3.1, which is inspired by the work of Gourley and Bartuccelli [37]:

Lemma 3.1 *For any given equilibrium of (3.11), the number of roots of the characteristic equation (3.17) in the right half of the complex plane equals*

$$\lim_{R\to\infty}\left(\frac{3}{2} - \frac{1}{\pi}\arg\Phi(iR, n^2, q)\right). \tag{3.18}$$

Proof By a well-known result in complex variable theory, it is known that when the equation $\Phi(\phi, n^2, q) = 0$ has no roots on the imaginary axis, the number of roots of this equation in the right half of the complex plane is given by

$$\lim_{R\to\infty}\frac{1}{2\pi i}\int_{C(R)}\frac{\Phi'(\phi, n^2, q)}{\Phi(\phi, n^2, q)}\mathrm{d}\phi.$$

Here, prime denotes differentiation with respect to ϕ, and $C(R)$ is the contour $C(R_1)\bigcup C(R_2)$, traversed in the anticlockwise sense, where

$$C(R_1) = \{\phi = ik : k \in [-R, R]\},\ C(R_2) = \{\phi = Re^{i\theta} : \theta \in [-\frac{1}{2}\pi, \frac{1}{2}\pi]\}.$$

This contour is completely contained in the right half plane $\mathrm{Re}\phi \geq 0$. And the number of roots of $\Phi(\phi, n^2, q)$ inside $C(R)$ is

$$\begin{aligned}&\frac{1}{2\pi i}\int_{C(R)}\frac{\Phi'(\phi, n^2, q)}{\Phi(\phi, n^2, q)}\mathrm{d}\phi\\ &\quad = \frac{1}{2\pi i}(\int_{C(R_1)}\frac{\Phi'(\phi, n^2, q)}{\Phi(\phi, n^2, q)}\mathrm{d}\phi + \int_{C(R_2)}\frac{\Phi'(\phi, n^2, q)}{\Phi(\phi, n^2, q)}\mathrm{d}\phi).\end{aligned}$$

There are two steps to prove the (3.18). The first step is proof that

$$\frac{1}{2\pi i}\int_{C(R_1)}\frac{\Phi'(\phi, n^2, q)}{\Phi(\phi, n^2, q)}\mathrm{d}\phi = -\frac{1}{\pi}arg\Phi(iR, n^2, q). \tag{3.19}$$

The second one is proof that

$$\lim_{R\to\infty}\frac{1}{2\pi i}\int_{C(R_2)}\frac{\Phi'(\phi,n^2,q)}{\Phi(\phi,n^2,q)}\mathrm{d}\phi=\frac{3}{2}. \tag{3.20}$$

For point ϕ on contour $C(R_1)$, the complex conjugate of $\bar{m}$ is

$$\overline{\bar{m}(ik+dn^2)}=\int_0^\infty e^{-ikt}e^{dn^2t}m_i(t)\mathrm{d}t=\bar{m}(-ik+dn^2).$$

Thus,

$$\overline{\mathcal{B}_j(ik,n^2,q)}=\mathcal{B}_j(-ik,n^2,q),\qquad j=0,1,2. \tag{3.21}$$

It follows from (3.21),

$$\begin{aligned}\overline{\Phi(ik,n^2,q)}&=k^3i-\mathcal{B}_2(ik,n^2,q)k^2-\mathcal{B}_1(ik,n^2,q)ki+\mathcal{B}_0(ik,n^2,q)\\&=\Phi(-ik,n^2,q).\end{aligned}$$

That is $\overline{\Phi(\phi,n^2,q)}=\Phi(\bar{\phi},n^2,q)$ for any $\phi\in C(R_1)$. Based on these observations, it can be deduced that the roots of the equation $\Phi(\phi,n^2,q)=0$ manifest themselves in complex conjugate pairs. Specifically, within the region $C(R_1)$,

$$\rho e^{iarg(\Phi(-iR))}=\Phi(-iR)=\overline{\Phi(iR)}=\overline{\rho e^{iarg(\Phi(iR))}}=\rho e^{-iarg(\Phi(iR))},$$

where ρ is the module of Φ. Thus,

$$iarg(\Phi(-iR))=-iarg(\Phi(iR)).$$

The contour $C(R_1)$ is traced from $\phi=iR$ to $\phi=-iR$. Then using the argument principle of complex number,

$$\begin{aligned}\frac{1}{2\pi i}\int_{C(R_1)}\frac{\Phi'(\phi,n^2,q)}{\Phi(\phi,n^2,q)}\mathrm{d}\phi&=\frac{arg\Phi(-iR,n^2,q)-arg\Phi(iR,n^2,q)}{2\pi},\\&=-\frac{1}{\pi}arg\Phi(iR,n^2,q).\end{aligned}$$

This completes the proof of (3.19). Next, we prove (3.20).

For any point ϕ on the contour $C(R_2)$,

$$|\bar{m}(\phi+dn^2)|\le\int_0^\infty e^{-(\phi+dn^2)t}m(t)\mathrm{d}t\le\int_0^\infty m(t)\mathrm{d}t=1.$$

Also,

$$|\frac{\mathrm{d}\bar{m}(\phi+dn^2)}{\mathrm{d}\phi}| \le \int_0^{\infty} te^{-(\phi+dn^2)t}m(t)\mathrm{d}t \le \int_0^{\infty} tm(t)\mathrm{d}t = T_m.$$

Consequently, $\mathcal{B}_j(\phi, n^2, q)$ are bounded functions of ϕ for Re$\phi \ge 0$, with bounds K_j and

$$\mathcal{B}_j'(\phi, n^2, q) \le Q_j(n^2, q)T_f \triangleq K_j', \quad j = 0, 1, 2,$$

where $Q_j(q)$ are the parameter polynomials. Note that

$$\frac{1}{2\pi i}\int_{C(R_2)} \frac{2}{\phi}\mathrm{d}\phi = \frac{1}{2\pi i}\int_{-\frac{\pi}{2}}^{\frac{\pi}{2}} \frac{2}{Re^{i\theta}} iRe^{i\theta}\mathrm{d}\theta = 1.$$

Then

$$\begin{aligned}\frac{1}{2\pi i}\int_{C(R_2)} \frac{\Phi'(\phi, n^2, q)}{\Phi(\phi, n^2, q)}\mathrm{d}\phi - \frac{3}{2} &= \frac{1}{2\pi i}\int_{C(R_2)} (\frac{\Phi'(\phi, n^2, q)}{\Phi(\phi, n^2, q)} - \frac{3}{\phi})\mathrm{d}\phi, \\ &= \frac{1}{2\pi i}\int_{C(R_2)} (\frac{-\mathcal{B}_2\phi - 2\mathcal{B}_1 + \mathcal{B}_1'\phi + B_0'}{\Phi} - \frac{3\mathcal{B}_0}{\phi\Phi})\mathrm{d}\phi.\end{aligned}$$

In order to simplify notation, $\mathcal{B}_j(\phi, n^2, q)$ and $B_j'(\phi, n^2, q)$ are abbreviated as $\mathcal{B}_j$ and $\mathcal{B}_j'$, respectively. It follows from the bounds of $\mathcal{B}_j$ and $\mathcal{B}_j'$,

$$|\Phi(iR, n^2, q)| \ge R^3 - K_2R^2 - K_1R - K_0.$$

Then,

$$\begin{aligned}&\lim_{R\to\infty}\frac{1}{2\pi i}\int_{C(R_2)} |\frac{\Phi'(\phi, n^2, q)}{\Phi(\phi, n^2, q)}\mathrm{d}\phi - \frac{3}{2}| \\ &= \lim_{R\to\infty}\frac{1}{2\pi i}\int_{-\frac{\pi}{2}}^{\frac{\pi}{2}} |(\frac{-\mathcal{B}_2Re^{i\theta} - 2\mathcal{B}_1 + \mathcal{B}_1'Re^{i\theta} + B_0'}{\Phi} - \frac{3\mathcal{B}_0}{Re^{i\theta}\Phi})Re^{i\theta}|\mathrm{d}\theta, \\ &\le \lim_{R\to\infty}\frac{1}{2\pi i}\int_{-\frac{\pi}{2}}^{\frac{\pi}{2}} |(\frac{\mathcal{B}_2Re^{i\theta} + 2\mathcal{B}_1 + \mathcal{B}_1'Re^{i\theta} + B_0'}{\Phi} + \frac{3\mathcal{B}_0}{Re^{i\theta}\Phi})Re^{i\theta}|\mathrm{d}\theta, \\ &\le \lim_{R\to\infty}\frac{K_2R^2 + 2K_1R + K_1'R^2 + K_0'R + 3K_0}{2\pi(R^3 - K_2R^2 - K_1R - K_0)}\int_{-\frac{\pi}{2}}^{\frac{\pi}{2}} \mathrm{d}\theta\end{aligned}$$

$$= \lim_{R\to\infty} \frac{K_2R^2 + 2K_1R + K_1'R^2 + K_0'R + 3K_0}{2(R^3 - K_2R^2 - K_1R - K_0)} = 0.$$

Theorem 3.1 *All roots of the characteristic equation* (3.17) *have negative real part if* $\Phi(0, n^2, q) < 0$, $Im\Phi'(0, n^2, q) < 0$,, *and there exists a* $x_0 \in (0, \infty)$, *such that* $Re\Phi'(x_0, n^2, q) = 0$ *and* $Re\Phi(x_0, n^2, q) > 0$, *where the prime is the derivative with respect to R.*

Proof To get the local stability for the system (3.11), we need to prove that

$$\lim_{R\to\infty} arg\Phi(iR, n^2, q) = \frac{3\pi}{2}. \tag{3.22}$$

In view of (3.17), we have

$$\Phi(iR, n^2, q) = R^3 i - \mathcal{B}_2(n^2, q)R^2 + \mathcal{B}_1(iR, n^2, q)iR + \mathcal{B}_0(iR, n^2, q) = 0.$$

For large value of R,

$$\mathrm{Re}\Phi(iR, n^2, q) \sim -\mathcal{B}_2(n^2, q)R^2, \quad \mathrm{Im}\Phi(iR, n^2, q) \sim R^3.$$

Also, when $R = 0$,

$$\Phi(0, n^2, q) = \mathcal{B}_0(0, n^2, q).$$

The sign of $\Phi(0, n^2, q)$ determines whether the complex number $\Phi(iR, n^2, q)$ starts on the positive or negative real axis when $R = 0$. Similarly, the sign of $\mathcal{B}_2(n^2, q)$ determines whether $\Phi(iR, n^2, q)$ will end up in the first or second quadrant as R tends to infinity. Based on the assumption that $\Phi(0, n^2, q) < 0$, the complex number $\Phi(iR, n^2, q)$ can be classified into the following two cases:

(1) When $\mathcal{B}_2(n^2, q) < 0$, $\Phi(iR, n^2, q)$ starts on the negative real axis and ends up in the first quadrant.
(2) When $\mathcal{B}_2(n^2, q) > 0$, the existence of x_0 guarantees that $\mathrm{Re}(iR, n^2, q)$ changes its monotonicity at x_0, indicating that $\Phi(iR, n^2, q)$ starts on the negative real axis and ends up in the second quadrant. For the above two cases, we show some qualitative sketches of graphs that correspond to three different values of $\lim_{R\to\infty} \arg\Phi(iR, n^2, q)$ (Fig. 3.2).

If $\Phi(0, n^2, q) = 0$, Theorem 3.2 mentioned above does not apply, and Eq. (3.17) has 0 as one of its roots. The signs of the remaining roots can be determined as follows. Let us assume that 0 is a root of (3.17) with a first-order multiplicity. Then (3.17) can be expressed as

$$\Phi(\phi, n^2, q) = -\phi\hat{\Phi}(\phi, n^2, q) = -\phi(\phi^2 - \mathcal{B}_2(n^2, q)\phi - \hat{B}_1(\phi, n^2, q)),$$

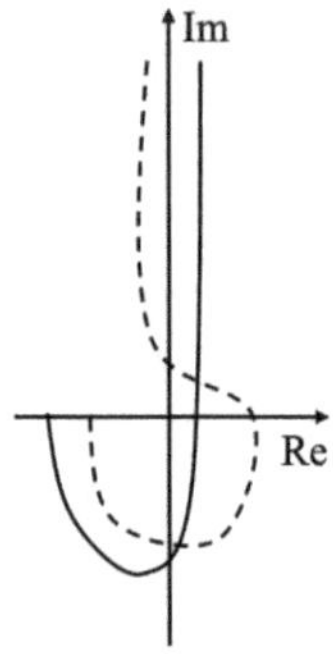

Fig. 3.2 Qualitative sketches for case I (solid line) and case II (dashed line)

where

$$\hat{B}_1(\phi, n^2, q) = \mathcal{B}_1(\phi, n^2, q) - \mathcal{B}_0(\phi, n^2, q)/\phi.$$

Notice that $\hat{B}_1(\phi, n^2, q)$ is analytic at 0 and $\hat{B}_1(0, n^2, q) \neq 0$. Thus the rest of the roots of (3.17) satisfy

$$\hat{\Phi}(\phi, n^2, q) = \phi^2 - \mathcal{B}_2(n^2, q)\phi - \hat{B}_1(\phi, n^2, q) = 0. \tag{3.23}$$

From [37], we have the following lemma.

Lemma 3.2 *The number of roots of the characteristic equation* (3.23) *in the right half complex plane equals*

$$\lim_{R\to\infty} \left(1 - \frac{1}{\pi}\arg\hat{\Phi}(iR, n^2, q)\right). \tag{3.24}$$

Theorem 3.2 *All the roots of* (3.23) *have negative real parts if*

$$\hat{\Phi}(0, n^2, q) > 0,\ \mathrm{Im}(\hat{\Phi}'(0, n^2, q)) > 0, \mathrm{Im}(\hat{\Phi}(iR, n^2, q)) \to \infty \text{ as } R \to \infty,$$

or

$$\hat{\Phi}(0, n^2, q) < 0, \mathrm{Im}(\hat{\Phi}'(0, n^2, q)) < 0,\ \mathrm{Im}(\hat{\Phi}(iR, n^2, q)) \to -\infty \text{ as } R \to \infty.$$

Proof Note that $\mathrm{Re}(\hat{\Phi}(iR, n^2, q)) \to -\infty$ as $R \to \infty$ and $\mathrm{Im}(\hat{\Phi}(iR, n^2, q)) \sim \mathcal{B}_2(n^2, q)R$ as $R \to \infty$. Hence, $\arg(\hat{\Phi}(iR, n^2, q))$ tends to either π or $-\pi$ as $R \to \infty$. The rest of the proof, which is similar to that of Theorem 3.1, shows that the given conditions guarantee

$$\lim_{R\to\infty} \arg\hat{\Phi}(iR, n^2, q) = \pi.$$

3.1.3 Local Stability of the NPZ Model

To investigate the linearized stability of (3.5), we set $N = N^* + u$, $P = P^* + v$, $Z = Z^* + w$, substitute into (3.5) and retain only linear terms in u, v and w, for all $x \in (0, l\pi)$, $t > 0$, model (3.5) can be rewritten as

$$\begin{cases} u_t(x,t) = du_{xx} - \mu P^* f'(N^*)u - \mu f(N^*)v \\ \qquad + \delta \int_{-\infty}^{t} \int_0^{l\pi} G(x,y,t-s) m_2(t-s) w(y,s) \mathrm{d}y \mathrm{d}s \\ \qquad + \lambda \int_{-\infty}^{t} \int_0^{l\pi} G(x,y,t-s) m_1(t-s) v(y,s) \mathrm{d}y \mathrm{d}s \\ \qquad + (1-r) g Z^* h'(P^*)) \int_{-\infty}^{t} \int_0^{l\pi} G(x,y,t-s) m_3(t-s) v(y,s) \mathrm{d}y \mathrm{d}s \\ \qquad + (1-r) g h(P^*)) \int_{-\infty}^{t} \int_0^{l\pi} G(x,y,t-s) m_3(t-s) w(y,s) \mathrm{d}y \mathrm{d}s, \\ v_t(x,t) = dv_{xx} + \mu P^* f'(N^*) u \\ \qquad + (\mu f(N^*) - g Z^* h'(P^*) - \lambda) v - g h(P^*) w, \\ w_t(x,t) = dw_{xx} + r g Z^* h'(P^*) v + (r g h(P^*) - \delta) w, \end{cases} \tag{3.25}$$

The Jacobian matrix of (3.5) at the equilibrium (N^*, P^*, Z^*) is

$$J = \begin{bmatrix} -b - \dfrac{dn^2}{l^2} & -\mu f(N^*) + \lambda \mathcal{M}_1 + (1-r)c\mathcal{M}_3 & \delta \mathcal{M}_2 + (1-r) g h(P^*) \mathcal{M}_3 \\ b & \mu f(N^*) - \lambda - c - \dfrac{dn^2}{l^2} & -g h(P^*) \\ 0 & r g h'(P^*) Z^* & r g h(P^*) - \delta - \dfrac{dn^2}{l^2} \end{bmatrix},$$

where $b = \mu P^* f'(N^*)$, $c = g h'(P^*) Z^*$. The characteristic equation is

$$\Phi(\phi, n^2, q) := -\phi^3 + \beta_2(n^2, q)\phi^2 + \beta_1(\phi, n^2, q)\phi + \beta_0(\phi, n^2, q) = 0, \tag{3.26}$$

where $\beta_{1,0}$ depends on $\mathcal{M}_{1,2,3}$.

To facilitate a comparison with the local stability of the closed NPZ model without diffusion, results from the previous literature are briefly reviewed [36].

Lemma 3.3 *Consider the model (3.5) with $d = 0$, and assume E_1 and E_2 exist. Then*

- *E_1 is stable if $N_T < f^{-1}\left(\frac{\lambda}{\mu}\right)$,*
- *E_2 is stable if $f^{-1}\left(\frac{\lambda}{\mu}\right) + \frac{2\lambda(1+T_1\lambda)}{\mu f'(N^*)} \leq N_T < f^{-1}\left(\frac{\lambda}{\mu}\right) + (1 + T_1\lambda)h^{-1}\left(\frac{\delta}{rg}\right)$.*

Furthermore, it has been demonstrated by Kloosterman et al. [36] that the local stability of E_3 can be achieved under a specific delay kernel $m_i(t) = \alpha e^{-\alpha t}$.

Lemma 3.4 *Consider the model (3.5) with $d = 0$, and assume E_3 exist, $m_i(t) = \alpha e^{-\alpha t}$, E_3 is locally stable if*

$$h'(P_3) \geq h(P_3)/P_3. \tag{3.27}$$

To provide a more comprehensive understanding, the local stability considering both the generalized delay function (3.1) and the absence of diffusion is investigated.

Lemma 3.5 *Consider the model (3.5) with $d = 0$, and assume E_3 exist; E_3 is locally stable if* (3.27) *is satisfied.*

Proof When $d = 0$, for the nontrivial equilibrium (N_3, P_3, Z_3), the Jacobian matrix is

$$J_0 = \begin{bmatrix} -G_{11} & G_{12} + \lambda\mathcal{M}_1(\phi) + G_{13}\mathcal{M}_3(\phi) & \delta\mathcal{M}_2(\phi) + G_{14}\mathcal{M}_3(\phi) \\ G_{11} & G_{21} & G_{22} \\ 0 & G_{31} & 0 \end{bmatrix},$$

with

$$G_{11} = \mu P_3 f'(N_3), G_{12} = -\mu f(N_3),$$

$$G_{13} = (1-r)gh'(P_3)Z_3, G_{14} = (1-r)gh(P_3),$$

$$G_{21} = \mu f(N_3) - \lambda - gh'(P_3)Z_3, G_{22} = -gh(P_3), G_{31} = rgh'(P_3)Z_3.$$

The characteristic equation of the linearization around (N_3, P_3, Z_3) is then

$$\Phi_0(\phi, 0, q) := -\phi^3 + \hat{Y}_2(0, q)\phi^2 + \hat{Y}_1(\phi, 0, q)\phi + \hat{Y}_0(\phi, 0, q) = 0, \tag{3.28}$$

where

$$\hat{Y}_2(0, q) = G_{21} - G_{11},$$

$$\hat{Y}_1(\phi, 0, q) = G_{11}(\lambda \mathcal{M}_1(\phi, 0) + G_{13}\mathcal{M}_3(\phi, 0) + G_{12} + G_{21}) + G_{22}G_{31} < 0,$$

$$\hat{Y}_0(\phi, 0, q) = G_{11}G_{31}(G_{22} + \delta \mathcal{M}_2(\phi, 0) + G_{14}\mathcal{M}_3(\phi, 0)).$$

Since $\Phi_0(0, 0, q) = \hat{Y}_0(0, 0, q) = G_{11}G_{31}(G_{22} + \delta + G_{14}) = 0$, we have that 0 is a root of (3.28), therefore, (3.28) can be written as

$$\Phi_0(\phi, 0, q) = \phi\hat{\Phi}_0(\phi, 0, q) = 0.$$

Notice that $\hat{\Phi}_0(0, 0, q) = \hat{Y}_1(0, 0, q) < 0$.

When $\phi = iR$, we have

$$\hat{\Phi}_0(iR, 0, q) = R^2 + \hat{Y}_2(0, q)Ri + \bar{Y}_1(iR, 0, q), \tag{3.29}$$

where $\bar{Y}_1(iR, n^2, q)) = \hat{Y}_1(iR, n^2, q) - \hat{Y}_0(iR, n^2, q)/iR$. Then, under the assumption (3.27), we have

$$\hat{Y}_2(0, q) = \frac{\delta Z_3}{rP_3}(1 - \frac{h'(P_3)}{h(P_3)}P_3) - \mu P_3 f'(N_3) < 0,$$

which implies that $\lim\limits_{R\to\infty} \operatorname{Im}(\hat{\Phi}_0(iR, n^2, q)) = -\infty$. Besides, it's clear that

$$\operatorname{Im}\Phi_0'(0, 0, q) < 0.$$

Thus, E_3 is locally stable based on Theorem 3.2.

In the following, the local stability of the equilibria of model (3.5) is considered when $d > 0$.

Theorem 3.3 *Assume that* $\mu f(N_T) - \lambda < 0$,*, then the trivial equilibrium* $E_1(N_T, 0, 0)$ *of (3.5) is locally stable.*

Proof For the trivial solution $(N_T, 0, 0)$, the Jacobian matrix is

$$J_1 = \begin{bmatrix} -\frac{dn^2}{l^2} & -\mu f(N_T) + \lambda \mathcal{M}_1(\phi) & \delta \mathcal{M}_2(\phi) \\ 0 & \mu f(N_T) - \lambda - \frac{dn^2}{l^2} & 0 \\ 0 & 0 & -\delta - \frac{dn^2}{l^2} \end{bmatrix}.$$

The characteristic equation of the linearization around $(N_T, 0, 0)$ is then

$$\Phi_1(\phi, n^2, q) := (-\frac{dn^2}{l^2} - \phi)(-\delta - \frac{dn^2}{l^2} - \phi)(\mu f(N_T) - \lambda - \frac{dn^2}{l^2} - \phi) = 0. \tag{3.30}$$

It is clear that the roots of (3.30) are

$$-\frac{dn^2}{l^2},\ -\delta - \frac{dn^2}{l^2},\ \mu f(N_T) - \lambda - \frac{dn^2}{l^2}.$$

Thus, under the conditions of the theorem, all roots are negative if $n > 0$. If $n = 0$, there is one zero root, and the rest are negative.

Theorem 3.4 *Assume that $\mu > \lambda$ and $N_T > f^{-1}\left(\frac{\lambda}{\mu}\right)$ are satisfied and*

$$rgh(P_2) - \delta < 0, \tag{3.31}$$

then the trivial equilibrium $E_2(N_2, P_2, 0)$ of (3.5) is locally stable.

Proof For the trivial equilibrium $(N_2, P_2, 0)$, the Jacobian matrix is

$$J_2 = \begin{bmatrix} -\mu P_2 f'(N_2) - \frac{dn^2}{l^2} & -\lambda + \lambda \mathcal{M}_1(\phi) & \delta \mathcal{M}_2(\phi) + (1-r)gh(P_2)\mathcal{M}_3(\phi) \\ \mu P_2 f'(N_2) & -\frac{dn^2}{l^2} & -gh(P_2) \\ 0 & 0 & rgh(P_2) - \delta - \frac{dn^2}{l^2} \end{bmatrix}.$$

The characteristic equation of the linearization around $E_2(N_2, P_2, 0)$ is then

$$\Phi_2(\phi, n^2, q) := -\phi^3 + C_2(n^2, q)\phi^2 + C_1(\phi, n^2, q)\phi + C_0(\phi, n^2, q) = 0, \tag{3.32}$$

where

$$C_2(n^2, q) = D_{33} - D_{11} - \frac{3dn^2}{l^2},$$

$$C_0(\phi, n^2, q) = (D_{33} - \frac{dn^2}{l^2})[\lambda D_{11}(1 - \mathcal{M}_1) + \frac{dn^2}{l^2}(D_{11} + \frac{dn^2}{l^2})],$$

$$C_1(\phi, n^2, q) = \lambda D_{11}(\mathcal{M}_1 - 1) + D_{11}D_{33} + 2(D_{33} - D_{11})\frac{dn^2}{l^2} - \frac{3d^2n^4}{l^4},$$

with

$$D_{11} = \mu P_2 f'(N_2),\ D_{13} = (1-r)gh(P_2),\ D_{23} = -gh(P_2),\ D_{33} = rgh(P_2) - \delta.$$

When the wave number n takes on the value of zero, the roots of Eq. (3.32) exhibit the characteristics outlined in Lemma 3.3. However, it is necessary to extend our analysis to the case where the wave number surpasses zero, specifically focusing on the scenario where $n > 0$.

By Eq. (3.14), we have $\mathcal{M}_1 = C_1(R) - \mathcal{S}_1(R)i$ and

$$\Phi_2(iR, n^2, q) = \mathrm{Re}\Phi_2(iR, n^2, q) + \mathrm{Im}\Phi_2(iR, n^2, q)i, \tag{3.33}$$

where

$$\begin{aligned}
\mathrm{Re}\Phi_2(iR, n^2, q) =& -(D_{33} - D_{11} - \frac{3dn^2}{l^2})R^2 + \lambda D_{11}\mathcal{S}_1(R)R + o(1),\\
\mathrm{Im}\Phi_2(iR, n^2, q) =& R^3 + (D_{11}(\lambda C_1(R) - \lambda + D_{33})\\
&+ 2(D_{33} - D_{11})\frac{dn^2}{l^2} - \frac{3d^2n^4}{l^4})R\\
&+ \lambda D_{11}(D_{33} - \frac{dn^2}{l^2})\mathcal{S}_1(R).
\end{aligned}$$

One easily concludes from (3.33) that, for large values of R,

$$\mathrm{Re}\Phi_2(iR, n^2, q) \sim constant\ R^2, \quad \mathrm{Im}\Phi_2(iR, n^2, q) \sim R^3.$$

Also, when $R = 0$,

$$\Phi_2(0, n^2, q) = (rgh(P_2) - \delta - \frac{dn^2}{l^2})(\mu P_2 f'(N_2)(\lambda - \lambda C(0) + \frac{dn^2}{l^2}) + \frac{d^2n^4}{l^4}).$$

Since $\lambda - \lambda C(0) \geq 0$, it is obvious that under the assumption (3.31),

$$\Phi_2(0, n^2, q) \leq \frac{dn^2}{l^2}(rgh(P_2) - \delta - \frac{dn^2}{l^2})(\mu P_2 f'(N_2) + \frac{dn^2}{l^2}) < 0.$$

In the view of (3.33),

$$\mathrm{Im}\Phi_2'(iR, n^2, q) = 3R^2 + W_{20}C_1'(R)R + W_{21}\mathcal{S}_1'(R) + W_{22},$$

where

$$\begin{aligned}
W_{20} &= \lambda D_{11},\ W_{21} = \lambda D_{11}(D_{33} - \frac{dn^2}{l^2}),\\
W_{22} &= D_{11}(\lambda C_1(R) - \lambda + D_{33}) + 2(D_{33} - D_{11})\frac{dn^2}{l^2} - \frac{3d^2n^4}{l^4}.
\end{aligned}$$

It is noteworthy that, under the assumption (3.31), D_{11} is greater than zero, while D_{33} is less than zero. Consequently, it follows that W_{20} is positive, whereas both W_{22} and W_{21} are negative. Then, for every wave number $n > 0$

$$\mathrm{Im}\Phi_2'(0, n^2, q) = W_{22} + W_{21}\mathcal{S}_1'(0) < W_{22} + W_{21}T_{f_1} < 0.$$

Further, it can be observed that

$$\lim_{R\to\infty} \mathrm{Im}\Phi_2'(iR, n^2, q) \to \infty, \lim_{R\to\infty} \mathrm{Re}\Phi_2(iR, n^2, q) \to \infty.$$

Therefore, based on Theorem 3.1 and Lemma 3.3, $E_2(N_2, P_2, 0)$ is locally stable.

Theorem 3.5 *Assume that* $rg > \delta$*,* $Z_3 > 0$*, and assume (3.27), then the nontrivial equilibrium* $E_3(N_3, P_3, Z_3)$ *of model (3.5) is locally stable.*

Proof For the nontrivial equilibrium $E_3(N_3, P_3, Z_3)$, the Jacobian matrix is

$$J_3 = \begin{bmatrix} -G_{11} - \frac{dn^2}{l^2} & G_{12} + \lambda\mathcal{M}_1(\phi) + G_{13}\mathcal{M}_3(\phi) & \delta\mathcal{M}_2(\phi) + G_{14}\mathcal{M}_3(\phi) \\ G_{11} & G_{21} - \frac{dn^2}{l^2} & G_{22} \\ 0 & G_{31} & -\frac{dn^2}{l^2} \end{bmatrix},$$

The characteristic equation of the linearization around (N_3, P_3, Z_3) is then

$$\Phi_3(\phi, n^2, q) := -\phi^3 + Y_2(n^2, q)\phi^2 + Y_1(\phi, n^2, q)\phi + Y_0(\phi, n^2, q) = 0, \tag{3.34}$$

where

$$\begin{aligned} Y_2(n^2, q) =& \hat{Y}_2(0, q) - 3\frac{dn^2}{l^2}, \\ Y_1(\phi, n^2, q) =& \hat{Y}_1(\phi, 0, q) + \frac{2dn^2}{l^2}\hat{Y}_2(0, q) - \frac{3d^2n^4}{l^4}, \\ Y_0(\phi, n^2, q) =& \hat{Y}_0(\phi, 0, q) + \frac{dn^2}{l^2}(G_{22}G_{31} + G_{11}(G_{12} + G_{21} + \lambda\mathcal{M}_1 + G_{13}\mathcal{M}_3)) \\ &+ \frac{d^2n^4}{l^4}\hat{Y}_2(0, q) - \frac{d^3n^6}{l^6}. \end{aligned}$$

For $n = 0$, the roots of (3.34) are as described in Lemma 3.5. Now consider $n > 0$. When $\phi = iR$, we have

$$\Phi_3(iR, n^2, q) = \mathrm{Re}\Phi_3(iR, n^2, q) + \mathrm{Im}\Phi_3(iR, n^2, q)i, \tag{3.35}$$

where

$$\begin{aligned}\mathrm{Re}\Phi_3(iR, n^2, q) &= -Y_2(n^2, q)R^2 + o(1)R + o(1),\\ \mathrm{Im}\Phi_3(iR, n^2, q) &= R^3 + W_{31}R + W_{30},\end{aligned}$$

with

$$\begin{aligned}W_{31} =& G_{22}G_{31} + \frac{2dn^2}{l^2}\hat{Y}_2(0, q) - \frac{3d^2n^4}{l^4} + G_{11}(G_{12} + G_{21}\\ &+ G_{13}C_3(R) + \lambda C_1(R)),\\ W_{30} =& -G_{11}(\frac{dn^2}{l^2}(\lambda S_1(R) + G_{13}S_3(R)) + G_{31}(\lambda S_2(R) + G_{14}S_3(R))).\end{aligned}$$

One easily concludes from (3.33) that, for large values of R,

$$\mathrm{Re}\Phi_3(iR, n^2, q) \sim constant\ R^2,\ \ \mathrm{Im}\Phi_3(iR, n^2, q) \sim R^3.$$

Also, when $R = 0$,

$$\begin{aligned}\Phi_3(0, n^2, q) =& G_{11}G_{31}(G_{22} + \delta C_2(R) + G_{14}C_3(R)) + \frac{dn^2}{l^2}(G_{22}G_{31} + G_{11}(G_{12}\\ &+ G_{21} + \lambda C_1(R) + G_{13}C_3)(R)) + \frac{d^2n^4}{l^4}\hat{Y}_2(0, q) - \frac{d^3n^6}{l^6}.\end{aligned}$$

By calculation, it can be deduced that

$$G_{12}+G_{21}+\lambda C_1(R)+G_{13}C_3(R) < 0,\ \ G_{22}+\delta C_2(R)+G_{14}C_3(R) \leq 0. \tag{3.36}$$

According to (3.36) and (3.27), $\Phi_3(0, n^2, q) < 0$. Besides, in the view of (3.35),

$$\mathrm{Im}\Phi_3'(iR, n^2, q) = 3R^2 + W_{40}R + W_{41} + W_{42},$$

where

$$\begin{aligned}W_{40} =& G_{11}(\lambda C_1'(R) + G_{13}C_3'(R)) > 0,\\ W_{41} =& -G_{11}(\frac{dn^2}{l^2}(\lambda S_1'(R) + G_{13}S_3'(R)) + G_{31}(\lambda S_2'(R) + G_{14}S_3'(R))) < 0,\\ W_{42} =& G_{22}G_{31} + \frac{2dn^2}{l^2}\hat{Y}_2(0, q) - \frac{3d^2n^4}{l^4} + G_{11}(G_{12} + G_{21} + G_{13}C_3(R)\\ &+ \lambda C_1(R)) < 0.\end{aligned}$$

Then, for wave number $n = 1, 2, \ldots$,

$$\mathrm{Im}\Phi_3'(0, n^2, q) = W_{42} + W_{41} < 0,$$

and

$$\lim_{R\to\infty} \mathrm{Im}\Phi_3'(iR, n^2, q) = \infty.$$

Besides,

$$\lim_{R\to\infty} \mathrm{Re}\Phi_3(iR, n^2, q) = \infty.$$

Based on Theorem 3.1 and Lemma 3.5, $E_3(N_3, P_3, Z_3)$ is locally stable.

3.1.4 Global Behavior of the NPZ Model

In this section, the global stability of the uniform equilibria in the system (3.5) is investigated. The focus is on the case where

$$m_i(t) = \frac{1}{\tau}e^{-\frac{t}{\tau}},\ i = 1, 2, 3.$$

By utilizing this specific delay kernel, the system (3.5) can be reformulated as a reaction-diffusion system without delay terms. This reformulation proves advantageous for conducting numerical simulations.

It is useful to define the following component, which can be thought of the detritus in the system

$$\begin{aligned} Q(x,t) = \int_{-\infty}^{t}\int_{0}^{l\pi} & G(x, y, t-s)e^{-\frac{t-s}{\tau}}(\lambda P(y, s) \\ & + \delta Z(y, s) + (1-r)gZ(y, s)h(P(y, s)))\mathrm{d}y\mathrm{d}s. \end{aligned}$$

It is then easy to see that system (3.5) is equivalent to the following system, for $t > 0, x \in (0, l\pi)$,

$$\begin{cases} N_t(x,t) = dN_{xx}(x,t) - \mu P(x,t)f(N(x,t)) + \dfrac{1}{\tau}Q(x,t), \\ P_t(x,t) = dP_{xx}(x,t) + \mu P(x,t)f(N(x,t)) - gZ(x,t)h(P(x,t)) - \lambda P(x,t), \\ Z_t(x,t) = dZ_{xx}(x,t) + rgZ(x,t)h(P(x,t)) - \delta Z(x,t), \\ Q_t(x,t) = dQ_{xx}(x,t) + \lambda P(x,t) + \delta Z(x,t) + (1-r)gZ(x,t)h(P(x,t) \\ \qquad\qquad - \dfrac{1}{\tau}Q(x,t), \end{cases} \tag{3.37}$$

with homogeneous Neumann boundary conditions and the initial conditions for $t \in [-\infty, 0]$, $x \in [0, l\pi]$,

$$N(x,0) = N_0(x) \geq 0,\ P(x,0) = P_0(x) \geq 0,\ Z(x,0) = Z_0(x) > 0,$$

$$Q(x,0) = \int_{-\infty}^{0}\int_{0}^{l\pi} G(x,y,-s)e^{\frac{s}{\tau}}(\lambda P(y,s) + \delta Z(y,s) + (1-r)gZ(y,s)h(P(y,s)))\mathrm{d}y\mathrm{d}s \geq 0.$$

Adding the equations in (3.37) and integrating x on the interval $(0, l\pi)$ yield

$$\int_0^{l\pi}(P_t(x,t) + N_t(x,t) + Z_t(x,t) + Q_t(x,t))\mathrm{d}x = \frac{\partial}{\partial t}\int_0^{l\pi}(P(x,t) + N(x,t) + Z(x,t) + Q(x,t))\mathrm{d}x = 0,$$

which leads to the conservation law

$$\int_0^{l\pi}(P(x,t)+N(x,t) + Z(x,t) + Q(x,t))\mathrm{d}x = \tilde{N}_T. \tag{3.38}$$

Thus, for a spatially homogeneous equilibrium solution, (N^*, P^*, Z^*, Q^*), the conservation law (3.38) can be written as

$$N_T = \frac{\tilde{N}_T}{l\pi} = P^* + N^* + Z^* + Q^*. \tag{3.39}$$

When $d = 0.002$, the conservation laws are examined under various spatial conditions in Fig. 3.3.

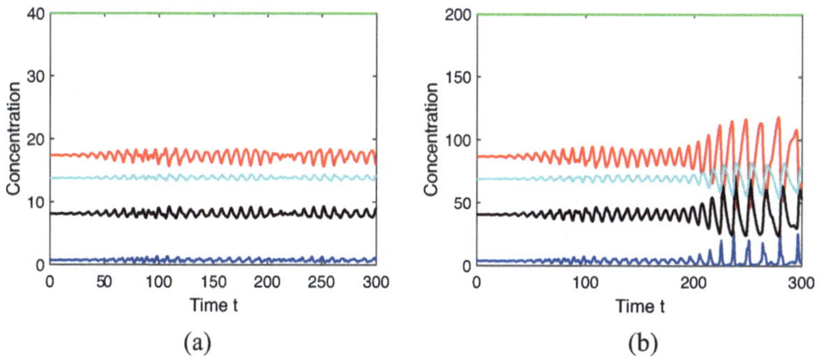

Fig. 3.3 The conservation laws of model (3.37) in $[0, l\pi]$ and $[0, 5l\pi]$

Using the conservation law (3.39), the equilibria of (3.37) can be determined as follows:

$$E_1^*(N_1^*, P_1^*, Z_1^*, Q_1^*) = (N_T, 0, 0, 0),$$

$$E_2^*(N_2^*, P_2^*, Z_2^*, Q_2^*) = (f^{-1}(\frac{\lambda}{\mu}), \frac{N_T - N_2^*}{1+\lambda\tau}, 0, \lambda\tau P_2^*),$$

$$E_3^*(N_3^*, P_3^*, Z_3^*, Q_3^*) = (N_3^*, h^{-1}(\frac{\delta}{rg}), \frac{(\mu f(N_3^*) - \lambda)P_3^* r}{\delta}, \tau\mu P_3^* f(N_3^*)),$$

where N_3^* satisfies the implicit expression

$$N_T = N_3^* + P_3^* + Z_3^* + \tau\mu P_3^* f(N_3^*). \tag{3.40}$$

It should be noted that E_2^* exists if and only if conditions $\mu > \lambda$ and $N_T > f^{-1}(\frac{\lambda}{\mu})$ hold, while E_3^* exists if and only if condition $rg > \delta$ is satisfied.

To establish the existence of a classical solution to (3.37), the method of upper and lower solutions for mixed quasi-monotone functions are employed. First, let's review the definitions of mixed monotonicity and upper/lower solutions [38]. The initial-boundary value problem for the reaction-diffusion system is given by

$$\begin{cases} \frac{\partial u_i}{\partial t} + L_i u_i = f_i(x, t, u_1, \ldots, u_N), & (x,t) \in \Omega \times (0, T], \\ B_i u_i = g_i(x, t), & (x, t) \in (0, T] \times \partial\Omega, \\ u_i(0, x) = u_i(x) & x \in \Omega. \end{cases} \tag{3.41}$$

where L_i is a uniformly elliptic operator on Ω and

$$L_i = -\sum_{i,j=1}^{n} a_{ij}(x)\frac{\partial^2}{\partial x_i y_i} + \sum_{i,j=1}^{n} b_i(x)\frac{\partial}{\partial x_i},$$

Definition 3.1 Let $\mathbf{f} = (f_1, \ldots, f_N)$ be a vector function. It is said that mixed quasimonotonicity is exhibited by $\mathbf{f}$ if, for each i, there exist nonnegative integers α_i and β_i satisfying $\alpha_i + \beta_i = N - 1$, such that $f_i(\cdot, u_i, [u]\alpha_i, [u]\beta_i)$ is monotonically nondecreasing with respect to each component in $[u]\alpha_i$, and monotonically nonincreasing with respect to each component in $[u]\beta_i$.

Definition 3.2 Let $\mathbf{f} = (f_1, \ldots, f_N)$ be a vector function with quasi-monotonicity. Then, we call $\tilde{u} = (\tilde{u}_1, \ldots, \tilde{u}_N) \in C^{1,0}((0, T] \times \overline{\Omega}) \cap C^{2,1}((0, T] \times \overline{\Omega})$ and $u = (\hat{u}_1, \ldots, \hat{u}_N) \in C^{1,0}((0, T] \times \overline{\Omega}) \cap C^{2,1}((0, T] \times \overline{\Omega})$ a pair of upper and lower solutions of (3.41) if $\tilde{u} \geq \hat{u} > 0$ and

$$\begin{cases} \mathcal{L}_i \tilde{u}_i \geq f_i(t, x, \tilde{u}_i, [\tilde{u}]_{\alpha_i}, [\hat{u}]_{\beta_i}), & (x, t) \in (0, T] \times \overline{\Omega}, \\ \mathcal{L}_i \hat{u}_i \leq f_i(t, x, \hat{u}_i, [\hat{u}]_{\alpha_i}, [\tilde{u}]_{\beta_i}), & (x, t) \in (0, T] \times \overline{\Omega}, \\ B_i(\tilde{u}_i) \geq 0 \geq B_i(\hat{u}_i), & (x, t) \in (0, T] \times \partial\Omega, \\ \tilde{u}_i(0, x) \geq u_i(x) \geq \hat{u}_i(0, x), & x \in \Omega. \end{cases}$$

Now, we focus on the global behavior of model (3.37).

Theorem 3.6 *Model (3.37) has a unique solution* $(N(x,t), P(x,t), Z(x,t), Q(x,t))$ *such that for* $t \in (0, \infty)$ *and* $x \in [0, l\pi]$, $0 < N(x,t) < \tilde{N}_T$, $0 < P(x,t) < \tilde{N}_T$, $0 < Z(x,t) < \tilde{N}_T$, $0 < Q(x,t) < \tilde{N}_T$.

Proof Define

$$g_1(x,t) = \frac{1}{\tau} Q - \mu P f(N), g_2(x,t) = \mu P f(N) - g Z h(P) - \lambda P,$$

$$g_3(x,t) = r g Z h(P) - \delta Z, g_4(x,t) = \lambda P + \delta Z + (1-r) g Z h(P) - \frac{1}{\tau} Q.$$

Since, in $\overline{\mathbb{R}_+^4} = \{N \geq 0, P \geq 0, Z \geq 0, Q \geq 0\}$,

$$\frac{\partial g_1}{\partial P} = -\mu f(N) < 0, \quad \frac{\partial g_1}{\partial Z} = 0, \quad \frac{\partial g_1}{\partial Q} = \frac{1}{\tau} > 0,$$

$$\frac{\partial g_2}{\partial N} = \mu P f'(N) > 0, \quad \frac{\partial g_2}{\partial Z} = -g h(P) < 0, \quad \frac{\partial g_2}{\partial Q} = 0,$$

$$\frac{\partial g_3}{\partial N} = 0, \quad \frac{\partial g_3}{\partial P} = r g Z h'(P) > 0, \quad \frac{\partial g_3}{\partial Q} = 0,$$

$$\frac{\partial g_4}{\partial N} = 0, \quad \frac{\partial g_4}{\partial P} = \lambda + (1-r) g Z h'(P) > 0, \quad \frac{\partial g_4}{\partial Z} = \delta + (1-r) g h(P) > 0,$$

and then (3.37) is a mixed quasi-monotone system. Let

$$(\hat{N}(x,t), \hat{P}(x,t), \hat{Z}(x,t), \hat{Q}(x,t)) = (0, 0, 0, 0)$$

and

$$(\tilde{N}(x,t), \tilde{P}(x,t), \tilde{Z}(x,t), \tilde{Q}(x,t)) = (\tilde{N}^*(t), \tilde{P}^*(t), \tilde{Z}^*(t), \tilde{Q}^*(t)),$$

where $(\tilde{N}^*(t), \tilde{P}^*(t), \tilde{Z}^*(t), \tilde{Q}^*(t))$ is the unique solution of

$$\begin{cases} \dfrac{\mathrm{d}N}{\mathrm{d}t} = \dfrac{1}{\tau}Q, \\ \dfrac{\mathrm{d}P}{\mathrm{d}t} = (\mu f(N) - \lambda)P, \\ \dfrac{\mathrm{d}Z}{\mathrm{d}t} = (rgh(P) - \delta)Z, \\ \dfrac{\mathrm{d}Q}{\mathrm{d}t} = \delta Z + \lambda P + (1-r)gZh(P) - \dfrac{1}{\tau}Q, \\ N(0) = \sup\limits_{[0,l\pi]} N_0(x),\ P(0) = \sup\limits_{[0,l\pi]} P_0(x), \\ Z(0) = \sup\limits_{[0,l\pi]} Z_0(x),\ Q(0) = \sup\limits_{[0,l\pi]} Q_0(x). \end{cases} \tag{3.42}$$

Then $(\hat{N}(x,t), \hat{P}(x,t), \hat{Z}(x,t), \hat{Q}(x,t))$ and $(\tilde{N}(x,t), \tilde{P}(x,t), \tilde{Z}(x,t), \tilde{Q}(x,t))$ are the lower solution and upper solution to (3.37), since

$$\begin{cases} \dfrac{\partial \tilde{N}(x,t)}{\partial t} - \tilde{N}_{xx} - g_1(\tilde{N}(x,t), \hat{P}(x,t), \hat{Z}(x,t), \tilde{Q}(x,t)) = \mu f(N^*(t)) > 0, \\ \dfrac{\partial \tilde{P}(x,t)}{\partial t} - \tilde{P}_{xx} - g_2(\tilde{N}(x,t), \tilde{P}(x,t), \hat{Z}(x,t), \hat{Q}(x,t)) \\ \qquad = gZ^*(t)h(P^*(t)) > 0, \\ \dfrac{\partial \tilde{Z}(x,t)}{\partial t} - \tilde{Z}_{xx} - g_3(\hat{N}(x,t), \tilde{P}(x,t), \tilde{Z}(x,t), \hat{Q}(x,t)) = 0, \\ \dfrac{\partial \tilde{Q}(x,t)}{\partial t} - \tilde{Q}_{xx} - g_4(\hat{N}(x,t), \tilde{P}(x,t), \tilde{Z}(x,t), \tilde{Q}(x,t)) = 0, \end{cases}$$

and

$$\begin{cases} \dfrac{\partial \hat{N}(x,t)}{\partial t} - d\hat{N}_{xx} - g_1(\hat{N}(x,t), \tilde{P}(x,t), \tilde{Z}(x,t), \hat{Q}(x,t)) = 0, \\ \dfrac{\partial \hat{P}(x,t)}{\partial t} - d\hat{P}_{xx} - g_2(\hat{N}(x,t), \hat{P}(x,t), \tilde{Z}(x,t), \tilde{Q}(x,t)) = 0, \\ \dfrac{\partial \hat{Z}(x,t)}{\partial t} - d\hat{Z}_{xx} - g_3(\tilde{N}(x,t), \hat{P}(x,t), \hat{Z}(x,t), \tilde{Q}(x,t)) = 0, \\ \dfrac{\partial \hat{Q}(x,t)}{\partial t} - d\hat{Q}_{xx} - g_4(\tilde{N}(x,t), \hat{P}(x,t), \hat{Z}(x,t), \hat{Q}(x,t)) = 0. \end{cases}$$

Also, the boundary conditions and the initial conditions are satisfied. For any (N_a, P_a, Z_a, Q_a) and (N_b, P_b, Z_b, Q_b) in X_T, it is clear that

$$|g_i(N_a, P_a, Z_a, Q_a) - g_i(N_b, P_b, Z_b, Q_b)|$$
$$\leq K_1(|N_a - N_b| + |P_a - P_b| + |Z_a - Z_b| + |Q_a - Q_b|),$$

where $K_i = K_i(x, t), i = 1, 2, 3, 4$. Therefore, Lemma 2.1 in [38] shows that (3.37) has a unique globally defined solution $(N(x, t), P(x, t), Z(x, t), Q(x, t))$, which satisfies

$$0 \leq N(x, t) \leq \tilde{N}^*(t), 0 \leq P(x, t) \leq \tilde{P}^*(t),$$

$$0 \leq Z(x, t) \leq \tilde{Z}^*(t), 0 \leq Q(x, t) \leq \tilde{Q}^*(t), \ t \geq 0.$$

The application of the strong maximum principle implies that $N(x, t)$, $P(x, t)$, $Z(x, t)$, and $Q(x, t)$ are all greater than zero when $t > 0$ for all $x \in [0, l\pi]$. Moreover, based on the positivity of solutions, the conservation law (3.38) reveals that $N(x, t)$, $P(x, t)$, $Z(x, t)$, and $Q(x, t)$ are bounded above by the total biomass, denoted as $\tilde{N}_T$.

Remark 3.1 Consider the case where the delay kernel in model (3.5) are changed to higher order gamma distributions:

$$m_i(t) = \frac{t^{k-1}}{(k-1)!\tau^k} e^{-\frac{t}{\tau}},$$

with $k > 1$ an integer. The model then becomes (3.37) with the additional equations

$$Q_{j,t}(x, t) = dQ_{j,xx}(x, t) + \frac{1}{\tau}(Q_{j-1}(x, t) - Q_j(x, t)), \ j = 1, \ldots, k-1.$$

The proof of Theorem 3.6 can be easily extended to this case.

Conditions for the global stability of E_1^* are now presented.

Theorem 3.7 $E_1^* = (N_T, 0, 0, 0)$ *is global stable if*

$$\mu f(N_T) - \lambda < 0. \tag{3.43}$$

Proof First, it is shown that under the assumption (3.43), the inequality $rgh(P(x, t)) - \delta < 0$ holds true. By employing the system (3.37) and referring to Theorem 3.6, it can be observed that $P(x, t)$ satisfies the following equation:

$$\begin{cases} P_t(x, t) < dP_{xx} + P(\mu f(N_T) - \lambda), & t > 0, x \in (0, l\pi), \\ \dfrac{\partial P}{\partial x} = 0, & t > 0, \ \ x = 0, l\pi. \end{cases} \tag{3.44}$$

By utilizing the comparison principle for parabolic equations, it can be readily verified that $P(x,t) < \check{P}(t)$, where $\check{P}(t)$ denotes the solution (spatially homogeneous) of

$$\begin{cases} P_t(x,t) = dP_{xx} + (\mu f(N_T) - \lambda)P, & t > 0, x \in (0, l\pi), \\ \dfrac{\partial P}{\partial x} = 0, & t > 0, x = 0, l\pi, \\ P(0) = \sup P_0(x) & x \in [0, l\pi], \end{cases} \tag{3.45}$$

that is,

$$\check{P}(t) = P(0)e^{[\mu f(N_T)-\lambda]t} > 0.$$

Based on Theorem 3.6, since $P(x,t) > 0$, we can deduce that $\lim_{t\to\infty} P(x,t) = 0$ for $x \in [0, l\pi]$. It follows that there is a function, $\omega_1(t) = rgh\check{P}(t)$, such that

$$\lim_{t\to\infty, x\in[0,l\pi]} \omega_1(x,t) = 0,$$

and

$$\begin{cases} Z_t(x,t) < dZ_{xx} + \omega_1 - \delta Z, & t > 0, x \in (0, l\pi), \\ \dfrac{\partial Z}{\partial x} = 0, & t > 0, x = 0, l\pi. \end{cases} \tag{3.46}$$

Then from the comparison principle, $Z(x,t) < \check{Z}(t)$, where $\check{Z}(t)$ is the solutions of

$$\begin{cases} Z_t(x,t) = dZ_{xx} + \omega_1 - \delta Z, & t > 0, x \in (0, l\pi), \\ \dfrac{\partial Z}{\partial x} = 0, & t > 0, x = 0, l\pi, \\ Z(0) = \sup Z_0(x) & x \in [0, l\pi], \end{cases} \tag{3.47}$$

that is,

$$\check{Z}(t) = e^{-\delta t}Z(0) + e^{-\delta t}\int_0^t e^{\delta u}\omega_1(u)\mathrm{d}u.$$

It is evident that in the case where $\int_0^t e^{\delta u}\omega_1(u)\mathrm{d}u$, the final term tends to zero as $t \to \infty$. Additionally, since $Z(x,t) > 0$, it follows that $\lim_{t\to\infty} Z(x,t) = 0$ for $x \in [0, l\pi]$. Similarly, there exists a function $\omega_2(x,t) = \lambda\check{P}(t) + (\delta + (1-r)g)\check{Z}(t)$, such that

$$\lim_{t\to\infty, x\in[0,l\pi]} \omega_2(x,t) = 0,$$

and

$$\begin{cases} Q_t(x,t) < dQ_{xx} + \omega_2 - \frac{1}{\tau}Q, & t > 0, x \in (0, l\pi), \\ \frac{\partial Q}{\partial x} = 0, & t > 0, x = 0, l\pi, \end{cases} \tag{3.48}$$

then from the comparison principle, $Q(x,t) < \check{Q}(t)$. Here, $\check{Q}(t)$ is the solutions of

$$\begin{cases} Q_t(x,t) = dQ_{xx} + \omega_2 - \frac{1}{\tau}Q, & t > 0, x \in (0, l\pi), \\ \frac{\partial Q}{\partial x} = 0, & t > 0, x = 0, l\pi, \\ Q(0) = \sup Q_0(x) & x \in [0, l\pi], \end{cases} \tag{3.49}$$

where

$$\check{Q}(t) = e^{-\frac{1}{\tau}t}Q(0) + e^{-\frac{1}{\tau}t}\int_0^t e^{\frac{1}{\tau}u}\omega_2(u)\mathrm{d}u.$$

As $t \to \infty$,

$$e^{-\frac{1}{\tau}t}\int_0^t e^{\frac{1}{\tau}u}\omega_2(u)\mathrm{d}u \to 0.$$

Since $Q(x,t) > 0$, it follows that

$$\lim_{t\to\infty} Q(x,t) = 0, \quad x \in [0, l\pi].$$

Considering the conservation law (3.39), it can be concluded that

$$\lim_{t\to\infty} N(x,t) = N_T, \quad x \in [0, l\pi].$$

Consequently, E_1^* is globally stable.

The Hopf bifurcation near E_3^* is examined in the subsequent analysis, utilizing the Hopf bifurcation theorem for PDEs [39, 40]. In this particular context, d is chosen as the bifurcation parameter.

For the nontrivial solution E_3^*, the Jacobian matrix is

$$J_3^* = \begin{bmatrix} -I_{11} - \dfrac{dn^2}{l^2} & I_{12} & 0 & \frac{1}{\tau} \\ I_{11} & I_{22} - \dfrac{dn^2}{l^2} & I_{23} & 0 \\ 0 & I_{32} & -\dfrac{dn^2}{l^2} & 0 \\ 0 & I_{42} & I_{43} & -\frac{1}{\tau} - \dfrac{dn^2}{l^2} \end{bmatrix},$$

where

$$\begin{aligned} I_{11} &= \mu P_3^* f'(N_3^*) > 0, \ I_{12} = -\mu f(N_3^*) < 0, \\ I_{22} &= \mu f(N_3^*) - \lambda - g Z_3^* h'(P_3^*), I_{23} = -\frac{\delta}{r} < 0, \\ I_{32} &= r g Z_3^* h'(P_3^*) > 0, \ I_{42} = \lambda + (1-r) g Z_3^* h'(P_3^*) > 0, I_{43} = \frac{\delta}{r} > 0. \end{aligned}$$

The characteristic equation is

$$\Phi_3^*(\phi, n^2, q) =: (\frac{dn^2}{l^2} + \phi)(\phi^3 + \phi^2 e_2 + \phi e_1 + e_0) = 0, \tag{3.50}$$

where

$$\begin{aligned} c_1 &= \frac{I_{32}}{r}(\delta + I_{11}) + I_{11}\lambda + \frac{e_4}{\tau}, \ c_0 = \frac{I_{32}}{r}(\delta(I_{11} + \frac{1}{\tau}) + I_{11}\frac{r}{\tau}), \\ e_2 &= \frac{3dn^2}{l^2} + \frac{1}{\tau} + e_4, \\ e_1 &= c_1 + \frac{3d^2n^4}{l^4} + \frac{2dn^2}{l^2}(\frac{1}{\tau} + e_4), \ e_0 = c_0 + \frac{dn^2}{l^2}(c_1 + \frac{dn^2}{l^2}(\frac{dn^2}{l^2} + \frac{1}{\tau} + e_4)), \\ e_4 &= \mu P_3^* f'(N_3^*) - \mu f(N_3^*) + g Z_3^* h'(P_3^*) + \lambda. \end{aligned}$$

Therefore, we have the following theorem.

Theorem 3.8 *Assume the equilibrium point E_3^* exists, if*

$$P_3^* \geq f(N_3^*)/f'(N_3^*),$$

E_3^* undergoes a Hopf bifurcation at each $d > 0$ and $n \geq 1$ satisfying

$$\frac{8d^3n^6}{l^6} + \frac{8d^2n^4}{l^4}(e_4 + \frac{1}{\tau}) + \frac{2dn^2}{l^2}(c_1 + (e_4 + \frac{1}{\tau})^2) + c_1(e_4 + \frac{1}{\tau}) - c_0 = 0.$$

Proof The inequality $P_3^* \geq f(N_3^*)/f'(N_3^*)$ ensures that e_2 and e_1 are positive. For a type II response function $f(N) = \frac{N}{N+k}$, this condition can be satisfied when $h^{-1}(\frac{\delta}{rg}) \geq \frac{N_3^*(N_3^*+k)}{k}$. Additionally, it can be demonstrated that $e_2e_1 = e_0$ under the condition that the bifurcation parameter d satisfies the equation

$$\frac{8d^3n^6}{l^6}+\frac{8d^2n^4}{l^4}(e_4+\frac{1}{\tau})+\frac{2dn^2}{l^2}(c_1+(e_4+\frac{1}{\tau})^2)+c_1(e_4+\frac{1}{\tau})-c_0=0. \quad (3.51)$$

Thus, if (3.51) is satisfied, Eq. (3.50) becomes

$$\Phi_3^*(\phi, n^2, q) = (\frac{dn^2}{l^2} + \phi)(\phi^2 + e_1)(\phi + e_2). \quad (3.52)$$

Then the roots of (3.52) are $\phi_{1,2} = \pm\sqrt{e_1}i$, $\phi_3 = -\frac{dn^2}{l^2} < 0$, $\phi_4 = -e_2 < 0$.

The transversality condition is derived below. Note that $\Phi_3^*(\phi, d) = \Phi_3^*(\phi + \frac{dn^2}{l^2})$. If $d = 0$, let ϕ^* be a root of $\Phi_3^*(\phi) = 0$. Then $\Phi_3^*(\phi+\frac{dn^2}{l^2}) = 0$ if $\phi+\frac{dn^2}{l^2} = \phi^*$, that actually is $\phi = \phi^* - \frac{dn^2}{l^2}$. This means that $\mathrm{Re}(\phi) = \mathrm{Re}(\phi^*) - \frac{dn^2}{l^2}$. So $\frac{\mathrm{dRe}(\phi)}{\mathrm{d}d} = -\frac{n^2}{l^2} < 0$ for $n = 1, 2 \ldots$

To make Theorem 3.8 clearer, we fixed the parameters as given in Table 3.1 and calculated the bifurcation values of the diffusion rate d_i, and the corresponding period $\rho_i = \rho_1 = \ldots = \rho_n = 2\pi/\phi_i = 2\pi/\sqrt{e_1}_{n=1}$, and frequency $1/\rho_i$. These are shown in Table 3.2.

For values of $N_T = 0.9$, there are prolonged steady states for model (3.37), as depicted in Fig. 3.4.

Diffusion-induced Hopf bifurcation of the unstable equilibrium point E_3^* is possible for specific values of d given in Table 3.2 for $N_T = 2$. It should be emphasized that as d decreases through each critical Hopf value, the real part of a pair of complex conjugate eigenvalues of E_3^* increases. Hence, for $d > d_1$, there

Table 3.1 Parameter meanings and values used for all computations

Parameter	Value
d	Varies
l	2
μ	5.9 day^{-1}
λ	0.017 day^{-1}
g	7 day^{-1}
r	0.7
δ	0.17 day^{-1}
τ	7 day

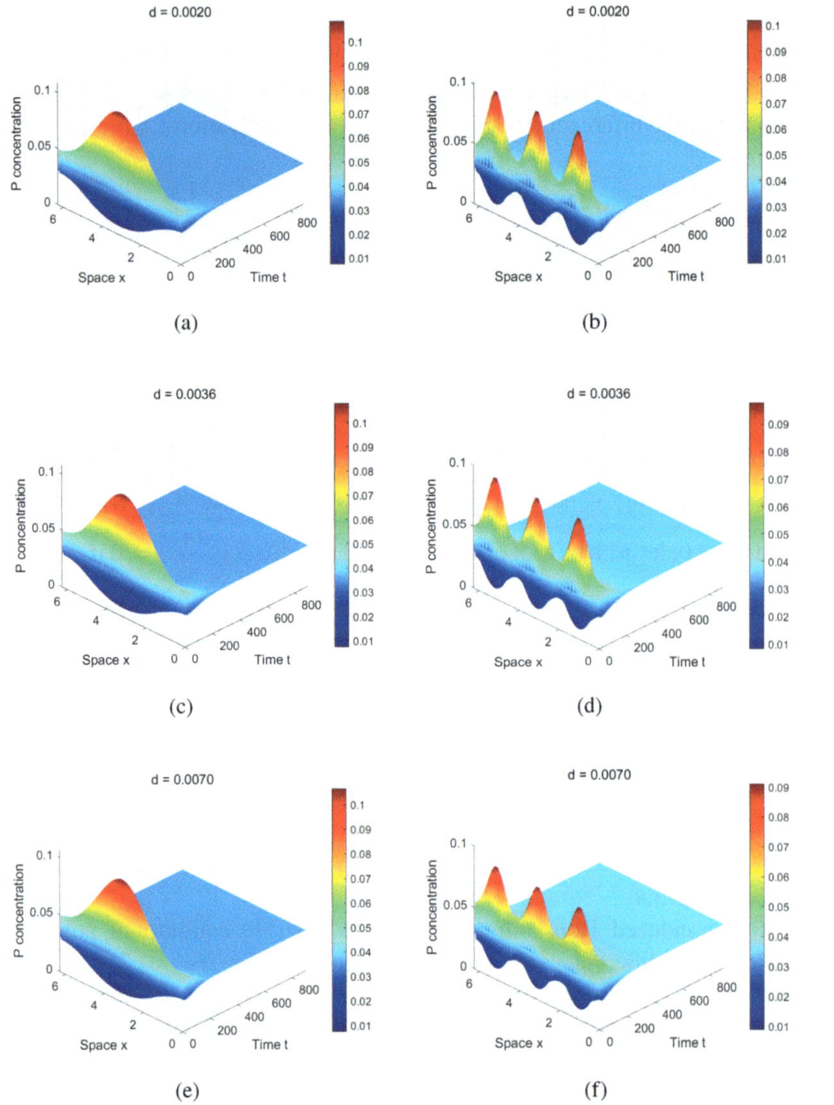

Fig. 3.4 Solutions under different perturbations with $N_T = 0.9$. The initial values are $N_0 = 0.166855 + 0.1\cos x\ (\cos 3x)$, $P_0 = 0.0359$, $Z_0 = 0.2093$, $Q_0 = 0.4879$

is one pair with a positive real part; for $d_2 < d < d_1$, there are two pairs, and so on. This implication will be explored further in the remaining part of this section.

Figure 3.5 illustrates bistability for $N_T = 2$. For $d = 0.007$, an inhomogeneous solution with regular oscillations in both time t and space x is observed in Fig. 3.5e. However, when perturbation equals $0.1 * \cos x$ in the initial displacement of the nutrients, the solution appears to be homogeneous (Fig. 3.5g). When the value of d

Table 3.2 Diffusion rate, period, and frequency of the Hopf bifurcation with parameter set in Table 3.1 for $N_T = 2$

Wave number i	Diffusion rate d_i	Period ρ_i	Frequency $1/\rho_i$
1	$0.032l^2$	8.0252	0.1246
2	$0.008l^2$	8.0252	0.1246
...	...	...	...
n	$\frac{0.032l^2}{n^2}$	8.0252	0.1246

is decreased to 0.0036, a transient is observed in the $0.1 * \cos x$ initial perturbation, where the solution transitions from inhomogeneous to homogeneous (Fig. 3.5c). Meanwhile, in the case of perturbation equal to $0.1 * \cos 3x$, the inhomogeneity disappears more rapidly (Fig. 3.5d). As d decreases to 0.002, all the solutions are significantly perturbed (Fig. 3.5a and b). To better understand these dynamics, contour plots of each solution at the end time, ranging from 700 to 900, are presented in Fig. 3.6.

The dynamical behavior of model (3.37) under random perturbation $0.1 * randn$, with initial values $N_0 = 1$ and $N_0 = 2$, is illustrated in Figs. 3.7 and 3.8. Here $randn$ is a random scalar obtained from a standard normal distribution. In both cases, $N(x, t)$ undergoes spatially and temporally varying perturbations, exhibiting an irregular and nonuniform spatial state.

3.1.5 Conclusion

This section presents a NPZ model with spatiotemporal distributed delays. The influence of diffusion rates and total biomass on the planktonic ecosystem is discussed. A general local stability theorem for three-dimensional reaction-diffusion systems with nonlocal delays is introduced, followed by an investigation of the local stability of equilibrium solutions in the NPZ model under generalized delay kernels. The results indicate that the local stability of equilibrium solutions depends on the total biomass of the system, the average delay, and the functional response of zooplankton grazing. By comparing the local stability conditions for $d = 0$ and $d \neq 0$, it is observed that the diffusion rate d does not affect the local stability of equilibrium solutions. Furthermore, the computation of Hopf bifurcations reveals that for $d \neq 0$, all the real parts of the eigenvalues are smaller than those corresponding to $d = 0$. Theoretical analysis and numerical simulations demonstrate that diffusion enhances the stability of the planktonic ecosystem. However, when an adequate amount of nutrients is supplied, the system exhibits various types of bistability, including uniform and nonuniform solutions, sustained and transient oscillations.

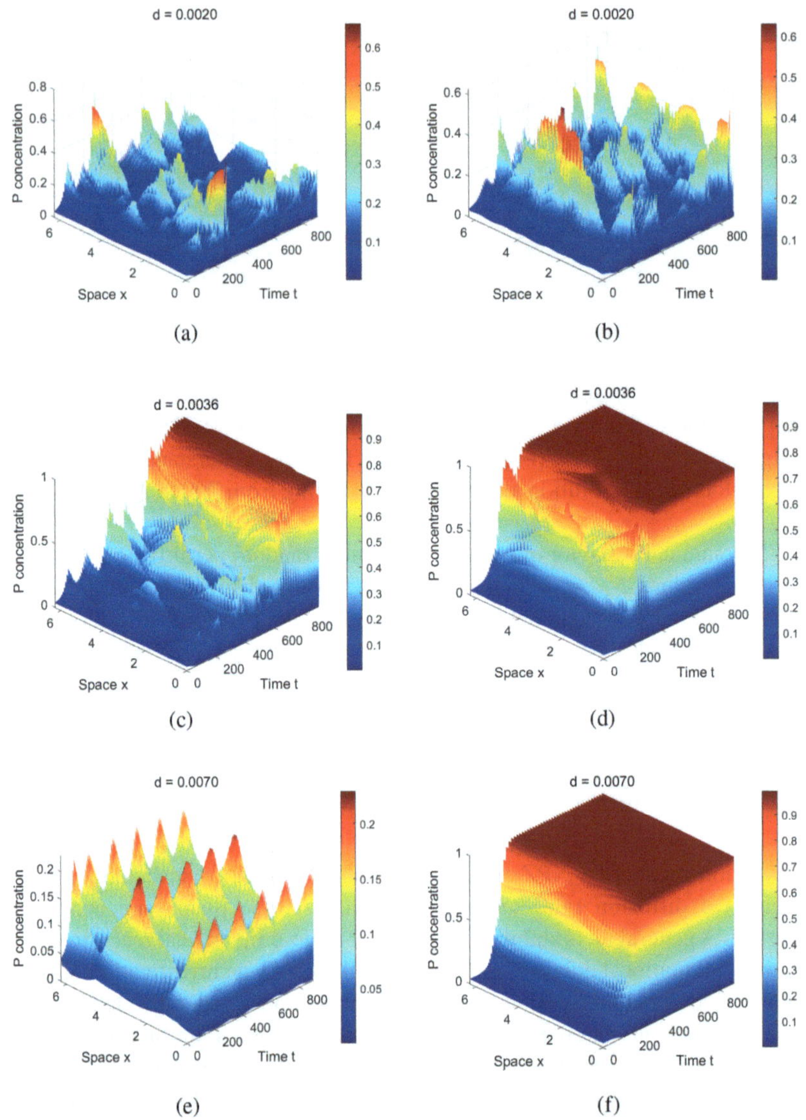

Fig. 3.5 Bistability under different perturbations for $N_T = 2$. The initial values are $N_0 = 0.866855 + 0.1\cos x\ (\cos 3x)$, $P_0 = 0.0359$, $Z_0 = 0.4093$, and $Q_0 = 0.6879$.

3.2 Dynamics of a Diffusive Predator-Prey Model with Nonlocal Delays

The relationship between predators and prey is one of the fundamental and widely observed relationships in the natural world, and it is a classic problem in population

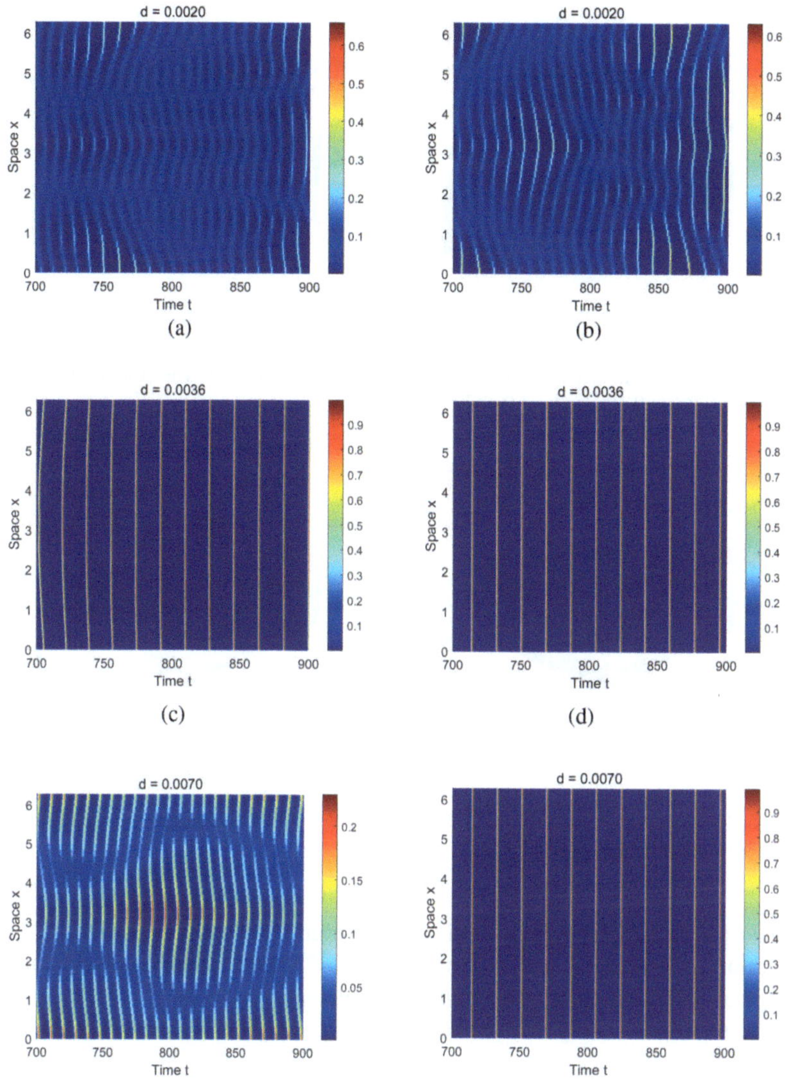

(a) (b) (c) (d) (e) (f)

Fig. 3.6 Spatiotemporal views of solutions in Fig. 3.5. The initial values are $N_0 = 0.866855 + 0.1\cos x\ (\cos 3x)$, $P_0 = 0.0359$, $Z_0 = 0.4093$, and $Q_0 = 0.6879$

ecology [41]. During the long-term development and evolution of organisms in nature, both intraspecific and interspecific relationships have emerged based on food, resources, and spatial interactions. Examples of such relationships include competition, mutualism, predation, and parasitism. Predation-prey dynamics models, hereafter referred to as predation models, effectively describe, predict, and

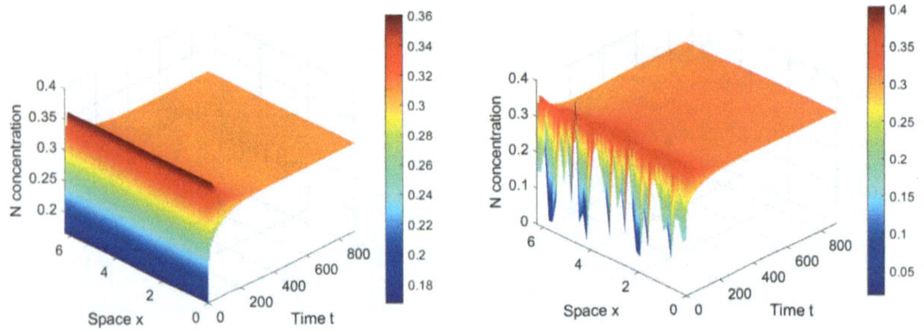

(a) (b)

Fig. 3.7 Solutions with or without random perturbation when $N_0 = 1$. The other initial values are $P_0 = 0.0359$, $Z_0 = 0.4093$, $Q_0 = 0.6879$

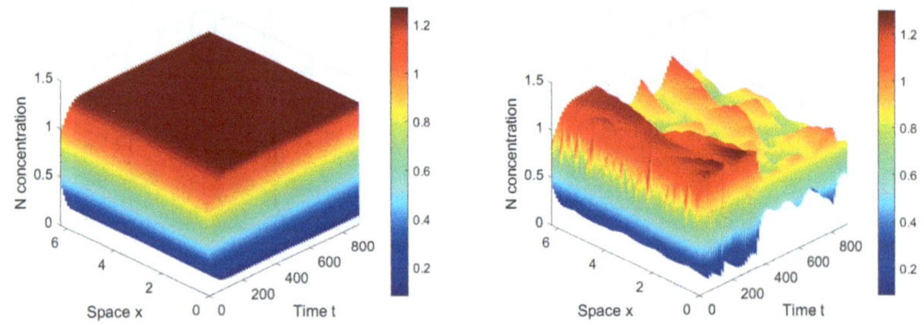

(a) (b)

Fig. 3.8 Solutions under random perturbation when $N_0 = 2$. The other initial values are $P_0 = 0.0359$, $Z_0 = 0.4093$, $Q_0 = 0.6879$

regulate the spatiotemporal dynamics and evolutionary trends of species' mutual adaptation, competition, and coevolution. They provide theoretical guidance for ecological conservation, management, and development.

In 1963, Rosenzweig and MacArthur proposed the Rosenzweig-MacArthur model [42], given by

$$\begin{cases} \dfrac{du}{dt} = g(u)(f(u) - v), \\ \dfrac{dv}{dt} = (g(u) - d)v, \end{cases} \tag{3.53}$$

where u and v represent the densities of predators and prey, respectively. d is the death rate of predators ($d > 0$), and $g(u)$ represents the functional response of predators. $g(u)f(u)$ represents the net growth rate of prey in the absence of predators. The functional response function represents the capturing rate, which indicates the mutual constraints between populations. It can be categorized as either prey-dependent or predator-dependent based on the density of different populations. Common prey-dependent functional responses include Holling types I, II, and III [43, 44], while predator-dependent functional responses include Beddington-DeAngelis [45, 46], Crowley-Martin [47], and ratio-dependent [48].

3.2.1 Rosenzweig-MacArthur Model with Nonlocal Delays

In recent years, increasing evidence has shown that environmental heterogeneity and individual mobility have a significant impact on population ecological balance [49]. When survival conditions are limited, animals actively disperse in their habitats to seek more resources for survival and mating opportunities. Time delays in population activities during specific stages such as incubation and gestation periods contribute to density variations that are delayed in time. Therefore, it is more realistic to consider models with time delays when studying population dynamics. In this section, we extend the model (3.53) by introducing spatiotemporal delays (3.1) and diffuions [50], given by

$$\begin{cases} u_t(x,t) = d_1 u_{xx} + g(u)f(u) - \tau_1 * g(u)v, & x \in \Omega, t > 0, \\ v_t(x,t) = d_2 v_{xx} + \tau_2 * g(u)v - dv, & x \in \Omega, t > 0, \\ \dfrac{\partial u}{\partial \nu} = \dfrac{\partial v}{\partial \nu} = 0, & x \in \partial\Omega, \quad t > 0, \\ u(x,0) = u_0(x) \geq 0, \ v(x,0) = v_0(x) \geq 0, & x \in \bar{\Omega}, t \in [-\infty, 0]. \end{cases} \tag{3.54}$$

Here, Ω represents a bounded region with a smooth boundary $\partial\Omega$, and $u(x,t)$ and $v(x,t)$ denote the densities of predators and prey at position x and time t. For simplicity, we consider $\Omega = [0, \pi]$. d_1 and d_2 represent the diffusion coefficients of the populations. Furthermore, the functions $f(u)$ and $g(u)$ satisfy the following conditions:

1. The function $f \in C^1(\bar{\mathbb{R}}^+)$ satisfies $f(0) > 0$. There exists a constant $K > 0$ such that for any $u > 0$ with $u \neq K$, $f(u)(u - K) < 0$ and $f(K) = 0$. Besides, there exists $\bar{\lambda} \in (0, K)$ such that $f'(u) > 0$ for $u \in [0, \bar{\lambda})$ and $f'(u) < 0$ for $u \in (\bar{\lambda}, K]$.
2. The function $g \in C^1(\bar{\mathbb{R}}^+)$ satisfies $g(0) = 0$. For $u > 0$, $g(u) > 0$, and for $u \geq 0$, $g'(u) > 0$. There exists a unique $\lambda > 0$ such that $g(\lambda) = d$.

Equation (3.3) shows that nonlocal delays have no effect on the spatially uniform equilibrium point. By simple calculations, it can be determined that the model (3.54) has three nonnegative constant equilibrium solutions: $(0, 0)$, $(K, 0)$, and (λ, v_λ), where λ is the unique solution satisfying $g(\lambda) = d$ and $v_\lambda = f(\lambda)$. The nontrivial equilibrium solution (λ, v_λ) exists if and only if $0 < \lambda < K$.

Since the spatial domain $(0, \pi)$ satisfies homogeneous Neumann boundary conditions, the solutions (u, v) of the model (3.54) can be written as

$$(u, v) = (c_1, c_2)e^{\phi t} \cos nx, \quad n = 0, 1, 2, \ldots \tag{3.55}$$

In this case, Eq. (3.1) can be written as

$$(\tau_i * u)(x, t) = \int_{-\infty}^{t} \int_{0}^{\pi} G(x, y, t - s)m_i(t - s)e^{\phi s} \cos nydyds.$$

After algebraic manipulation,

$$(\tau_i * u)(x, t) = \bar{\tau}_i(\phi + dn^2)e^{\phi t} \cos nx.$$

Here, $\bar{\tau}_i$ represents the Laplace transform of m_i. It is worth noting that for any kernel function m_i, we have

$$\bar{\tau}_i(iR + dn^2) = \int_0^{\infty} m_i(t)e^{-(iR+dn^2)t}dt = C_i(R, n) - iS_i(R, n), \tag{3.56}$$

where

$$C_i(R, n) = \int_0^{\infty} m_i(t)e^{-dn^2t} \cos Rtdt, \quad S_i(R, n) = \int_0^{\infty} m_i(t)e^{-dn^2t} \sin Rtdt.$$

For $i \in \mathbb{N}$ and $n \geq 0$,

$$|C_i(R, n)| \leq 1, \quad |S_i(R, n)| \leq 1.$$

For $R = 0$ and $n = 0$,

$$C_i(0, 0) = 1, \quad Si(0, 0) = 0.$$

Furthermore,

$$|Si(R, n)| \leq \int_0^{\infty} m_i(t)e^{-dn^2t}|\sin Rt|dt \leq \int_0^{\infty} m_i(t)e^{-dn^2t}Rtdt \leq RTi, \tag{3.57}$$

and

$$\left.\frac{dSi(R, n)}{dR}\right| R = 0 = \int_0^{\infty} m_i(t)te^{-dn^2t}dt \leq Ti. \tag{3.58}$$

In $X = \left\{(\psi, \varphi) \in H^2(\Omega) \times H^2(\Omega) : \frac{\partial \psi}{\partial \nu} = \frac{\partial \varphi}{\partial \nu} = 0\right\}$, the linear equation of model (3.54) at the equilibrium point $e^*(u^*, v^*)$ can be written as

$$\begin{pmatrix} \psi_t \\ \varphi_t \end{pmatrix} = L\begin{pmatrix} \psi \\ \varphi \end{pmatrix} := J_{(u^*,v^*)}\begin{pmatrix} \psi \\ \varphi \end{pmatrix} + D\begin{pmatrix} \psi_{xx} \\ \varphi_{xx} \end{pmatrix},$$

where

$$D = \begin{pmatrix} d_1 & 0 \\ 0 & d_2 \end{pmatrix}, \quad J_{(u^*,v^*)} = \begin{pmatrix} A(u, v, \bar{\tau_1}) - d_1 n^2 & B(u, v, \bar{\tau_1}) \\ C(u, v, \bar{\tau_2}) & D(u, v, \bar{\tau_2}) - d_2 n^2 \end{pmatrix},$$

and

$$A(u, v, \bar{\tau_1}) = g'(u) f(u) - v g'(u) \bar{\tau_1}(\phi + d_1 n^2) + g(u) f'(u),$$

$$B(u, v, \bar{\tau_1}) = -g(u) \bar{\tau_1}(\phi + d_1 n^2),$$

$$C(u, v, \bar{\tau_2}) = v g'(u) \bar{\tau_2}(\phi + d_2 n^2),$$

$$D(u, v, \bar{\tau_2}) = g(u) \bar{\tau_2}(\phi + d_2 n^2) - d.$$

The characteristic equation at $e^*(u^*, v^*)$ is

$$\Phi(\phi, n^2, q) = \text{Det}\begin{pmatrix} A(u^*, v^*, \bar{\tau_1}) - \phi - d_1 n^2 & B(u^*, v^*, \bar{\tau_1}) \\ C(u^*, v^*, \bar{\tau_2}) & D(u^*, v^*, \bar{\tau_2}) - \phi - d_2 n^2 \end{pmatrix} = 0,$$

Here, q represents the parameter set composed of all parameters, and ϕ represents the eigenvalues.

3.2.2 *Local Stability of the Rosenzweig-MacArthur Model*

In the subsequent discussion, it is defined that for wave numbers $n = 1, 2, 3, \ldots$, the equilibrium point is considered locally stable if all the roots of the characteristic equation $\Phi(\phi, n^2, q) = 0$ lie in the left half plane. Similarly, for wave numbers $n = 0, 1, 2, 3, \ldots$, the equilibrium point is regarded as locally asymptotically stable if all the roots of the characteristic equation $\Phi(\phi, n^2, q) = 0$ lie in the left half plane.

The local stability results of the model (3.54) without time delay effects, as presented in the literature [51], are reviewed.

Lemma 3.6 *For the model* (3.54), *with the assumption that* $d, d_1, d_2 > 0$*:*

1. *If* $\lambda > 0$*, then* $(0, 0)$ *is unstable.*
2. *If* $\lambda > K$*, then* $(K, 0)$ *is locally asymptotically stable. If* $\lambda < K$*, then it is unstable.*
3. *If* $\bar{\lambda} < \lambda < K$*, then* (λ, v_λ) *is locally asymptotically stable. If* $0 < \lambda < \bar{\lambda}$*, then it is unstable.*

Here, $\bar{\lambda} \in (0, K)$ *represents the unique root of the equation* $f'(\lambda) = 0$*.*

Next, the analysis of the local stability of the equilibrium solutions of the model (3.54) is carried out, based on a lemma derived from the literature [37].

Lemma 3.7 *The number of roots of the characteristic equation* (3.23) *in the right half plane is given by*

$$\lim_{R\to\infty}\left(1 - \frac{1}{\pi}\arg\Phi(iR, n^2, q)\right).$$

Lemma 3.8 *Assume that*

$$\Phi(0, n^2, q) > 0, \quad \lim_{R\to\infty}\operatorname{Im}(\Phi(iR, n^2, q)) = \infty, \; \operatorname{Im}(\Phi'(0, n^2, q)) > 0,$$

then all the roots of the characteristic equation (3.23) *have negative real parts.*

Next, the case of $n \geq 1$ is considerd.

Theorem 3.9 *In the model* (3.54), *assuming* $d, d_1, d_2 > 0$ *and* $n \geq 1$,

1. *If* $f(0)g'(0) < d_1$*, then* $(0, 0)$ *is locally stable.*
2. *If* $\lambda > K$*, then* $(K, 0)$ *is locally stable. If* $\lambda < K$ *and* $g(K) < d_2 + d$*, then* $(K, 0)$ *is locally stable.*
3. *If* $0 < \lambda < \bar{\lambda}$ *and* $0 < v_\lambda g'(\lambda) + df'(\lambda) < d_1$*, then* (λ, v_λ) *is locally stable. If* $\bar{\lambda} < \lambda < K$ *and* $0 < v_\lambda g'(\lambda) < d_1$*, then* (λ, v_λ) *is locally stable.*

Proof

(1) The Jacobian matrix of model (3.54) at the equilibrium (0, 0) is given by

$$J_{(0,0)} = \begin{pmatrix} f(0)g'(0) - d_1 n^2 & 0 \\ 0 & -d - d_2 n^2 \end{pmatrix}.$$

The characteristic equation is

$$\Phi_1(\phi, n^2, q) := (f(0)g'(0) - d_1 n^2 - \phi)(-d - d_2 n^2 - \phi) = 0. \tag{3.59}$$

The roots of the characteristic equation (3.59) are given by

$$f(0)g'(0) - d_1 n^2, \; -d - d_2 n^2.$$

Thus, when $n \geq 1$ and $f(0)g'(0) < d_1$, all the roots have negative real parts. Therefore, (0, 0) is locally stable.

(2) The Jacobian matrix of model (3.54) at the equilibrium $(K, 0)$ is given by

$$J_{(K,0)} = \begin{pmatrix} f'(K)g(K) - d_1n^2 & -g(K)\bar{\tau}_1(\phi + d_1n^2) \\ 0 & -d + g(K)\bar{\tau}_2(\phi + d_2n^2) - d_2n^2 \end{pmatrix}.$$

The characteristic equation is

$$\Phi_2(\phi, n^2, q) := (d_1n^2+\phi-f'(K)g(K))(d_2n^2+\phi+d-g(K)\bar{\tau}_2(\phi+d_2n^2)) = 0.$$

When $\phi = iR$ and $R \to \infty$,

$$\mathrm{Re}\Phi_2(iR, n^2, q) \sim -R^2,$$

$$\mathrm{Im}\Phi_2(iR, n^2, q) \sim (d - C_1(R, n)g(K) - f'(K)g(K) + (d_1 + d_2)n^2)R.$$

When $R = 0$,

$$\Phi_2(0, n^2, q) = (d_1n^2 - f'(K)g(K))(d - C_1(0, n)g(K) + d_2n^2).$$

When $\lambda > K$, it can be deduced that $f'(K) < 0$ and $d - C_1g(K) > d - g(K) > 0$. Hence,

$$\Phi_2(0, n^2, q) > (d_1n^2 - f'(K)g(K))(d - g(K) + d_2n^2) > 0,$$

$$\mathrm{Im}\Phi_2'(0, n^2, q) = (d - C_1(0, n)g(K) - f'(K)g(K) + (d_1 + d_2)n^2) > 0,$$

and

$$\lim_{R\to\infty} \mathrm{Im}\Phi_2(iR, n^2, q) \to \infty.$$

According to Lemma 3.8, it is concluded that the characteristic equation has no roots in the right half plane. Therefore, $(K, 0)$ is locally stable.

When $\lambda < K$, we have $d - g(K) < 0$, $f'(K) < 0$, and

$$d_1n^2 - f'(K)g(K) > 0.$$

Based on the assumption $n \geq 1$ and $g(K) < d_2 + d$, $d + d_2n^2 - g(K) > 0$. From the previous analysis,

$$\Phi_2(0, n^2, q) > 0, \ \lim_{R\to\infty} \mathrm{Im}\Phi_2(iR, n^2, q) = \infty, \ \mathrm{Im}\Phi_2'(0, n^2, q) > 0.$$

According to Lemma 3.8, the theorem holds.

(3) The Jacobian matrix of model (3.54) at the equilibrium(λ, v_λ) is given by

$$J_{(\lambda,v_\lambda)} = \begin{pmatrix} A_1(\lambda) - B_1(\lambda)\bar{\tau}_1(\phi + d_1n^2) - d_1n^2 & -d\bar{\tau}_1(\phi + d_1n^2) \\ B_1(\lambda)\bar{\tau}_2(\phi + d_2n^2) & -d + d\bar{\tau}_2(\phi + d_2n^2) - d_2n^2 \end{pmatrix},$$

here

$$A_1(\lambda) = v_\lambda g'(\lambda) + df'(\lambda), \qquad B_1(\lambda) = v_\lambda g'(\lambda) > 0.$$

The characteristic equation is

$$\begin{aligned}\Phi_3(\phi, n^2, q) :=&(d + \phi + d_2n^2 - d\bar{\tau}_2(\phi + d_2n^2)) \\ &\times (\phi + d_1n^2 + B_1\bar{\tau}_1(\phi + d_1n^2) - A_1) \\ &+ B_1d\bar{\tau}_1(\phi + d_1n^2)\bar{\tau}_2(\phi + d_2n^2) = 0.\end{aligned}$$

when $\phi = iR$, $R \to \infty$,

$$\mathrm{Re}\Phi_3(iR, n^2, q) \sim - R^2,$$

$$\mathrm{Im}\Phi_3(iR, n^2, q) \sim (B_1C_1(R, n) - A_1 + (1 - C_2(R, n))d + (d_1 + d_2)n^2)R.$$

when $R = 0$,

$$\begin{aligned}\Phi_3(0, n^2, q) = &\, dB_1C_1(0, n)C_2(0, n) \\ &- (A_1 - B_1C_1(0, n) - d_1n^2)((1 - C_2(0, n))d + d_2n^2),\end{aligned}$$

$$\mathrm{Im}\Phi_3'(0, n^2, q) = B_1C_1(0, n) - A_1 + (1 - C_2(0, n))d + (d_1 + d_2)n^2.$$

Two cases are considered in the below.

When $0 < \lambda < \bar{\lambda}$, it follows from $f'(\lambda) > 0$ that $\mathrm{Im}\Phi_3'(0, 0, q) < 0$. Therefore, when $n = 0$, the equilibrium (λ, v_λ) is unstable.

When $\bar{\lambda} < \lambda < K$, $f'(\lambda) < 0$. By assuming $A_1 < 0$, it can be deduced that

$$\lim_{R\to\infty} \mathrm{Im}\Phi_3(iR, 0, q) \to \infty \text{ and } \mathrm{Im}\Phi_3'(0, 0, q) > 0.$$

Lemma 3.8 and Theorem 3.9 lead to the conclusion that the equilibrium (λ, v_λ) is locally asymptotically stable.

3.2.3 Global Behavior of the Rosenzweig-MacArthur Model

Next, the global dynamical behavior of the model (3.54) in the presence of time delay only during the process of prey provisioning for predators is considered.

Let

$$m_1(t) = \delta(t), m_2(t) = \frac{1}{\tau_2} e^{-\frac{t}{\tau_2}}, \tag{3.60}$$

here, $m_1(t)$ represents the Dirac delta function, and $m_2(t)$ represents a weak delay kernel function.

Define

$$q(x,t) = \int_{-\infty}^{t} \int_{0}^{\pi} G_2(x, y, t-s) e^{-\frac{t-s}{\tau_2}} g(u(y,s))v(y,s) dy ds, \tag{3.61}$$

Under the kernel function (3.60), model (3.54) can be expressed as

$$\begin{cases} u_t(x,t) = d_1 u_{xx} + g(u)(f(u) - v), & x \in (0,\pi), t > 0, \\ v_t(x,t) = d_2 v_{xx} + \dfrac{1}{\tau_2} q - dv, & x \in (0,\pi), t > 0, \\ q_t(x,t) = d_2 q_{xx} + g(u)v - \dfrac{1}{\tau_2} q, & x \in (0,\pi), t > 0, \\ \dfrac{\partial u}{\partial x} = \dfrac{\partial v}{\partial x} = \dfrac{\partial q}{\partial x} = 0, & x = 0, \pi, t > 0, \end{cases} \tag{3.62}$$

with intial conditions, for $t \in [-\infty, 0]$ and $x \in [0, \pi]$,

$$u(x,0) = u_0(x) \geq 0, v(x,0) = v_0(x) \geq 0,$$

$$q(x,0) = \int_{-\infty}^{0} \int_{0}^{\pi} G_2(x, y, t-s) e^{\frac{s}{\tau_2}} g(u(x,s))v(x,s) dy ds \geq 0.$$

The model (3.62) exhibits three equilibria, namely,

$$(0,0,0), (K,0,0), \text{ and } (\lambda, v_\lambda, \tau_2 v_\lambda g(\lambda)).$$

The existence of solutions for the system (3.62) will be proven using the upper and lower solution method based on mixed monotonicity.

Theorem 3.10 *Assume $d, d_1, d_2 > 0$, and $t \in (0, \infty)$, $x \in [0, \pi]$.*

1. *Model (3.62) has a unique solution $(u(x,t), v(x,t), q(x,t))$ such that $u(x,t) > 0$, $v(x,t) > 0$, and $q(x,t) > 0$.*

2. *For any solution* $u(x,t)$ *of model* (3.62), *it holds that* $\lim_{t\to\infty} \sup u(x,t) \leq K$. *Moreover, when* $d_1 = d_2$,

$$\int_0^{\pi} v(x,t)dt \leq (K + \frac{g(K)f(\bar{\lambda})}{d})\pi.$$

Proof

1. Define

$$m_1(x,t) = g(u)(f(u) - v),$$

$$m_2(x,t) = \frac{1}{\tau_2}q - dv,$$

$$m_3(x,t) = g(u)v - \frac{1}{\tau_2}q.$$

In $\overline{R_+^3} = \{u \geq 0, v \geq 0, q \geq 0\}$

$$\frac{\partial m_1}{\partial v} = -g(u) < 0, \frac{\partial m_1}{\partial q} = 0,$$

$$\frac{\partial m_2}{\partial u} = 0, \frac{\partial m_2}{\partial q} = \frac{1}{\tau_2} > 0,$$

$$\frac{\partial m_3}{\partial u} = g'(u)v > 0, \frac{\partial m_3}{\partial v} = g(u) > 0.$$

So the model (3.62) is quasi-monotone. Let $(\hat{u}(x,t), \hat{v}(x,t), \hat{q}(x,t)) = (0,0,0)$ and $(\tilde{u}(x,t), \tilde{v}(x,t), \tilde{q}(x,t)) = (\tilde{u}^(t), \tilde{v}^(t), \tilde{q}^(t))$, where $(\tilde{u}^(t), \tilde{v}^(t), \tilde{q}^(t))$ is a solution of Eq. (3.63).

$$\begin{cases} \frac{du}{dt} = f(u)g(u), \\ \frac{dv}{dt} = \frac{1}{\tau_2}q, \\ \frac{dq}{dt} = g(u)v, \\ u(0) = \sup\limits_{[0,\pi]} u_0(x), v(0) = \sup\limits_{[0,\pi]} v_0(x), q(0) = \sup\limits_{[0,\pi]} q_0(x). \end{cases} \tag{3.63}$$

Since

$$\begin{cases} \dfrac{\partial \hat{u}(x,t)}{\partial t} - \Delta \hat{u}(x,t) - m_1(\hat{u}(x,t), \tilde{v}(x,t), \hat{q}(x,t)) = 0, \\ \dfrac{\partial \hat{v}(x,t)}{\partial t} - \Delta \hat{v}(x,t) - m_2(\hat{v}(x,t), \tilde{u}(x,t), \hat{q}(x,t)) = 0, \\ \dfrac{\partial \hat{q}(x,t)}{\partial t} - \Delta \hat{q}(x,t) - m_3(\hat{q}(x,t), \hat{u}(x,t), \hat{v}(x,t)) = 0, \end{cases}$$

and

$$\begin{cases} \dfrac{\partial \tilde{u}(x,t)}{\partial t} - \Delta \tilde{u}(x,t) - m_1(\tilde{u}(x,t), \hat{v}(x,t), \tilde{q}(x,t)) = 0, \\ \dfrac{\partial \tilde{v}(x,t)}{\partial t} - \Delta \tilde{v}(x,t) - m_2(\tilde{v}(x,t), \hat{u}(x,t), \tilde{q}(x,t)) = d\tilde{v}^*(t) \geq 0, \\ \dfrac{\partial \tilde{q}(x,t)}{\partial t} - \Delta \tilde{q}(x,t) - m_3(\tilde{q}(x,t), \tilde{u}(x,t), \tilde{v}(x,t)) = \dfrac{1}{\tau_2}\tilde{q}^*(t) \geq 0, \end{cases}$$

Therefore, $(\hat{u}(x,t), \hat{v}(x,t), \hat{q}(x,t))$ and $(\tilde{u}(x,t), \tilde{v}(x,t), \tilde{q}(x,t))$ are upper and lower solutions of the model (3.62). Thus, the model (3.37) has a unique solution $(u(x,t), v(x,t), q(x,t))$ that satisfies

$$0 \leq u(x,t) \leq \tilde{u}^*(t), 0 \leq v(x,t) \leq \tilde{v}^*(t), 0 \leq q(x,t) \leq \tilde{q}^*(t),\ t \geq 0.$$

By applying the strong maximum principle, it follows that when $t > 0$ and $x \in [0, \pi]$,

$$u(x,t), v(x,t), q(x,t) > 0.$$

2. Based on the above discussion, it can be concluded that for $t > 0$, we have $u(x,t) \leq u^*(t)$. Since the property of f and g, as $t \to \infty$, it follows that $u^*(t) \to K$. Hence, for any $\varepsilon > 0$, there exists $T > 0$ such that within X_T, we have $u(x,t) \leq K + \varepsilon$. When $d_1 = d_2$,

$$r_t = u_t + q_t = d_1(u_{xx} + q_{xx}) + f(u)g(u) - \frac{1}{\tau_2}q.$$

According to the comparison principle for parabolic equations,

$$r(x,t) \leq u^*(t) \leq K + \varepsilon.$$

Define

$$\int_0^\pi u(x,t)dx = U(t), \int_0^\pi v(x,t)dx = V(t),$$

$$\int_0^\pi q(x,t)dx = Q(t), \int_0^\pi r(x,t)dx = R(t).$$

According to the Neumann boundary condition characteristics, it can be inferred that

$$\begin{aligned}\frac{d(U+V+Q)}{dt} &= \int_0^\pi d_1(u_{xx}+v_{xx}+q_{xx})dx + \int_0^\pi g(u)(f(u)-v)dx \\ &\quad + \int_0^\pi g(u)vdx - d\int_0^\pi vdx, \\ &= \int_0^\pi g(u)f(u)dx - dV.\end{aligned}$$

and

$$\begin{aligned}(U+V+Q)_t &= -d(U+V+Q) + dR + \int_0^\pi g(u)f(u)dx, \\ &\le -d(U+V+Q) + \pi(d(K+\varepsilon) + g(K+\varepsilon)f(\bar{\lambda})).\end{aligned}$$

When integrating both sides of the equation,

$$\int_0^\pi v(x,t)dx = V(t) < U(t)+V(t)+Q(t) \le \frac{\pi}{d}(d(K+\varepsilon)+g(K+\varepsilon)f(\bar{\lambda})).$$

Thus, any solution $v(x,t)$ satisfies the a priori estimate in L^1:

$$K_1 = \frac{\pi(d(K+\varepsilon)+g(K+\varepsilon)f(\bar{\lambda}))}{d}.$$

The global stability of the semi-trivial equilibrium $(K, 0, 0)$ and the positive equilibrium $(\lambda, v_\lambda, g(\lambda)v_\lambda\tau_2)$ is considered as follows:

Theorem 3.11 *Assume that* $d, d_1, d_2 > 0$.

1. *When* $\lambda_0 < \lambda \le K$, *where* $\lambda_0 \in (\bar{\lambda}, K)$ *and* $f(\lambda) = f(0)$, *the equilibrium solution* $(\lambda, v_\lambda, g(\lambda)v_\lambda\tau_2)$ *is globally asymptotically stable.*
2. *When* $\lambda > K$ *(i.e.,* $d > g(K)$*), the equilibrium solution* $(K, 0, 0)$ *is globally asymptotically stable.*

Proof

1. When $\lambda_0 < \lambda \le K$, $(\lambda, v_\lambda, g(\lambda)v_\lambda\tau_2)$ is the unique positive equilibrium solution of model (3.62). A Lyapunov function is conducted as follows:

$$V(x,t) = \int_0^\pi \int_\lambda^u \frac{g(\zeta)-d}{g(\zeta)} d\zeta dx + \int_0^\pi \int_{v_\lambda+q}^{v+q} \frac{\eta - q - v_\lambda}{\eta - q} d\eta dx.$$

Based on the calculations, it can be determined that

$$\begin{aligned}\frac{dV}{dt} &= \int_0^\pi \frac{g(u)-d}{g(u)} u_t dx + \int_0^\pi \frac{v-v_\lambda}{v}(v_t+q_t)dx,\\ &\le \int_0^\pi (g(u)-g(\lambda))(f(u)-f(\lambda))dx - d_1 d\int_0^\pi \frac{g'(u)}{g^2(u)}|\nabla u|^2 dx\\ &- d_2 v_\lambda \int_0^\pi \frac{|\nabla v|^2}{v^2} dx.\end{aligned}$$

By the definition of λ_0 and condition (ii), it is concluded that $V_t \le 0$. Therefore, $(u(x,t), v(x,t), q(x,t))$ is globally stable.

2. When $\lambda > K$, a similar Lyapunov function is conducted as

$$V(x,t) = \int_0^\pi \int_K^u \frac{g(\zeta)-g(K)}{g(\zeta)} d\zeta dx + \int_0^\pi (v+q)dx.$$

Based on the calculations, it is concluded that

$$\begin{aligned}\frac{dV}{dt} &= \int_0^\pi \frac{g(u)-g(K)}{g(u)} u_t dx + \int_0^\pi (v_t+q_t)dx,\\ &\le - d_1 g(K)\int_0^\pi \frac{g'(u)}{g^2(u)}|\nabla u|^2 dx\\ &+ \int_0^\pi (g(u)-g(K))f(u) + (g(K)-d)v\\ &- d_2 v_\lambda \int_0^\pi \frac{|\nabla v||\nabla q|}{v^2} dx.\end{aligned}$$

From $g(K) < d$, it can be deduced that $V_t \le 0$. Therefore, $(K, 0, 0)$ is globally stable.

Next, we apply the reaction-diffusion equation's Hopf bifurcation theory [40] to study the Hopf bifurcation of model (3.62), with d_1 selected as the bifurcation parameter.

The Jacobian matrix of model (3.62) at the equilibrium $(\lambda, v_\lambda, v_\lambda\tau_2 g(\lambda))$ is given by

$$J_{(\lambda,v_\lambda,v_\lambda\tau_2 g(\lambda))} = \begin{pmatrix} A_1(\lambda) - d_1 n^2 - d & 0 \\ 0 & -d - d_2 n^2 & \frac{1}{\tau_2} \\ B_1(\lambda) & d & -\frac{1}{\tau_2} - d_2 n^2 \end{pmatrix}.$$

The characteristic equation is

$$\Phi_4(\phi, n^2, q) = \phi^3 + b_2\phi^2 + b_1\phi + b_0 = 0, \tag{3.64}$$

here

$$b_0 = d_2 n^2(\gamma_2 - d_2 n^2)(d_1 n^2 - A_1) + \frac{dB_1}{\tau_2},$$

$$b_1 = n^2(d_2(\gamma_1 + d_1 n^2) + (d_1 + d_2)(d + \frac{1}{\tau_2})) - A_1\gamma_2,$$

$$b_2 = d + (d_1 + 2d_2)n^2 + \frac{1}{\tau_2} - A_1,$$

where $\gamma_1 = (d_1 + d_2)n^2 > 0$, $\gamma_2 = d + 2d_2 n^2 + \frac{1}{\tau_2} > 0$.

Theorem 3.12 *Assuming* $d, d_1, d_2 > 0$, $n \geq 1$ *and* $v_\lambda g'(\lambda) + df'(\lambda) < 0$, *model* (3.62) *have a Hopf bifurcation at the equilibrium* $(\lambda, v_\lambda, v_\lambda\tau_2 d)$ *when the bifurcation parameter* d_1 *satisfies the condition*

$$\tau_2\gamma_2(A_1(A_1 - \gamma_2 - 2d_1 n^2) + \gamma_1(d + \gamma_1 + \frac{1}{\tau_2})) = dv_\lambda g'(\lambda). \tag{3.65}$$

Proof Given the assumption $v_\lambda g'(\lambda) + df'(\lambda) < 0$, it follows that b_2, b_1, and b_0 are all positive. This observation indicates that the existence of a Hopf bifurcation point is restricted to the interval $[0, \bar{\lambda})$. Additionally, when the condition in Eq. (3.65) is satisfied by the parameter d_1, the equality $b_2 b_1 = b_0$ holds. Consequently, Eq. (3.64) can be expressed as

$$\Phi_4(\phi, n^2, q) = (\phi^2 + b_1)(\phi + b_2). \tag{3.66}$$

Since $A_1 < 0$, the roots of Eq. (3.66) are

$$\phi_{1,2} = \pm\sqrt{b_1}i, \phi_3 = -b_2 < 0.$$

Now the transversality condition for the Hopf bifurcation is driven.

$$\frac{d\phi}{dd_1} = -\frac{d\Phi_4}{dd_1}/\frac{d\Phi_4}{d\phi} = -\frac{\zeta_2\phi^2 + \zeta_1\phi + \zeta_0}{3\phi^2 + 2b_2\phi + b_1},$$

here

$$\zeta_2 = n^2 > 0, \zeta_1 = n^2(d + 2d_2n^2 + \frac{1}{\tau_2}) > 0, \zeta_0 = d_2n^4(d + d_2n^2 + \frac{1}{\tau_2}) > 0.$$

When $b_2b_1 = b_0$ and $n \geq 1$,

$$\begin{aligned} Re(\frac{d\phi}{dd_1}) &= \frac{-b_1\zeta_2 + \zeta_0 - \zeta_1 b_2}{2(b_1 + b_2^2)}, \\ &= \frac{-n^2(d_2^2n^4 + b_1 + (d + \gamma_1 + \frac{1}{\tau_2} - A_1)\gamma_2)}{2(b_1 + b_2^2)}, \\ &< 0. \end{aligned}$$

Thus, for model (3.62), a Hopf bifurcation occurs at $(\lambda, v_\lambda, v_\lambda\tau_2 d)$.

Subsequently, the focus will be on the dynamics of a predator-prey model with time delays that occur exclusively in the predator's feeding process. The kernel function is defined as follows:

$$m_1(t) = \frac{1}{\tau_1}e^{-\frac{t}{\tau_1}}, \; m_2(t) = \delta(t).$$

Then (3.54) can be written as

$$\begin{cases} u_t(x,t) = d_1u_{xx} + g(u)f(u) - \dfrac{1}{\tau_1}p, & x \in (0,\pi), t > 0, \\ v_t(x,t) = d_2v_{xx} + g(u)v - dv, & x \in (0,\pi), t > 0, \\ p_t(x,t) = d_1p_{xx} + g(u)v - \dfrac{1}{\tau_1}p, & x \in (0,\pi), t > 0, \\ \dfrac{\partial u}{\partial x} = \dfrac{\partial v}{\partial x} = \dfrac{\partial p}{\partial x} = 0, & x = 0, \pi, t > 0, \\ u(x,0) = u_0(x) \geq 0, v(x,0) = v_0(x) \geq 0, & x \in [0,\pi], \\ p(x,0) = \int_{-\infty}^{0}\int_0^{\pi} G_1(x,y,-s) & \\ \qquad \times e^{\frac{s}{\tau_1}} g(u(y,s))v(y,s)dyds \geq 0, & t \in [-\infty, 0]. \end{cases} \tag{3.67}$$

Model (3.67) has three equilibria: $(0,0,0)$, $(K,0,0)$, and $(\lambda, v_\lambda, \lambda v_\lambda \tau_1)$.

The occurrence of a Hopf bifurcation for model (3.67) at $(\lambda, v_\lambda, \lambda v_\lambda \tau_1)$ is considered, where d_1 is the bifurcation parameter.

The Jacobian matrix of model (3.67) at $(\lambda, v_\lambda, \lambda v_\lambda \tau_1)$ is given by

$$J_{(\lambda, v_\lambda, \lambda v_\lambda \tau_1)} = \begin{pmatrix} A_1(\lambda) - d_1 n^2 & 0 & -\dfrac{1}{\tau_1} \\ B_1(\lambda) & -d_2 n^2 & 0 \\ B_1(\lambda) & -d & -\dfrac{1}{\tau_1} - d_1 n^2 \end{pmatrix}.$$

The characteristic equation is

$$\Phi_5(\phi, n^2, q) = \phi^3 + \phi^2 a_2 + \phi a_1 + a_0 = 0, \tag{3.68}$$

here

$$a_0 = d_2 n^2 (d_1 n^2 + \frac{1}{\tau_1})(d_1 n^2 - A_1) + \frac{B_1}{\tau_1}(d_2 n^2 - d),$$

$$a_1 = d_1(d_1 + 2d_2)n^4 + (B_1 + (d_1 + d_2)n^2)\frac{1}{\tau_1} - A_1((d_1 + d_2)n^2 + \frac{1}{\tau_1}),$$

$$a_2 = (2d_1 + d_2)n^2 + \frac{1}{\tau_1} - A_1.$$

Let

$$A_1 = v_\lambda g'(\lambda) + df'(\lambda) < 0, d < d_2, \tag{3.69}$$

It implies that $a_2, a_1, a_0 > 0$. Furthermore, when $a_2 a_1 = a_0$, the model (3.68) can be expressed as

$$\Phi_5(\phi, n^2, q) = (\phi^2 + a_1)(\phi + a_2). \tag{3.70}$$

In this case, the characteristic roots are given by

$$\phi_{1,2} = \pm\sqrt{a_1} i, \ \phi_3 = -a_2 < 0.$$

For convenience, let

$$A_2 = A_1 - 2d_1 n^2 - \frac{1}{\tau_1} < 0.$$

By performing the necessary calculations under the assumption (3.69), the following results are obtained:

$$a_2a_1 - a_0 = A_1\tau_1(A_2 - (d_1 + d_2)n^2)((d_1 + d_2)n^2 + \frac{1}{\tau_1}) - B_1(A_2 - d) > 0,$$

which leads to a contradiction. Therefore, when $d, d_1, d_2 > 0$, $d_2 > d$, and $v_\lambda g'(\lambda) + df'(\lambda) < 0$, the model (3.67) does not undergo a Hopf bifurcation at the point $(\lambda, v_\lambda, \lambda v_\lambda \tau_1)$.

3.2.4 Conclusion

This subsection presents an exploration into diffusive predator-prey models featuring nonlocal delays, aiming to investigate the impact of delays and population diffusion rates on the dynamic behavior of ecological systems. Through a comparative analysis of the local stability conditions between the instantaneous response model and the delayed response model, it is discerned that nonlocal delays do not exert an influence on the local asymptotic stability of equilibrium solutions. Furthermore, the text delves into the examination of solution existence, uniqueness, and boundedness, focusing on specific time delay kernels. The discussion extends to encompass global stability of equilibrium points and the occurrence of Hopf bifurcations. Notably, two distinctive scenarios are examined by selecting distinct kernel functions: (1) a delay solely in the predator's prey capture process, and (2) a delay exclusively in the predator's feeding process. The findings unveil that under identical conditions, the model in scenario 1 can undergo a Hopf bifurcation in proximity to the positive equilibrium, whereas the model in scenario 2 does not exhibit a Hopf bifurcation phenomenon.

Chapter 4
Physiological Modeling with Dynamic Systems

Diabetes is a chronic lifelong disease that can lead to chronic tissue damage, neuropathy, and functional impairments in the body. Specific complications include diabetic nephropathy, diabetic foot, diabetic cerebrovascular disease, diabetic ketoacidosis, diabetic retinopathy, and diabetic heart disease. According to the World Health Organization (WHO) and the International Diabetes Federation (IDF), in 2019, approximately 415 million people worldwide between the ages of 20 and 79 had diabetes, with a prevalence rate of 9.3%. The average growth rate was 51%, and it is estimated that by 2045, there will be 700 million people with diabetes globally [52]. With the development of the world economy and population aging, the high disability and mortality rates associated with diabetes have become one of the fastest-growing major social issues in the twenty-first century [53, 54]

Diabetes is caused by defects in insulin secretion or impaired biological action, leading to disturbances in glucose and lipid metabolism. Insulin, as the only protein hormone in the body that lowers blood glucose, is synthesized in response to plasma glucose concentration. When blood glucose levels rise, insulin synthesis accelerates in pancreatic β cells, promoting the oxidative breakdown of glucose and glycogen synthesis. Diabetes can be broadly categorized into three types: Type I diabetes, Type II diabetes, and gestational diabetes. Type I diabetes is characterized by a significant reduction or absence of insulin secretion due to a decrease or loss of pancreatic β cells, predominantly affecting children and adolescents. Type II diabetes is characterized by decreased insulin sensitivity (insulin resistance) and reduced insulin secretion (or relative deficiency) due to pancreatic β-cell dysfunction, primarily affecting individuals over the age of 50. Gestational diabetes results from abnormal glucose metabolism during pregnancy.

J. Ren, Y. Tao, *Modeling in Life Sciences and Ecology*, Springer Asia Pacific Mathematics Series 4, https://doi.org/10.1007/978-981-95-1038-2_4

4.1 Dynamics of a Glucose Regulatory Model

The precise regulation of glucose in the human body is a crucial characteristic of internal balance, contributing to the optimization of energy release in muscles and organs. The rhythmic variations in blood glucose levels are influenced by both internal and external stimuli, and these variations have been observed physiologically in humans and comprehensively understood through mathematical modeling. These mathematical models not only provide qualitative and quantitative analytical frameworks but also aid in predicting the dynamic processes of glucose regulation. The increasing utilization of wearable monitoring devices [55] has further enhanced their significance in informing clinical decisions, tracking the progression of conditions like diabetes [56], and devising optimal treatment strategies that incorporate natural circadian rhythms [57]. In this chapter, we elucidate the dynamical behavior of the human blood glucose regulatory system under the influence of periodic perturbations and in the absence of perturbations [58]. This encompasses various complex dynamical phenomena, such as periodic solutions of different periods, bifurcations, chaos, and bistability.

4.1.1 Glucose-Insulin-β-Cell System

Topp et al. [59] proposed a glucose-insulin-β-cell regulatory system, which can be described by the following three-dimensional system of ordinary differential equations:

$$\begin{cases} \dot{G} = a - (b + cI)G, \\ \dot{I} = \dfrac{d\beta G^2}{e + G^2} - fI, \\ \dot{\beta} = (-g + hG - g_1 G^2)\beta. \end{cases} \tag{4.1}$$

Here, G, I, and β represent the concentrations of blood glucose, insulin, and β-cell mass, respectively. a represents the hepatic glucose production rate, b is the glucose consumption rate independent of insulin, c represents insulin sensitivity, d is the insulin secretion rate, e is the half-saturation constant, f represents insulin clearance rate, h represents glucose tolerance, and g and g_1 are the rates of β-cell apoptosis and necrosis, respectively.

Through calculations, it can be concluded that model (4.1) has three equilibria:

$$E_0(\frac{a}{b}, 0, 0), \ \ E_1(G_1, I_1, \beta_1), \ \ E_2(G_2, I_2, \beta_2),$$

where

$$G_{1,2} = \frac{h \pm \sqrt{h^2 - 4g_1 g}}{2g_1}, \quad I_i = \frac{a}{cG_i} - \frac{b}{c}, \quad \beta_i = \frac{f I_i (e + G_i^2)}{d G_i^2}.$$

The Jacobi matrix of model (4.1) at any point (G, I, β) is

$$J(G, I, \beta) = \begin{pmatrix} -(b + cI) & -cG & 0 \\ \dfrac{2deG\beta}{(e + G^2)^2} & -f & \dfrac{dG^2}{e + G^2} \\ (h - 2g_1 G)\beta & 0 & hG - g_1 G^2 - g \end{pmatrix}.$$

The characteristic equation at E_i is

$$H(\lambda)_{i=1,2} = \lambda^3 + \delta(G_i)\lambda^2 + \zeta(G_i)\lambda + \theta(G_i) = 0, \tag{4.2}$$

where

$$\begin{aligned} \delta(G_i) &= f + \frac{a}{G_i}, \\ \zeta(G_i) &= \frac{af}{G_i} + \frac{2ef(a - bG_i)}{G_i(e + G_i^2)}, \\ \theta(G_i) &= f(bG_i - a)(2g_1 G_i - h). \end{aligned}$$

4.1.2 Bifurcation Analysis of the $GI\beta$ System

Theorem 4.1 *For model (4.1), the trivial equilibrium $E_0\left(\frac{a}{b}, 0, 0\right)$ is characterized by the following properties:*

- *It is a saddle-node of codimension 1 when $\frac{hb}{a} - \frac{g_1 b^2}{a^2} - g = 0$ and $\frac{2ag_1}{b} - h \neq 0$.*
- *It is a semi-hyperbolic point of codimension 2 when $\frac{hb}{a} - \frac{g_1 b^2}{a^2} - g = \frac{2ag_1}{b} - h = 0$.*

Proof For

$$\frac{hb}{a} - \frac{g_1 b^2}{a^2} - g = 0,$$

the eigenvalues of the Jacobian matrix of model (4.1) at E_0 are $-b, 0, -f$.

Let

$$T = (T_{ij})_{3\times 3} = (V_1, V_2, V_3)$$

be the transformation matrix, where V_1, V_2, V_3 are the corresponding eigenvectors of the Jacobian matrix. By performing the transformation

$$x = G - \frac{a}{b},\ y = I,\ \text{and}\ z = \beta,$$

the equilibrium E_0 can be brought to the origin.

Let

$$\begin{pmatrix} x \\ y \\ z \end{pmatrix} = \begin{pmatrix} 1 & 1 & 1 \\ 0 & -\frac{b^2}{ac} & \frac{b(f-b)}{ac} \\ 0 & \frac{-fb^2(a^2+eb^2)}{cda^3} & 0 \end{pmatrix} \begin{pmatrix} u \\ v \\ w \end{pmatrix},$$

model (4.1) becomes

$$\begin{cases} \dot{u} = -bu + \sum\limits_{i,j,k\in N}^{i+j+k=2} m_{ijk}u^i v^j w^k + O(|u, v, w|^3), \\ \dot{v} = \sum\limits_{i,j,k\in N}^{i+j+k=2,3} l_{ijk}u^i v^j w^k + O(|u, v, w|^4), \\ \dot{w} = -fw + \sum\limits_{i,j,k\in N}^{i+j+k=2} n_{ijk}u^i v^j w^k + O(|u, v, w|^3), \end{cases} \tag{4.3}$$

where

$$l_{110} = l_{011} = l_{020} = h - \frac{2ag_1}{b}, \qquad l_{030} = -g_1.$$

Then the center manifold can be computed for $v \sim 0$, indicating the existence of a center manifold,

$$u = \frac{m_{020}}{b}v^2 + O(v^3), \qquad w = \frac{n_{020}}{f}v^2 + O(v^3). \tag{4.4}$$

If $l_{020} = 0$, the reduced model on the one-dimensional center manifold (4.4) for model (4.3) is given by

$$\begin{aligned} \dot{v} &= (l_{030} + l_{110}\frac{m_{200}}{b} + l_{011}\frac{n_{200}}{f})v^3 + O(v^4), \\ &= -g_1 v^3 + O(v^4). \end{aligned}$$

The coefficient of v^3 is well-defined and does not vanish. Therefore, E is a semi-hyperbolic point of codimension 2.

If $l_{020} \neq 0$, it can be deduced that the model on the center manifold is topologically equivalent to

$$\dot{v} = \pm v^2 + O(v^3). \tag{4.5}$$

Therefore, E is a saddle node of codimension 1.

In the following contexts, the bifurcations of the nontrivial equilibrium will be discussed.

Theorem 4.2 *Model (4.1) undergoes a Hopf bifurcation at E_1 if the condition*

$$\delta(G_1)\zeta(G_1) = \theta(G_1)$$

is satisfied.

Proof Considering g_1 as a bifurcation parameter, substituting $\delta(G_1)\zeta(G_1) = \theta(G_1)$ into (4.2) yields

$$(\lambda^2 + \zeta)(\lambda + \delta) = 0. \tag{4.6}$$

The three roots of (4.6) are

$$\lambda_{1,2} = \pm\sqrt{\zeta(G_1)}i, \qquad \lambda_3 = \delta(G_1).$$

It can be deduced that

$$\frac{\partial \lambda_{1,2}}{\partial g_1} = \frac{2fg_1G_1(a - bG_1)}{3\lambda^2 + 2\delta\lambda + \zeta},$$

and for $\zeta > 0$,

$$(\mathrm{Re}\frac{\partial \lambda_{1,2}}{\partial g_1}) \mid_{\delta\zeta=\theta} = \frac{bfg_1G_1}{\zeta^2 + \zeta\delta^2} \neq 0,$$

$$(\mathrm{Im}\frac{\partial \lambda_{1,2}}{\partial g_1}) \mid_{\delta\zeta=\theta} = \frac{\sqrt{b}afg_1G_1}{\zeta^2 + \zeta\delta^2} \neq 0.$$

Similarly, considering a as a bifurcation parameter, and for $\zeta > 0$, the following equation is obtained:

$$(\mathrm{Re}\frac{\partial \lambda_{1,2}}{\partial a}) \mid_{\delta\zeta=\theta} = \frac{bf(bG_1 - a)}{2(\zeta^2 + \zeta\delta^2)} \neq 0,$$

$$(\mathrm{Im}\frac{\partial \lambda_{1,2}}{\partial a}) \mid_{\delta\zeta=\theta} = \frac{\sqrt{b}af(bG_1 - a)}{2(\zeta^2 + \zeta\delta^2)} \neq 0.$$

Hence, model (4.1) undergoes Hopf bifurcation at E_1.

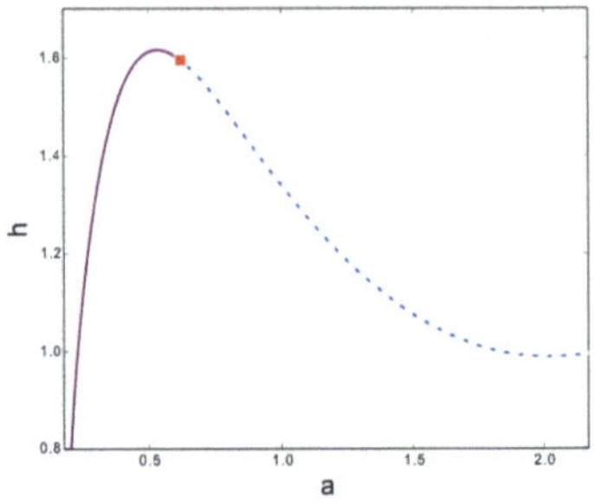

Fig. 4.1 Hopf bifurcation curve with $a = 6.13,\ b = 5.9,\ c = 1,\ d = 4,\ e = 0.1,\ f = 1,\ g = 0.99,\ h = 1.83,\ g_1 = 0.72$. The solid curve represents the subcritical Hopf bifurcation curve, while the dashed curve represents the supercritical Hopf bifurcation curve. The square denotes the degenerated Hopf point

The Hopf bifurcation curve for the given parameters $a = 6.13,\ b = 5.9,\ c = 1,\ d = 4,\ e = 0.1,\ f = 1,\ g = 0.99,\ h = 1.83,\ g_1 = 0.72$ is shown in Fig. 4.1. The solid curve represents the subcritical Hopf bifurcation curve, and the dashed curve represents the supercritical Hopf bifurcation curve. The square on the curve denotes the degenerate Hopf point.

In the following, the proof of the saddle-node bifurcation of codimension 1 for the model is presented. The two nontrivial equilibria, E_1 and E_2, collide at $\bar{E}(\bar{G}, \bar{I}, \bar{\beta})$, where

$$\bar{G} = \frac{h}{2g_1},\quad \bar{I} = \frac{2ag_1 - bh}{ch},\quad \bar{\beta} = \frac{4efg_1^2(2ag_1 - bh)}{cdh}.$$

Theorem 4.3 *For model (4.1), the nontrivial equilibrium $\bar{E}(\bar{G}, \bar{I}, \bar{\beta})$ is a saddle node of codimension 1 when $h^2 - 4g_1g = 0$.*

Proof For $h^2 - 4g_1g = 0$, the corresponding characteristic equation is

$$H(\lambda) = \lambda^3 + \delta(G_i)\lambda^2 + \zeta(G_i)\lambda,$$

with a zero eigenvalue.

By utilizing the transformation $x = G - \bar{G}$, $y = I - \bar{I}$, and $z = \beta - \bar{\beta}$, model (4.1) can be expressed as

$$\begin{pmatrix} \dot{x} \\ \dot{y} \\ \dot{z} \end{pmatrix} = J(\bar{E}) \begin{pmatrix} x \\ y \\ z \end{pmatrix} + \begin{pmatrix} a_{110}xy \\ b_{200}y^2 + b_{101}xz + b_{300}x^3 + b_{201}x^2z \\ c_{200}z^2 + c_{101}xz + c_{201}x^2z \end{pmatrix} + O(|x, y, z|)^4, \tag{4.7}$$

where

$$a_{110} = -c, b_{101} = \frac{2ed\bar{G}}{(e+\bar{G}^2)^2}, b_{200} = \frac{ed\bar{\beta}(e-3\bar{G}^2)}{(e+\bar{G}^2)^3}, c_{101} = h - 2g_1\bar{G},$$

$$c_{201} = -\frac{g_1}{3}, b_{300} = -\frac{2ed\bar{\beta}\bar{G}}{(e+\bar{G}^2)^3}, b_{201} = \frac{ed(e-3\bar{G}^2)}{(e+\bar{G}^2)^3}, c_{200} = -g_1\bar{\beta}.$$

Let $q_0 \in \mathbb{R}^3$ be proper eigenvector for $Jq_0 = 0$ and $p_0 \in \mathbb{R}^3$ be corresponding adjoint eigenvector for $J^T p_0 = 0$. To achieve the required normalization $\langle p_0, q_0\rangle = 1$, one can choose

$$q_0 = \begin{pmatrix} \frac{n}{s} \\ \frac{(b+c\bar{I})n}{cs\bar{G}} \\ 1 \end{pmatrix}, \qquad p_0 = \begin{pmatrix} 0 \\ 0 \\ 1 \end{pmatrix},$$

where

$$n = \frac{-d\bar{G}^2}{e+\bar{G}^2}, \qquad s = f(b+c\bar{I}) + \frac{2cde\bar{\beta}\bar{G}^2}{(e+\bar{G}^2)^2}.$$

According to the center manifold theorem, the restriction of model (4.7) to the center manifold takes the form

$$\dot{v} = \frac{1}{2}\langle p, B(q_0, q_0)\rangle v^2 + O(v^3),$$

where

$$B(q_0, q_0) = 2\frac{n}{s}\begin{pmatrix} a_{110}\frac{n}{s} \\ b_{200}\frac{n}{s} + b_{101} \\ c_{200}\frac{n}{s} \end{pmatrix}, \quad \langle p_0, B(q_0, q_0)\rangle = -2g_1\bar{\beta}\frac{n^2}{s^2} \neq 0.$$

Therefore, $\bar{E}$ is a nondegenerate saddle node of codimension 1, and the model (4.7) restricted to the center manifold is topologically equivalent to the model (4.5).

Next, the zero-Hopf bifurcation of codimension 2 and the Bogdanov-Takens bifurcation of codimension 2 and 3 will be studied.

Theorem 4.4 *For model (4.1), the nontrivial equilibrium $\bar{E}$ exhibits the following properties:*

- *It is a zero-Hopf bifurcation point of codimension 2 when $\delta(\bar{G}) = \theta(\bar{G}) = 0$ and $F(0) \neq 0$;*

- *it is a zero-Hopf bifurcation point of codimension at least 3 when* $\delta(\bar{G}) = \theta(\bar{G}) = F(0) = 0$*;*

where

$$F(0) = \frac{-mc\bar{G}}{[(b+c\bar{I})^2 + \omega^2 + mc\bar{G}]}(\frac{1}{\bar{G}} - \frac{ed\bar{\beta}(e-3\bar{G}^2)}{(e+\bar{G}^2)^3} - \frac{g_1\bar{\beta}n(b+c\bar{I})}{cs\bar{G}}).$$

Proof By selecting a and c as bifurcation parameters, denoted further by $\alpha = (a, c)$, the characteristic equation corresponding to the coinciding equilibria, where $\delta(\bar{G}) = \theta(\bar{G}) = 0$, is given by

$$H(\lambda) = \lambda^3 + \zeta(G_i)\lambda = 0,$$

and its eigenvalues are 0 and $\pm\,\omega i$, where $\omega^2 = \dfrac{af}{\bar{G}} + \dfrac{2ef(a-b\bar{G})}{\bar{G}(e+\bar{G}^2)}$.

Let $q_1, p_1 \in \mathbb{C}^3$ are corresponding eigenvector and adjoint eigenvector given by

$$Jq_1 = i\omega, \quad J^T p_1 = -i\omega,$$

such that

$$\langle p_0, q_0\rangle = 1,\ \langle p_1, q_1\rangle = 1,\ \langle p_0, q_1\rangle = 0,\ \langle p_1, q_0\rangle = 0,$$

where

$$q_1 = \begin{pmatrix} 1 \\ \dfrac{-(i\omega + b + c\bar{I})}{c\bar{G}} \\ 0 \end{pmatrix},$$

$$p_1 = \frac{-mc\bar{G}}{(b+c\bar{I})^2 + \omega^2 + mc\bar{G}} \begin{pmatrix} -1 \\ \dfrac{-i\omega + b + c\bar{I}}{m} \\ \dfrac{n}{s} - \dfrac{n(b+c\bar{I})(b+c\bar{I}-i\omega)}{smc\bar{G}} \end{pmatrix},$$

with $m = \dfrac{2ed\bar{G}\bar{\beta}}{(e+\bar{G}^2)^2}$.

By employing the transformation

$$u = \langle p_0(\alpha), x\rangle, z = \langle p_1(\alpha), x\rangle,$$

where $x = uq_0+zq_1+\bar{z}\bar{q}_1$, the model (4.7) can be further expressed in the following form:

$$\begin{cases} \dot{u} = g(u, z, \bar{z}, \alpha), \\ \dot{z} = \lambda z + h(u, z, \bar{z}, \alpha), \end{cases} \tag{4.8}$$

where

$$g(u, z, \bar{z}, \alpha) =\langle p_0, F(uq_0 + zq_1 + \bar{z}\bar{q}_1, \alpha)\rangle,$$

$$h(u, z, \bar{z}, \alpha) =\langle p_1, F(uq_0 + zq_1 + \bar{z}\bar{q}_1, \alpha)\rangle.$$

Moreover, these functions are smooth with respect to their variables, and their Taylor expansions in u, z, and $\bar{z}$ start with quadratic terms:

$$g(u, z, \bar{z}, \alpha) = \sum_{i+j+k\geq 2} \frac{1}{i!j!k!} g_{ijk}(\alpha)u^i z^j \bar{z}^k,$$

$$h(u, z, \bar{z}, \alpha) = \sum_{i+j+k\geq 2} \frac{1}{i!j!k!} h_{ijk}(\alpha)u^i z^j \bar{z}^k.$$

By calculation, one can get

$$g_{200} = c_{200}\frac{n^2}{s^2} + c_{101}\frac{n}{s},\ g_{020} = c_{200},\ g_{002} = c_{200},\ g_{101} = 2c_{200}\frac{n}{s} + c_{101},$$

$$g_{110} = 2c_{200}\frac{n}{s} + c_{101},\ g_{011} = 2c_{200},\ h_{011} = 2(\frac{(b + c\bar{I})}{\bar{G}} + m_1) + 2m_2 i,$$

$$h_{020} = (\frac{b + c\bar{I}}{\bar{G}} + m_1) + (\frac{\omega}{\bar{G}} + m_2)i,\ h_{002} = (\frac{b + c\bar{I}}{\bar{G}} + m_1) + (\frac{-\omega}{\bar{G}} + m_2)i,$$

$$h_{101} = 2m_1\frac{n}{s} + m_3 + (\frac{-n\omega}{s\bar{G}} + \frac{2m_2 n}{s} + m_4)i,$$

$$h_{110} = 2m_1\frac{n}{s} + m_3 + (\frac{n\omega}{s\bar{G}} + \frac{2m_2 n}{s} + m_4)i,$$

$$h_{200} = \frac{-n^2(b + c\bar{I})}{s^2\bar{G}} + (m_1\frac{n}{s} + m_3)\frac{n}{s} + (m_2\frac{n}{s} + m_4)\frac{n}{s}i,$$

where

$$m_1 = \frac{b_{200}(b + c\bar{I})}{m} + c_{200}\frac{nmc\bar{G} - n(b + c\bar{I})^2}{mcs\bar{G}},$$

$$m_2 = \frac{\omega}{m}(-b_{200} + \frac{c_{200}n(b+c\bar{I})}{cs\bar{G}}),$$

$$m_3 = \frac{b_{101}(b+c\bar{I})}{m} + c_{101}\frac{nmc\bar{G} - n(b+c\bar{I})^2}{mcs\bar{G}},$$

$$m_4 = \frac{\omega}{m}(-b_{101} + \frac{c_{101}n(b+c\bar{I})}{cs\bar{G}}).$$

Under the assumption $g_{200} \neq 0$ and $g_{011} \neq 0$, the changes are introduced

$$v = u - \frac{1}{2}\frac{g_{020}}{2\omega i}z^2 + \frac{1}{2}\frac{g_{002}}{2\omega i}\bar{z}^2 - \frac{g_{110}}{\omega i}uz + \frac{g_{101}}{\omega i}u\bar{z},$$

$$w = z + \frac{1}{2}\frac{h_{200}}{\omega i}u^2 - \frac{1}{2}\frac{h_{020}}{\omega i}baru^2 + \frac{1}{2}\frac{h_{002}}{3\omega i}\bar{z}^2 + \frac{h_{101}}{2\omega i}u\bar{z} + \frac{h_{011}}{\omega i}z\bar{z},$$

and performed a linear scaling of the variables and time

$$u = v + \frac{1}{2}r_3(\alpha)v^2,$$

$$z = w + K(\alpha)vw,$$

$$dt = (1 + r_1 v + r_2|w|^2)d\tau,$$

where

$$r_1 = \frac{\text{Im}g_{110}h_{200}}{g_{200}\omega}, \quad K = \frac{r_2\omega i + H_{021}}{g_{011}},$$

$$r_3 = 2\text{Re}K - r_1 - \frac{2g_{110}\text{Im}h_{011} + g_{020}\text{Im}h_{010}}{g_{011}\omega},$$

with

$$H_{021} = \frac{i}{2\omega}(h_{011}h_{020} - \frac{1}{2}h_{101}g_{020} - 2|h_{011}|^2 - \frac{1}{3}|h_{011}|^2).$$

Then, model (4.8) becomes

$$\begin{cases} \dot{u} = \frac{1}{2}g_{200}u^2 + g_{011}|z|^2 + O(||(u, z, \bar{z})||^4), \\ \dot{z} = i\omega z + D(0)uz + E(0)u^2 z + O(||(u, z, \bar{z})||^4), \end{cases} \tag{4.9}$$

where

$$D(0) = 2m_1\frac{n}{s} + m_3 + (\frac{n\omega}{s\bar{G}} + \frac{4m_2 n}{s} + 3m_4)i,$$

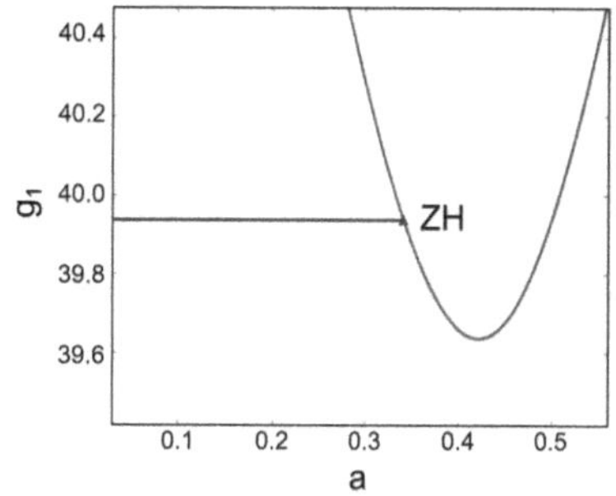

Fig. 4.2 Fold-Hopf bifurcation curve with $a = 0.34,\ b = 4,\ c = 2,\ d = 2,\ e = 100,\ f = 1,\ g = 7,\ h = 35.34,\ g_1 = 39.93$

$$F(0) = \frac{-mc\bar{G}}{[(b+c\bar{I})^2+\omega^2+mc\bar{G}]}\left(\frac{1}{\bar{G}} - \frac{ed\bar{\beta}(e-3\bar{G}^2)}{(e+\bar{G}^2)^3} - \frac{g_1\bar{\beta}n(b+c\bar{I})}{cs\bar{G}}\right).$$

If $F(0) = 0$, then $\bar{E}$ represents a degenerate zero-Hopf bifurcation point of codimension at least 3. Conversely, if $F(0) \neq 0$, the bifurcation can be further analyzed using linear scaling, and time-reparametrization

$$u = \frac{g_{200}}{F_0}u_1,\ z = \frac{g_{200}^3}{g_{011}F^2(0)}v_1,\ t = \frac{F(0)}{g_{200}^2}\tau,$$

model (4.9) is topologically equivalent to

$$\begin{cases} \dot{u_1} = u_1^2 \pm |u_1|^2 + O(||(u_1, v_1, \bar{v}_1)||^4), \\ \dot{v}_1 = \dfrac{E(0)}{g_{200}^2}i\omega u_1 + \dfrac{D(0)}{g_{200}}u_1v_1 + u_1^2v_1 + O(||(u_1, v_1, \bar{v}_1)||^4). \end{cases} \tag{4.10}$$

Hence, $\bar{E}$ is a nondegenerate zero-Hopf bifurcation point of codimension 2.

When $a = 0.34,\ b = 4,\ c = 2,\ d = 2,\ e = 100,\ f = 1,\ g = 7,\ h = 35.34,\ g_1 = 39.93$, model (4.1) exhibits a fold-Hopf bifurcation point in the parameter plane (a, g_1) as shown in Fig. 4.2. The dashed curve represents the fold bifurcation curve, and the solid curve corresponds to the Hopf bifurcation curve.

Theorem 4.5 *Suppose $\delta(\bar{G}) = \theta(\bar{G}) = 0$ and $\bar{M}_{11} \neq 0$. $\bar{E}$ is a Bogdanov-Takens point of codimension 2, and then model (4.1) localized at $\bar{E}$ is topologically equivalent to*

$$\begin{cases} \dot{u} = v, \\ \dot{v} = u^2 + sign(\bar{M}_{11})uv + O(|u, v|^3), \end{cases} \tag{4.11}$$

where

$$\bar{M}_{11} = \frac{g_1 f(e+\bar{G}^2)}{d\bar{G}^2} + \frac{cf[2(b+c\bar{I})-1]}{b+c\bar{I}} + \frac{(e-3\bar{G}^2)(2\bar{I}+d\bar{\beta}\bar{G})}{4e\bar{G}\bar{I}^2(e+\bar{G}^2)}.$$

Proof For $\delta(\bar{G}) = \theta(\bar{G}) = 0$, the characteristic equation corresponding to the coinciding equilibria $\bar{E}$ is given by

$$H(\lambda) = \lambda^3 + \zeta(\bar{G}_i)\lambda = 0.$$

The generalized eigenvectors corresponding to λ_1 and $\lambda_2 = 0$, as well as the eigenvector corresponding to $\lambda_3 = -f - b - c\bar{I}$, are given by

$$V_1 = \begin{pmatrix} -f \\ \dfrac{-2ef\bar{I}}{\bar{G}(e+\bar{G}^2)} \\ 0 \end{pmatrix}, \quad V_2 = \begin{pmatrix} \dfrac{f(1-b-c\bar{I})}{(b+c\bar{I})} \\ \dfrac{2ef\bar{I}}{\bar{G}(e+\bar{G}^2)} \\ \dfrac{-(e+\bar{G}^2)(f+b+c\bar{I})}{d\bar{G}^2(b+c\bar{I})} \end{pmatrix},$$

$$V_3 = \begin{pmatrix} b+c\bar{I} \\ \dfrac{-2ef\bar{I}}{\bar{G}(e+\bar{G}^2)} \\ 0 \end{pmatrix}.$$

Let $\bar{T} = (V_1, V_2, V_3)^T$, under the non-singular linear translation

$$\begin{pmatrix} x \\ y \\ z \end{pmatrix} = \bar{T} \begin{pmatrix} u \\ v \\ w \end{pmatrix},$$

model (4.1) becomes

$$\begin{cases} \dot{u} = -v + L_{20}u^2 + L_{11}uv + L_{02}v^2 + O(|u, v, w|^3), \\ \dot{v} = M_{20}u^2 + M_{11}uv + M_{02}v^2 + O(|u, v, w|^3), \\ \dot{w} = -(f+b+c\bar{I})w + O(|u, v, w|^2), \end{cases} \tag{4.12}$$

where

$$L_{20} = a_{110}\frac{f}{M} + b_{200}\frac{Nf^2}{Mm^2} + c_{200}\frac{n(MN-f)f^2}{m^2M^2},$$

$$L_{21} = \frac{Nf^2}{Mm^2}[3b_{300}\frac{f(N-1)}{mN} + b_{201}\frac{N}{M} + c_{201}\frac{(MN-f)}{N^2}],$$

$$L_{12} = \frac{(N-1)^2 f^2}{Mm^2N^2}[3b_{300}\frac{-f}{mN} + b_{201}\frac{M}{n} + c_{201}\frac{(MN-f)}{1-N}],$$

$$M_{02} = c_{101}\frac{f(N-1)}{mN}, \quad M_{21} = c_{201}\frac{f^2M}{m^2nN},$$

$$M_{20} = c_{200}\frac{nNf^2}{Mm^2}, \quad M_{11} = c_{200}\frac{2n(1-N)f^2}{Mm^2},$$

$$L_{11} = \frac{cf(2N-1)}{N} + \frac{f}{m}[\frac{b_{200}Mf(2-N)}{m} - \frac{b_{201}}{n}] + \frac{nf(MN-f)}{mNM^2}[c_{201}\frac{M}{n} + c_{200}\frac{f(N-2)}{m}],$$

with $M = f + b + c\bar{I}$, $N = b + c\bar{I}$, $m = \frac{2ef\bar{I}}{\bar{G}(e+G^2)}$. The other coefficients of the quadratic terms in model (4.12) are not presented here.

Based on the center manifold theorem, for sufficiently small ε_1 and ε_2, there exists a center manifold for model (4.12). Locally, this center manifold can be represented as follows:

$$W^c = \{(u, v, w)|w = H(u, v), |u| < \varepsilon_1, |v| < \varepsilon_2, H(0, 0) = 0, DH(0, 0) = 0\}.$$

Then model (4.12) restricted to the center manifold is given by

$$\begin{cases} \dot{u} = -v + L_{20}u^2 + L_{11}uv + L_{02}v^2 + O(|u, v, w|^3), \\ \dot{v} = M_{20}u^2 + M_{11}uv + M_{02}v^2 + O(|u, v, w|^3). \end{cases} \tag{4.13}$$

Using the near-identify transformation

$$\begin{cases} u = X, \\ v = -Y + L_{20}X^2 + M_{02}XY + O(|X, Y|^3), \end{cases}$$

and rewrite X, Y into u, v, it can be deduced that

$$\begin{cases} \dot{u} = v, \\ \dot{v} = \bar{M}_{20}u^2 + \bar{M}_{11}uv + O(|u, v|^3), \end{cases} \tag{4.14}$$

where $\bar{M}_{20} = -M_{20}$, $\bar{M}_{11} = -M_{02} - L_{11}$.

By calculation, one can get

$$\bar{M}_{20} = -\frac{f(b + c\bar{I})\bar{G}^2(e + \bar{G}^2)^2}{4g_1(f + b + c\bar{I})\bar{I}e^2} \neq 0,$$

$$\bar{M}_{11} = \frac{g_1 f(e+\bar{G}^2)}{d\bar{G}^2} + \frac{cf[2(b+c\bar{I})-1]}{b+c\bar{I}} + \frac{(e-3\bar{G}^2)(2\bar{I}+d\beta\bar{G})}{4e\bar{G}\bar{I}^2(e+\bar{G}^2)}.$$

Under the above assumptions, a change of coordinates and time can be made to preserve the orientation with respect to time

$$u \to |\frac{\bar{M}_{20}}{\bar{M}_{11}^2}|u, \qquad v \to |\frac{\bar{M}_{20}^2}{\bar{M}_{11}^3}|v, \qquad t \to |\frac{\bar{M}_{11}}{\bar{M}_{20}}|t,$$

model (4.1) topologically equivalent to the normal form (4.11).

Theorem 4.6 *Suppose that* $\zeta(\bar{G}) = \theta(\bar{G}) = \bar{M}_{11} = 0$ *and* $\bar{M}31 \neq 0$. *In this case,* $\bar{E}$ *represents a degenerate Bogdanov-Takens singularity of codimension 3. Moreover, the model (4.1) localized at* $\bar{E}$ *is topologically equivalent to the following system:*

$$\begin{cases} \dot{u} = v, \\ \dot{v} = u^2 + u^3 v + O(|u, v|^4)v, \end{cases} \tag{4.15}$$

where $\bar{M}_{31} = -(2L_{12}M_{20} + L_{21}M_{11} + L_{11}M_{21})$.

Proof The proof consists of seven steps.
Step 1: Expand model (4.7) up to order 4, and diagonalize the linear part using the transformation $\bar{T}$.
Step 2: Calculate the center manifold $Z = H(u, v)$ up to $O(|u, v|^4)$, and reduce the model to the center manifold. The reduced model is given by

$$\begin{cases} \dot{u} = -v + L_{20}u^2 + L_{02}v^2 + L_{11}u_1v_1 + L_{30}u^3 + L_{03}v^3 + L_{21}u^2v \\ \qquad + L_{12}v^2u + O(|u, v|^4), \\ \dot{v} = M_{02}v^2 + M_{02}v^2 + M_{03}v^3 + M_{21}u^2v + O(|u, v|^4). \end{cases} \tag{4.16}$$

Step 3: Reduction to a nonlinear oscillator.
Apply a near-identity transformation to obtain

$$\begin{cases} u_1 = u, \\ v_1 = -v + L_{20}u^2 + L_{02}v^2 + L_{11}u_1v_1 + L_{30}u^3 + L_{03}v^3 \\ \qquad + L_{21}u^2v + L_{12}v^2u + O(|u, v|^4), \end{cases}$$

which brings model (4.16) into

$$\begin{cases} \dot{u_1} = v_1, \\ \dot{v_1} = \sum\limits_{i=2,3,4} \bar{M}_{i0} u_1^i + \sum\limits_{i,j \in N}^{i+j=2,3,4} \bar{M}_{ij} u_1^i v_1^j + O(|u_1, v_1|^5), \end{cases} \tag{4.17}$$

where $\bar{M}_{20} = -M_{20}$ and $\bar{M}_{31} = -2L_{12}M_{20} - L_{21}M_{11} - L_{11}M_{21}$. The other coefficients of quadratic terms of model (4.17) are omitted here.

Step 4: Eliminating nonresonant terms in model (4.17).

Through a series of near-identity transformations, the model is transformed to

$$\begin{cases} u_{i-1} = u_i + u_{i-1}^i, \\ v_{i-1} = v_i, \end{cases}$$

model (4.17) is transformed to

$$\begin{cases} \dot{u_4} = v_4, \\ \dot{v_4} = \bar{M}_{20} u_4^2 + \bar{M}_{30} u_4^3 + \bar{M}_{40} u_4^4 + \bar{M}_{21} u_4^2 v_4 + \bar{M}_{31} u_4^3 v_4 + O(|u_4, v_4|^5). \end{cases} \tag{4.18}$$

Step 5: Removing $u_4^2 v_4$ terms from model (4.18).

Note that $\bar{M}20 \neq 0$. The expression $\bar{M}20u_4^2 + \bar{M}21u_4^2v_4$ can be simplified as $\bar{M}20u_4^2(1 + rv_4)$, where $r = \frac{\bar{M}21}{\bar{M}20}$. To simplify further, a change of variables is introduced: $u_4 = u_5$ and $v_4 = v_4(v_5) = v_5 + O(v_5^2)$. After making these substitutions, one can obtain

$$v_5 dv_5 = \frac{v_4}{1 + rv_4} dv_4. \tag{4.19}$$

That is

$$v_5^2 = \frac{2}{r}[v_4 - \frac{1}{r} ln(1 + rv_4)] = v_4^2 - \frac{2r}{3} v_4^3 + \frac{r^2}{2} v_4^4 + O(v_4^5).$$

Or, equivalently

$$\begin{aligned} v_5 =& v_4 - \frac{r}{3} v_4^2 - \frac{7r^2}{36} v_4^3 + O(v_4^4), \\ v_4 =& v_4(v_5) = v_5 + \frac{r}{3} v_5^2 + \frac{r^2}{36} v_5^3 + O(v_5^4). \end{aligned} \tag{4.20}$$

Model (4.18) can be transformed into the following form:

$$v_4 dv_4 = [Q(u_4, v_4) + O(|u_4, v_4|^5)]du_4, \tag{4.21}$$

where

$$Q(u_4, v_4) = \bar{M}_{20}u_4^2 + \bar{M}_{30}u_4^3 + \bar{M}_{40}u_4^4 + \bar{M}_{21}u_4^2 v_4 + \bar{M}_{31}u_4^3 v_4.$$

Now, dividing both sides of (4.21) by $1 + rv_4$ and utilizing the expression (4.19), one can obtain

$$v_5 dv_5 = \left\{ \frac{[Q(u_5, v_4(v_5)) - \bar{M}_{20}u_5^2[1 + rv_4(v_5)] + O(|u_5, v_4(v_5)|^5)}{1 + rv_4(v_5)} \right. \\ \left. + \bar{M}_{20}u_5^2 \right\} du_5,$$

then using the second equality of (4.20), we obtain

$$v_5 dv_5 = [\bar{M}_{20}u_5^2 + O(|u_5|^3) + \bar{M}_{31}u_5^3 v_5 + O(|u_5, v_5|^4)v_5]du_5,$$

which is equivalent to

$$\begin{cases} \dot{u_5} = v_5, \\ \dot{v_5} = \bar{M}_{20}u_5^2 + O(|u_5|^3) + \bar{M}_{31}u_5^3 v_5 + O(|u_5, v_5|^4)v_5. \end{cases} \tag{4.22}$$

Step 6: Changing $\bar{M}_{20}$ and $\bar{M}_{31}$ to be 1 in (4.22).

Under the assumption $\bar{M}20\bar{M}31 \neq 0$, the following changes of variables and time are made:

$$u_5 = \bar{M}_{20}^{\frac{1}{5}}\bar{M}_{31}^{-\frac{2}{5}}u_6, \quad v_5 = \bar{M}_{20}^{\frac{4}{5}}\bar{M}_{31}^{-\frac{3}{5}}v_6, \quad t = \bar{M}_{20}^{-\frac{3}{5}}\bar{M}_{31}^{\frac{1}{5}}\tau,$$

then model (4.22) becomes

$$\begin{cases} \dot{u_6} = v_6, \\ \dot{v_6} = u_6^2 + O(|u_6|^3) + u_6^3 v_6 + O(|u_6, v_6|^4)v_6. \end{cases} \tag{4.23}$$

Step 7: Eliminating the term $O(|u_6|^3)$ in model (4.23).

Let $\phi(u_6) = u_6^2 + O(u_6^3)$ and $\psi(u_6) = \int_0^{u_6} \phi(u_6)du_6$. By a change of coordinates and time

$$u_6 = (3\psi(u_6))^{\frac{1}{3}}, \quad v_6 = v_6, \quad \tau = (3\psi(u_6))^{-\frac{2}{3}}\phi(u_6)\tau,$$

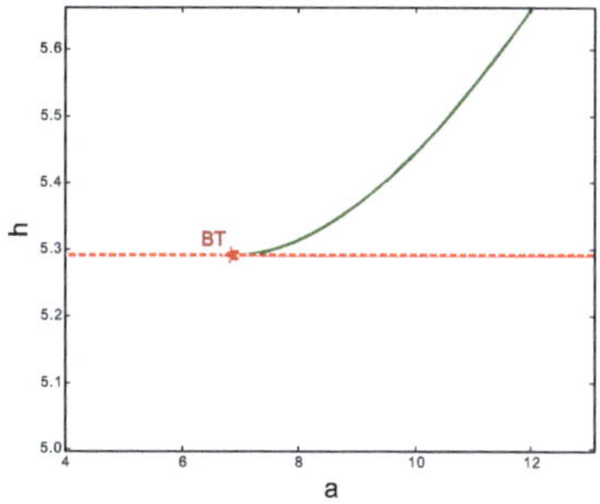

Fig. 4.3 Bogdanov-Takens bifurcation point with $a = 26.74,\ b = 4,\ c = 2,\ d = 2,\ e = 100,\ f = 1,\ g = 7,\ h = 8.24,\ g_1 = 1$

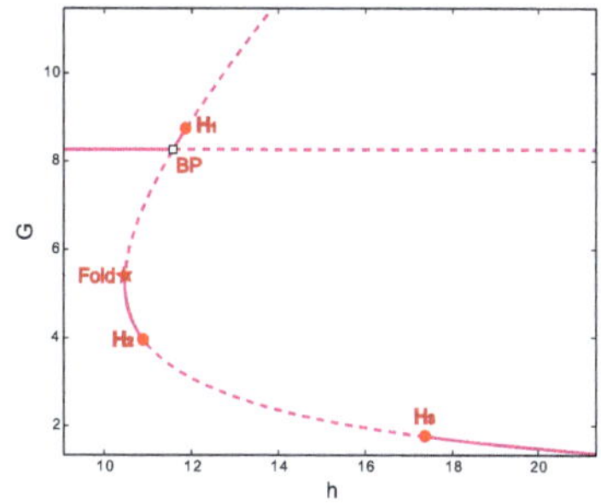

Fig. 4.4 Bifurcation diagram with $a = 16.5,\ b = 2,\ c = 1,\ d = 4,\ e = 3.6,\ f = 1,\ g = 27.4,\ h = 40,\ g_1 = 1$

and rewriting u_6, v_6 into u, v, one can get that

$$\begin{cases} \dot{u} = v, \\ \dot{v} = u^2 + u^3 v + O(|u, v|^4)v. \end{cases}$$

When $a = 26.74, b = 4, c = 2, d = 2, e = 100, f = 1, g = 7, h = 8.24, g_1 = 1$, model (4.1) exhibits a Bogdanov-Takens bifurcation point in the parameter plane (a, h) as shown in Fig. 4.3. The dashed curve represents the fold bifurcation curve, and the solid curve corresponds to the Hopf bifurcation curve.

To provide a more intuitive representation of the bifurcation results, Fig. 4.4 illustrates the bifurcation diagrams of model (4.1) for the parameter values $a = 16.5,\ b = 2,\ c = 1,\ d = 4,\ e = 3.6,\ f = 1,\ g = 27.4,\ h = 40,\ g_1 = 1$. The points $\text{H}_i (i = 1, 2, 3)$, Fold, and BP correspond to Hopf bifurcation points, fold bifurcation points, and Bogdanov-Takens bifurcation points, respectively. Solid lines represent stable equilibrium solutions, while dashed lines indicate unstable equilibrium solutions.

4.2 Dynamics of Perturbed Glucose Regulatory System

4.2.1 Approach to Investigation

Maintaining the balance of blood glucose homeostasis relies on intricate equilibrium and is influenced by factors such as insomnia [60], obesity [61], smoking [62], dietary habits [63], and time zone differences[64]. The internal circadian oscillator governs both physiological parameters and the cyclic external environment, both of which demonstrate evident circadian rhythmic patterns [64, 65].

Given the pivotal roles of hepatic glucose production and the glucose tolerance of β cells in glucose metabolism, we have employed the $GI\beta$ model (4.1) with four distinct mechanisms of periodicity. Specifically,

$$\begin{aligned}
&\text{mechanism 1: } a = a_0(1 + \varepsilon \sin 2\pi\omega t),\\
&\text{mechanism 2: } h = h_0(1 + \varepsilon \sin 2\pi\omega t),\\
&\text{mechanism 3: } a = a_0(1 + \varepsilon_1 \sin 2\pi\omega t) \text{ and } h = h_0(1 + \varepsilon_2 \sin 2\pi\omega t),\\
&\text{mechanism 4: } a = a_0(1 + \varepsilon_1 \sin 2\pi\omega t) \text{ and } h = h_0(1 + \varepsilon_2 \cos 2\pi\omega t).
\end{aligned} \tag{4.24}$$

Here, ε signifies the extent of periodicity, while a_0 and h_0 represent the mean values of variables a and h, respectively.

The bifurcation analysis of the model under periodic perturbation is conducted using a Poincaré map facilitated by a continuous technique. The introduction of periodic perturbations involves integrating a nonlinear oscillator's components to achieve the desired periodicity. For the model (4.1), the perturbation is set to a 1-day period. In this scenario, we adopt the following oscillator configuration:

$$\begin{cases}
\dot{v} = v + 2\pi w - v(v^2 + w^2),\\
\dot{w} = -2\pi v + w - w(v^2 + w^2),
\end{cases} \tag{4.25}$$

with an asymptotically stable periodic solution $v = \sin 2\pi t,\ w = \cos 2\pi t$. Under the periodicity mechanisms 1 and 2 outlined in (4.24), the original model (4.1) can be converted into four distinct autonomous five-dimensional models. As an illustration, considering the second mechanism (4.24), the resulting higher-dimensional autonomous model is

$$\begin{cases}
\dot{G} = a_0(1 + \varepsilon v) - (b + cI)G,\\
\dot{I} = \dfrac{d\beta G^2}{e + G^2} - fI,\\
\dot{\beta} = (-g + hG - g_1 G^2)\beta,\\
\dot{v} = v + 2\pi w - v(v^2 + w^2),\\
\dot{w} = -2\pi v + w - w(v^2 + w^2),
\end{cases} \tag{4.26}$$

and

$$\begin{cases} \dot{G} = a - (b + cI)G, \\ \dot{I} = \dfrac{d\beta G^2}{e + G^2} - fI, \\ \dot{\beta} = (-g + h_0(1 + \varepsilon v)G - g_1 G^2)\beta, \\ \dot{v} = v + 2\pi w - v(v^2 + w^2), \\ \dot{w} = -2\pi v + w - w(v^2 + w^2). \end{cases} \tag{4.27}$$

The equilibrium (G_a, I_a, β_a) of model (4.1) corresponds the periodic solution $(G_a, I_a, \beta_a, \sin 2\pi t, \cos 2\pi t)$ of the perturbed model (4.27). When $\varepsilon \neq 0$, we can calculate the bifurcation diagrams of model (4.27) by using the Poincaré map $\mathcal{P}$. For model (4.1), the first return map is

$$\mathcal{P} : (y_1(0),\ y_2(0),\ y_3(0),\ v(0),\ w(0)) \mapsto (y_1(1),\ y_2(1),\ y_3(1),\ v(1),\ w(1)).$$

The phase portrait of the mapping $\mathcal{P}$ comprises fixed points, regular and irregular invariant sets, as well as other trajectories. Specifically, we consider

Fixed points after the kth iteration of the mapping, corresponding to k-subharmonic solutions in the original continuous system. Closed regular invariant sets of the mapping, corresponding to quasiperiodic solutions (invariant tori) in the original continuous system. Irregular invariant sets of the mapping, corresponding to chaotic behavior in the original continuous system. For a periodically perturbed system, the bifurcations that may occur in its Poincaré map include Neimark-Sacker bifurcations (NS bifurcations), period-doubling bifurcations, flip bifurcations, and transcritical bifurcations, among others. We denote the curves corresponding to NS bifurcations, period-doubling bifurcations, flip bifurcations, and transcritical bifurcations as h^k, f^k, t^k, and r^k, respectively.

- NS bifurcation curve h^k
  On the curve h^k, the fixed points with a period of k for the mapping $\mathcal{P}$ have conjugate complex multipliers $\mu^k_{1,2} = \exp(\pm i\omega)$. Corresponding to the original continuous system, crossing h^k results in the bifurcation of a stable (unstable) quasiperiodic orbit of period k along with an unstable (stable) periodic orbit of period k.
- Period-doubling bifurcation curve f^k
  On the curve f^k, the fixed points with a period of k for the mapping $\mathcal{P}$ have the multiplier $\mu^k = -1$. Corresponding to the original continuous system, crossing f^k leads to the bifurcation of a saddle (non-saddle) periodic orbit of period $2k$ and a non-saddle (saddle) periodic orbit of period k.
- Flip bifurcation curve t^k

On the curve t^k, the fixed points with a period of k for the mapping $\mathcal{P}$ have the multiplier $\mu^k = 1$. Corresponding to the original continuous system, crossing t^k results in the disappearance of two periodic orbits of period k or the emergence of two new periodic orbits of period k.

- Transcritical bifurcation curve r^k

  On the curve r^k, the fixed points with a period of k for the mapping $\mathcal{P}$ have the multiplier $\mu^k = 1$. Corresponding to the original continuous system, crossing r^k leads to a change in stability between two periodic orbits of period k.

The aforementioned bifurcation curves can be obtained by the continuous extension of bifurcation points. Here are some symbols used in the bifurcation diagrams in the subsequent content:

- $\mu_1 = \mu_2 = 1$, $\mathcal{A}$ or $\mathcal{A}_i$, $i \in \mathbb{N}$: 1:1 resonance.
- $\mu_1 = \mu_2 = -1$, $\mathcal{B}$ or $\mathcal{B}_i$, $i \in \mathbb{N}$: 1:2 resonance.
- $\mu_{1,2} = \exp(\pm\frac{2\pi i}{3})$, C or C_i, $i \in \mathbb{N}$: 1:3 resonance.
- $\mu_{1,2} = \exp(\pm\frac{\pi i}{2})$, $\mathcal{D}$ or $\mathcal{D}_i$, $i \in \mathbb{N}$: 1:4 resonance.

4.2.2 Bifurcation Analysis with Synchronous Perturbations

Model (4.1) are categorized into four distinct scenarios, based on the specific types of Hopf bifurcations. This particular subsection focuses revolves around the investigation and presentation of bifurcation diagrams derived from the Poincaré map for the perturbed model. Both periodic mechanism 1 and mechanism 2 are considered.

Taking the parameter values as follows: $a_0 = 3$, $b = 6.788$, $c = 1.371$, $d = 4$, $e = 0.1697$, $f = 450.3$, $g = 1.948$, $h_0 = 2.459$, and $g_1 = 0.72$, the unperturbed model (4.1) experiences a subcritical Hopf bifurcation at $a_H = 0.999$ as the parameter a_0 changes. The numerically determined asymptotic period of the cycle is 2.093 when a_0 approaches a_H. The bifurcation diagrams for the perturbed model are illustrated on the (ε, a_0) plane in Fig. 4.5. In Fig. 4.5, the point denoted as H on the a_0-axis represents the subcritical Hopf bifurcation of the unperturbed model, serving as the starting point for the period-one torus bifurcation curve $h^{(1)}$. This curve concludes at the 1:2 resonance point $\mathcal{B}$. The curve f^1 intersects both point $\mathcal{B}$ and the period-two torus bifurcation h^2. Furthermore, two branches of fold bifurcation curves, t_1^2 and t_2^2, correspond to period two. An unstable period-one fixed point turns stable while crossing h^1 from region 2 to region 1, leading to the emergence of an unstable closed invariant curve. Subsequently, the attracting point loses its stability as it crosses f^1 into region 3, bounded by f^1, h^2, and $\varepsilon = 1$. Two pairs of period-two points appear while crossing t_1^2. These two saddle points exist only within the region enclosed by t_1^2 and $\varepsilon = 1$. Upon crossing h^2, these unstable points become stable within region 4, and they vanish upon crossing f^1 either from region 4 to 2 or from 3 to 1.

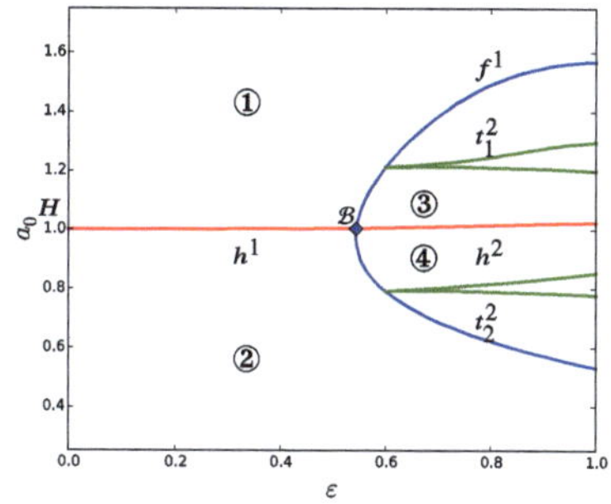

Fig. 4.5 Bifurcation diagrams of the perturbed model (4.26) with $a_0 = 3$, $b = 6.788, c = 1.371, d = 4$, $e = 0.1697,\ f = 450.3$, $g = 1.948, h_0 = 2.459$, and $g_1 = 0.72$

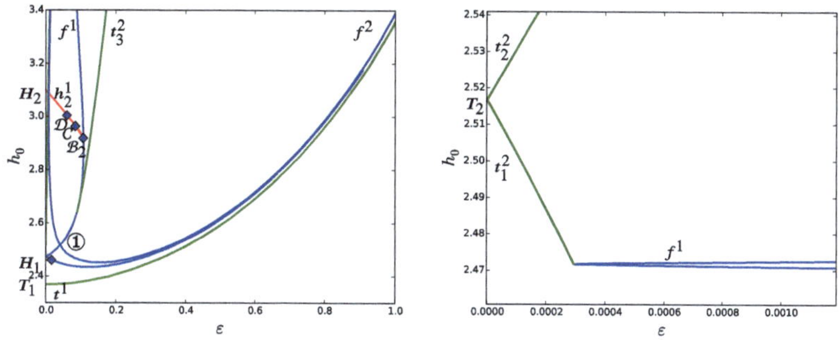

(a) (b)

Fig. 4.6 Bifurcation diagrams of the perturbed model (4.27) with $a_0 = 3, b = 6.788, c = 1.371, d = 4, e = 0.1697, f = 450.3, g = 1.948, h_0 = 2.459, g_1 = 0.72$

Taking the parameter values as in Fig. 4.5, for model (4.1), two subcritical Hopf bifurcations occur at $h_H^1 = 3.158$ and $h_H^2 = 2.459$ as the parameter h_0 varies. Within these parameter ranges, the unperturbed model demonstrates unstable limit cycles. The numerically determined asymptotic period of the cycle is 2.093 as h_0 approaches h_H^1. The other asymptotic period of the cycle is 0.897 as h_0 approaches h_H^2. The bifurcation diagrams for the perturbed model are illustrated on the (ε, h_0) plane in Fig. 4.6. In Fig. 4.6a, the points marked as H_1 and H_2 along the h_0-axis correspond to two instances of subcritical Hopf bifurcations within the unperturbed model (4.1). These points serve as the origins for the period-one torus bifurcation curves h_1^1 and h_2^1. The curve h_2^1 concludes at the 1:2 resonance point $\mathcal{B}_1$. The period-one flip curve $f_1^{(1)}$ passes through both $\mathcal{B}_1$ and $\mathcal{B}_2$. Another period-one torus bifurcation curve, h_1^1, intersects points C (a 1:3 resonance) and $\mathcal{D}$ (a 1:4 resonance), ultimately terminating at $\mathcal{B}_2$. Figure 4.6b provides insights into the scenarios at points T_1 and T_2, where the limit cycle manifests periods of 1 and 2, respectively. These points correspond to the starting positions of the period-one fold bifurcation t^1 and the two period-two fold bifurcations t_1^2 and t_2^2. Furthermore, a period-two flip bifurcation curve f^2 is displayed.

Two unstable period-one fixed points collide while traversing t^1, and they vanish as they cross t^1 downward. The stability of one of these points changes,

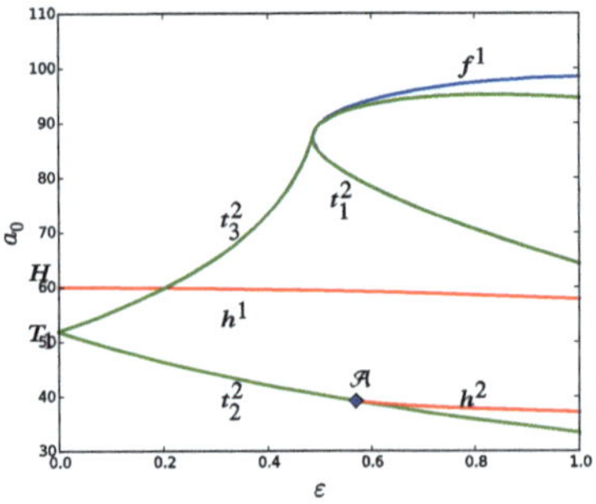

Fig. 4.7 Bifurcation diagrams of the perturbed model (4.26) with $a_0 = 60, b = 2, c = 1, d = 5, e = 10, f = 20, g = 0.6, h_0 = 0.177, g_1 = 2.14 \times 10^{-3}$

resulting in the emergence of an unstable closed invariant curve as it crosses h_2^1 upward. Subsequently, this point regains instability upon crossing h_1^1, leading to the appearance of a stable closed invariant curve. Moreover, it loses stability again within the region enclosed by f^1 and $\varepsilon = 1$. Two pairs of period-two fixed points (comprising two saddles and two non-saddle points) appear while crossing t_1^2 or t_2^2 to the right. These two saddle points disappear upon traversing the region of f^1 situated between t_1^2 and t_2^2. The remaining two points become unstable within the area bounded by f^2 and $\varepsilon = 1$, gradually dissipating through the section of f^1 located to the right of the intersections between t_1^2 and f^1. Moreover, while crossing f^1 (above the intersection of t_3^2 with f^1) to the right, two unstable points emerge and subsequently vanish above t_3^2.

By adopting the parameter values as follows: $a_0 = 60$, $b = 2$, $c = 1$, $d = 5$, $e = 10$, $r = 20$, $g = 0.6$, $h_0 = 0.177$, and $g_1 = 2.14 \times 10^{-3}$, the unperturbed model (4.1) manifests a supercritical Hopf bifurcation at $a_H = 60$ as a_0 is adjusted. The numerically derived asymptotic period of the cycle is 1.612 when a_0 converges toward a_H. Furthermore, the model (4.1) experiences a supercritical Hopf bifurcation at $h_H = 0.1754$ when the parameter h_0 undergoes modification. Within the context of these parameter values, the unperturbed model sustains oscillations around a stable limit cycle. The numerically determined asymptotic period is 1.621 as h_0 approaches h_H. The bifurcation diagrams portraying the behavior of the perturbed model are depicted on both the (ε, a_0) and (ε, h_0) planes in Fig. 4.7a and b.

In Fig. 4.7, the position denoted as H on the a_0-axis corresponds to the occurrence of a supercritical Hopf bifurcation within the unperturbed model (4.1). This point serves as the starting point for the period-one torus curve h_1. At the point T, the period of the unstable cycle is 2. Two branches of the period-two tangent bifurcation curve, t_2^2 and t_3^2, originate from T. Additionally, a period-two torus bifurcation curve h^2 originates from $\mathcal{A}$, which represents a 1:1 resonance point. As the unstable period-one fixed point crosses through h^1 upward, its stability changes, leading to the emergence of an unstable closed invariant curve. This point transforms into a saddle as it traverses the region enclosed by f^1 and $\varepsilon = 1$. Two pairs of period-two fixed points (comprising two saddle points and two attracting points)

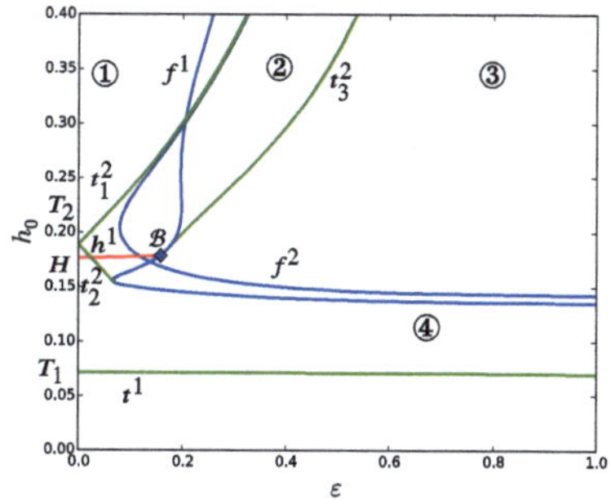

Fig. 4.8 Bifurcation diagrams of the perturbed model (4.27) with $a_0 = 60, b = 2, c = 1, d = 5, e = 10, f = 20, g = 0.6, h_0 = 0.177, g_1 = 2.14 \times 10^{-3}$

emerge while crossing t_1^2 upward. These two saddle points exist exclusively within the region bounded by f^1 and $\varepsilon = 1$. The two attracting points become unstable and disappear via t_2^2. Furthermore, two pairs of period-two points (comprising two saddle points and two non-saddle points) manifest upon crossing either t_2^2 or t_3^2 to the right. These two saddle points are limited to the region bounded by t_2^2, t_3^2, and f^1. The other two non-saddle points remain stable above h^2 and vanish through the portion of f^1 situated to the right of the intersections with t_3^2.

In Fig. 4.8, the location marked as H on the h_0-axis corresponds to the occurrence of a supercritical Hopf bifurcation within the unperturbed model (4.1), and this point is the root of the curve h^1, which terminates at $\mathcal{B}$, representing a 1:2 resonance. A period-one flip bifurcation curve f^1 traverses through $\mathcal{B}$. T_1 corresponds to the tangent bifurcation within the unperturbed model (4.1) and serves as the root of t^1. A branch of the period-two tangent bifurcation curve, t_1^2 and t_2^2, originates from T_2. The unstable period-one point gains stability as it crosses h^1 upward. Subsequently, it transforms into an unstable saddle while traversing the region bounded by the curve f^1 and $\varepsilon = 1$. The collision of the saddle point and the non-saddle point occurs on t^1, leading to their disappearance below t^1. Two pairs of unstable period-two fixed points (comprising two saddle points and two unstable points) emerge upon crossing t_2^2 upward and t_1^2 downward. The two saddle points vanish while traversing the region between t_2^2 and t_1^2 through f^1 to the right. The two unstable points become attractive within the region enclosed by f^2 and $\varepsilon = 1$, subsequently disappearing through f^1 transitions from region 2 to 1 and region 3 to 4.

4.2.3 *Bifurcation Analysis with Nonsynchronous Perturbations*

In this subsection, we investigate the final two mechanisms (3 and 4) described by Eq. (4.24), which introduce periodic and nonsynchronous perturbations. To comprehensively analyze the bifurcation diagrams, we refrain from constraining the ranges of ε_1 and ε_2. For each periodic mechanism, we explore two scenarios that correspond to the cases of the unperturbed model (4.1), characterized by a stable or unstable limit cycle.

The perturbed model associated with periodic mechanisms 3 and 4 is given by

$$\begin{cases} \dot{G} = a_0(1+\varepsilon_1 v) - (b+cI)G, \\ \dot{I} = \dfrac{d\beta G^2}{e+G^2} - fI, \\ \dot{\beta} = (-g + h_0(1+\varepsilon_2 v)G - g_1 G^2)\beta, \\ \dot{v} = v + 2\pi w - v(v^2+w^2), \\ \dot{w} = -2\pi v + w - w(v^2+w^2), \end{cases} \tag{4.28}$$

and

$$\begin{cases} \dot{G} = a_0(1+\varepsilon_1 v) - (b+cI)G, \\ \dot{I} = \dfrac{d\beta G^2}{e+G^2} - fI, \\ \dot{\beta} = (-g + h_0(1+\varepsilon_2 w)G - g_1 G^2)\beta, \\ \dot{v} = v + 2\pi w - v(v^2+w^2), \\ \dot{w} = -2\pi v + w - w(v^2+w^2). \end{cases} \tag{4.29}$$

In Fig. 4.9, we set the parameters as $a_0 = 3, b = 6.788, c = 1.371, d = 4, e = 0.1697, f = 450.3, g = 1.948, h_0 = 2.459, g_1 = 0.72$. Under these parameter values, the model exhibits unstable limit cycles. For the perturbed model (4.28), an unstable period-one solution emerges in region 1, transforming into a saddle upon crossing from region 1 to the area enclosed by the closed curve f^1. Additionally, within region 2 (situated between f^1 and t^2), a pair of stable period-two solutions manifests. Furthermore, when crossing from region 3 to region 2, two pairs of period-two solutions arise and subsequently vanish upon reaching t^2. The two saddles are solely present within the region bounded by the closed curve f^1, while the other two points transition from unstable to stable as they cross the curve f^2 but then become unstable again upon crossing from f^1 to the exterior.

By selecting $a_0 = 1, b = 5.9, c = 1, d = 4, e = 0.1, r = 1, g = 1.152, h = 1.83, g_1 = 0.72$, the bifurcation diagram on the $(\varepsilon_1, \varepsilon_2)$ plane is depicted in Fig. 4.10. Under these parameter choices, the unperturbed model (4.28) exhibits oscillations on a stable limit cycle. Three nontrivial periodic solutions emerge within the region enclosed by the closed curve t^1. Two nontrivial period-one solutions cease to exist upon crossing outside t^1. Furthermore, while crossing the curves h_1^1 or h_2^1, the stability of the third fixed point undergoes a change.

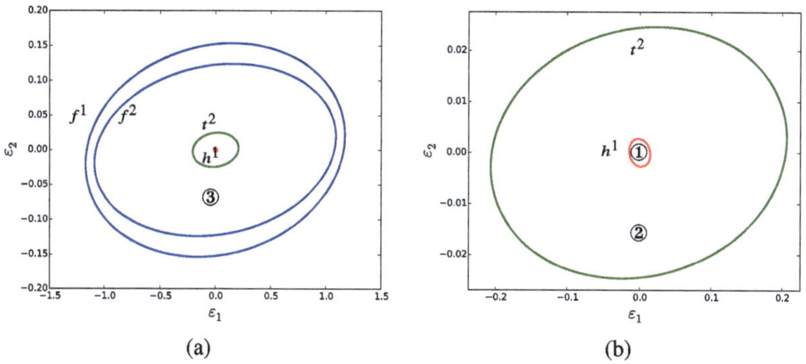

Fig. 4.9 Bifurcation diagrams of the perturbed model of (4.28) with $a = 59.278, b = 51.611, c = 1, d = 5, e = 10, f = 20, g = 0.6, h = 7.32 \times 10^{-2}, g_1 = 2.14 \times 10^{-3}$. (**b**) is the partial enlargement drawings of (**a**)

Fig. 4.10 Bifurcation diagrams of the perturbed model of (4.28) with $a_0 = 1, b = 5.9, c = 1, d = 4, e = 0.1, f = 1, g = 1.152, h = 1.83, g_1 = 0.72$

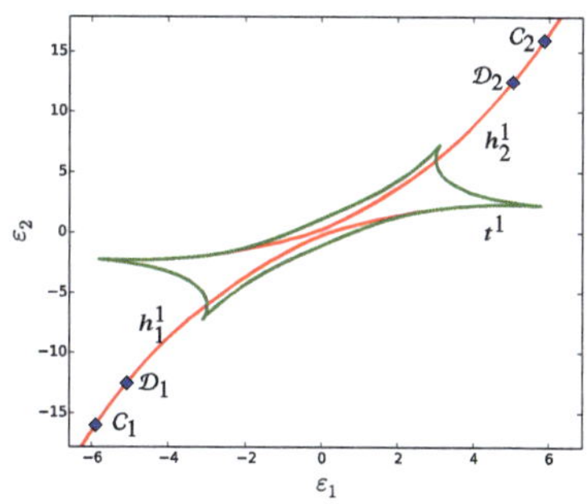

4.2.4 *Solutions and Chaos of the GIβ System*

Next, a subset of parameters to demonstrate the complex dynamical behaviors of chaos and quasiperiodic solutions under periodically perturbation. Consider h and ε as free parameters, while keeping the values of $a = 16.51, b = 2, c = 1, d = 4, e = 3.559, f = 1, g = 27.4255, g_1 = 1, \varepsilon = 0.4$ fixed, this parameter selection allows us to delve into the solutions of period two and period four for model (4.27) (Fig. 4.11).

Chaos is a dynamic behavior that is exceedingly complex and difficult to anticipate. It is identified by the system's evolution exhibiting increased sensitivity and a lack of regularity. Long-term forecasts are challenging, because even little variations in the starting circumstances might cause the system's paths to vary significantly. Chaos generally develops in nonlinear systems and appears as motion that appears erratic and unreliable. Chaos is frequently accompanied by dynamical characteristics including bifurcation processes and odd attractors. Between periodic solutions and chaos, quasiperiodic solutions reflect a dynamic behavior. Although the system's history is more irregular in quasiperiodic solutions, it still lacks the uniformity and predictability observed in periodic ones. Multiple frequencies in

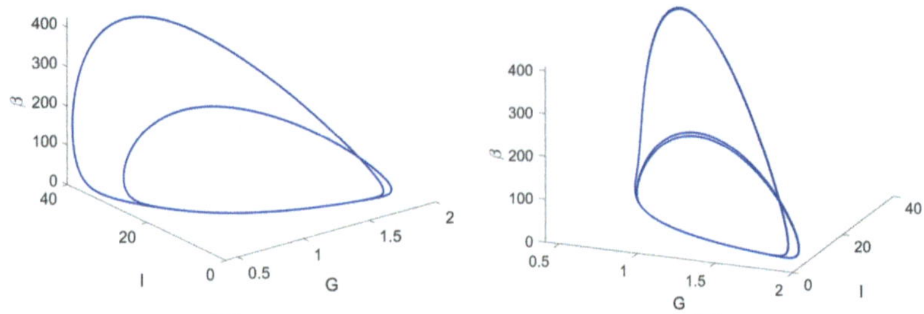

(a) (b)

Fig. 4.11 Phase portraits of different solutions of model (4.27). (**a**) Periodic solution with period two for $h = 27$. (**b**) Periodic solution with period four for $h = 26.27$

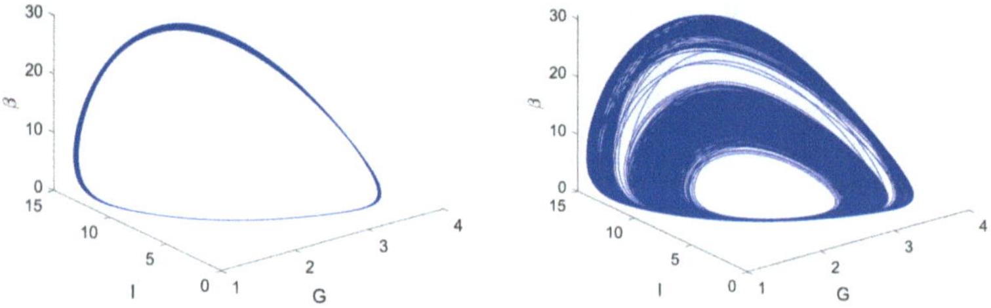

(a) (b)

Fig. 4.12 Phase portraits of different states of model (4.27). (**a**) Quasiperiodic solution for $\varepsilon = 0.04$. (**b**) Chaos for $\varepsilon = 0.155$

the system interact regularly to provide quasiperiodic solutions, which produce trajectory types that have a shape. We observe that under periodic perturbation mechanisms, even when the perturbation amplitude varies only slightly (when ε increase from 0.04 to 0.155), the state of (4.27) model can transition from chaos to quasiperiodic solutions. Here, we keep other parameters fixed: $a = 16.51$, $b = 2$, $c = 1$, $d = 4$, $e = 3.559$, $f = 1$, $g = 27.42$, $h = 15$, and $g_1 = 1$ (Fig. 4.12).

Bistability pertains to a phenomenon within a system where it possesses two distinct stable equilibrium states. Under such circumstances, the system is capable of transitioning between these two states without undergoing spontaneous departure from them. In instances of bistability, the stability characteristics of the system may exhibit variability across same parameter range, leading to a diverse array of behaviors exhibited by the system within those specific ranges. In our context, we maintain fixed values for the parameters, specifically $a = 16.51$, $b = 2$, $c = 1$, $d = 4$, $e = 3.559$, $f = 1$, $g = 27.42$, $h = 15$, $g_1 = 1$, and $\varepsilon = 0.155$. By selecting distinct initial values, $G_0 = 3$, $I_0 = 3$, $\beta_0 = 1.5$, and $G_0 = 3$, $I_0 = 0.5$, and $\beta_0 = 1.5$, we are able to observe the manifestation of bistability within the model (4.27). This phenomenon is depicted in Fig. 4.13.

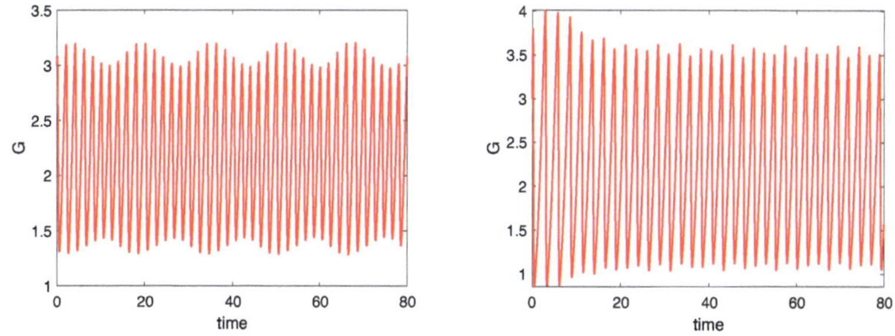

(a) (b)

Fig. 4.13 Bistable solutions of model (4.27). (**a**) Initial values for $G_0 = 3$, $I_0 = 3$, $\beta_0 = 1.5$, and (**b**) initial values for $G_0 = 3$, $I_0 = 0.5$, and $\beta_0 = 1.5$

4.2.5 Conclusion

In this chapter, we have conducted a comprehensive investigation into the bifurcations and associated dynamics of the blood glucose regulation system (4.1). Our analysis reveals that under specific parameter conditions, the system (4.1) can exhibit a variety of bifurcation phenomena, including one-dimensional Hopf bifurcations, degenerate Hopf bifurcations, fold bifurcations, and transcritical bifurcations. Additionally, two-dimensional cusp bifurcations, Bogdanov-Takens bifurcations, and three-dimensional Bogdanov-Takens bifurcations are also identified.

Moreover, due to the influence of the body's internal biological clock and external periodic variations, the blood glucose regulation system exhibits circadian rhythmicity. We have also investigated the dynamic behavior of system (4.1) under four different periodic mechanisms. The results demonstrate that these periodic mechanisms give rise to the appearance of multiple attractors, including periodic solutions, quasiperiodic solutions, and chaotic solutions, each exhibiting distinct periods and stability properties.

From a theoretical perspective, near these bifurcation points and state transition regions, the system might undergo abrupt transitions from one state to another. In clinical practice, these transitions can be understood as shifts between different physiological states. For instance, such transitions could include shifts from a normal state to a physiologically elevated state, from a normal state to a physiologically lowered state, from a physiologically lowered state to a pathologically lowered state, or from a physiologically elevated state to a pathologically elevated state. This study enhances our understanding of the complex dynamic behaviors within the blood glucose regulation system.

Chapter 5
Machine Learning in Clinical Medicine

At present, machine learning (ML) models have been extensively employed in various healthcare aspects, including disease prevention, diagnosis, treatment, and prognosis prediction—such as disease risk prediction, patient readmission prediction, death prediction, and patient care need prediction—delivering commendable results. The primary objectives of implementing ML in clinical medicine encompass (i) delivering precise predictions and judgments for predictive tasks and (ii) utilizing trained models to guide clinical practice and research.

Besides, due to the inherent complexity of ML models, the process that culminates in the final output often lacks transparency and interpretability. High interpretability in a model implies that end users can comprehend and interpret future predictions with greater ease. Thus, the interpretability of ML models is considered as crucial as achieving high predictive accuracy. Furthermore, in clinical applications, ML models are typically trained and evaluated for specific diseases within a limited context and heavily rely on individual statistical and ML expertise. In response to these challenges, several interpretable ML methods have emerged to assist healthcare professionals in understanding the foundational mechanisms through which ML models generate predictions.

5.1 Predicting Mortality Risk in Rheumatic Heart Disease Patients

Rheumatic heart disease (RHD) is a cardiac condition caused by rheumatic inflammation, primarily affecting heart valves and leading to stenosis and valve insufficiency, which can result in heart failure. RHD originates from rheumatic fever, triggered by an infection with group A beta-hemolytic streptococcus [66]. The recent Global Burden of Diseases report on RHD serves as a timely reminder of the notable global heterogeneity in RHD burden [67, 68]. In developed nations, the

J. Ren, Y. Tao, *Modeling in Life Sciences and Ecology*, Springer Asia Pacific Mathematics Series 4, https://doi.org/10.1007/978-981-95-1038-2_5

incidence of RHD has markedly decreased, attributed to improved living standards, widespread dissemination of healthcare knowledge, and extensive use of penicillins. However, within specific regions and among vulnerable groups, particularly in low- and middle-income countries like Africa, the Western Pacific, and India, the prevalence of RHD persists at elevated levels, primarily due to restricted healthcare resources and challenges associated with early diagnosis and treatment [69].

Predicting in-hospital mortality risk in RHD patients is crucial for optimizing treatment plans and clinical decision-making. Traditional statistical models using serum markers such as IL-8, IL-6, CXCL-1, and antistreptolysin often fall short in specificity and sensitivity due to their inability to handle complex nonlinear relationships [70, 71]. In contrast, recent advancements in machine learning (ML) have demonstrated superior performance in health assessments of critically ill patients [72]. ML algorithms, renowned for their efficiency in processing large datasets [73], have been successfully applied in prognostic assessments for various patient populations [74, 75], showing excellent performance in predicting outcomes. However, one major challenge in medical datasets, particularly in binary classification problems, is the imbalance between majority and minority classes, which can lead to frequent prediction errors [26, 75–77]. In RHD, errors in predicting the minority class, such as disease progression or death, have significant clinical impacts. Addressing this imbalance and achieving interpretable machine learning models remains an ongoing challenge in the field of RHD prognosis.

To address this, this chapter introduces a mortality risk prediction model for RHD patients based on an imbalanced dataset. Additionally, it integrates interpretable ML techniques to analyze the prioritization of key features associated with mortality risk, the determination of thresholds, and the understanding of how variations in these features influence patient mortality rates.

5.1.1 Data Preprocessing

5.1.1.1 Data Source

The freely accessible critical care database known as the Medical Information Mart for Intensive Care (MIMIC-IV) database version 4.1 was used, with the necessary permission granted (certificate number: 48369375). The MIMIC-IV database contains comprehensive clinical data pertaining to patients admitted to the Beth Israel Deaconess Medical Center between 2008 and 2019. It encompasses a vast dataset, including over 200,000 emergency department admissions, and more than 70,000 ICU stays.

The clinical data within MIMIC-IV encompass a wide range of information, including demographic characteristics, vital signs, results from imaging examinations, laboratory tests, a data dictionary, and documents containing codes from the International Classification of Diseases, Ninth and Tenth Revisions (ICD-9 and ICD-10, respectively). Additionally, it contains records of hourly physiologic data obtained from bedside monitors, which have undergone validation by ICU nurses. Crucially, it's important to note that all health information sourced from the MIMIC-IV database are anonymized, eliminating the need for informed consent from the patients involved. This database has received approval from the Institutional Review Boards of the Massachusetts Institute of Technology.

5.1.1.2 Data Extraction

After obtaining access to the data, clinical information related to patients with RHD was extracted from the MIMIC-IV database. Patients meeting specific criteria were selected for this study, including (1) initial ICU admission during their first hospitalization, (2) an ICU length of stay exceeding 24 hours, and (3) age over 18 years. After excluding 452 cases with a missing rate exceeding 80% for laboratory indicators and 507 cases with a total indicator missing rate exceeding 20%, 1,266 study samples remained. This cohort included 1,150 in-hospital survivors and 116 in-hospital deaths. Patients with RHD were identified using ICD-9 and ICD-10 codes. Each case contained 163 data items, covering demographic information, vital signs recorded within 24 hours of ICU admission, and laboratory results. The data extraction process is depicted in Fig. 5.1.

In Fig. 5.2, the distribution of inpatient status in the research cohort is presented, with 91% classified as survivors and 9% as deceased. Regarding gender distribution, females account for 49%, while males make up 51%. In terms of racial distribution,

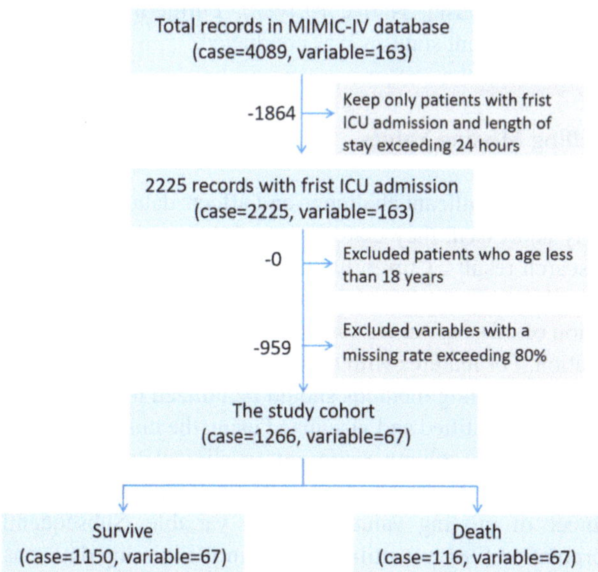

Fig. 5.1 The detailed process of data extraction

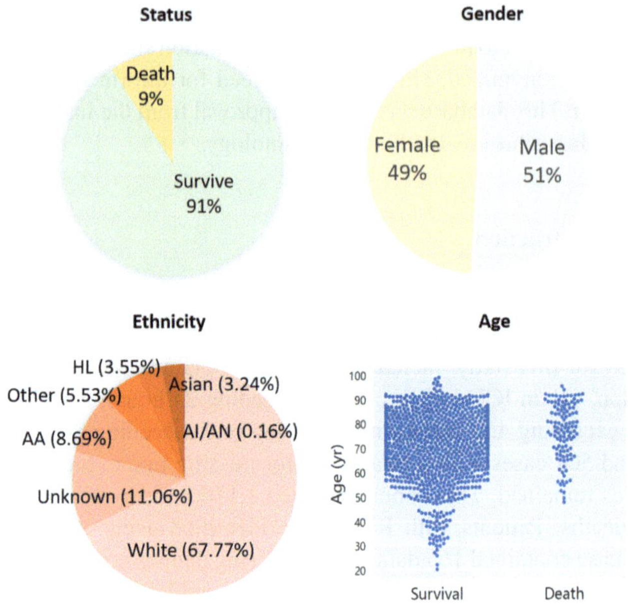

Fig. 5.2 The status, gender, ethnicity, and age distribution of the study cohort

the majority are White (67.77%), followed by African Americans (8.69%), other races (5.53%), Hispanic/Latinx (3.55%), Asians (3.24%), unknown race (11.06%), and American Indian/Alaska Native (0.16%). Furthermore, an analysis of age distribution across different statuses was conducted.

5.1.1.3 Handling Missing Values

Missing data poses a significant challenge and affects data quality. If missing values are not properly addressed, they can have a substantial impact on the accuracy of subsequent research results. Currently, the treatment of missing values is typically categorized into three approaches: deletion, filling, and no treatment. Features with a high proportion of missing values can be directly eliminated due to the substantial loss of information. For features with a small proportion of missing values that hold practical significance, filling methods should be utilized to ensure data integrity.

Missing data was identified and visualized using the missingno module in Python 3.9.12. In Fig. 5.3, each column represents a clinical variable, with white lines indicating missing data. The density of white lines within each column correlates with the number of missing values for that variable. Subsequently, K-nearest neighbor interpolation [78] was utilized to fill in the missing data, ensuring a more complete dataset for analysis.

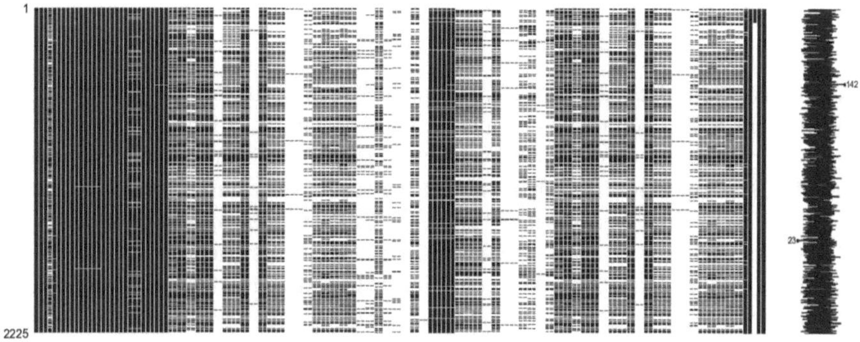

Fig. 5.3 Visualization for variable missingness

5.1.1.4 Feature Vectorization

Categorical data contains variable indicators that require vectorization before modeling and analysis. For instance, in the context of patient survival status, "0" indicates that the patient survived the ICU stay, while "1" indicates that the patient died during the ICU stay. Similarly, in the case of gender, "0" represents male and "1" represents female. For categorical variables like race, one-Hot coding [79] is employed for vectorization. This step enables the representation of categorical patient characteristics in a reasonable vectorized format.

5.1.1.5 Data Normalization

In a multi-indicator system, indicators often possess varying scales and quantitative levels due to their different nature and meaning. Directly using raw information for analysis can result in the attenuation of indicators with lower values and the overemphasis of those with higher values. Therefore, this section employs the normalization method to process the continuous indicators dimensionlessly, ensuring the reliability and validity of the analysis results. The calculation formula for normalization is as follows:

$$x_{ij}^* = \frac{x_{ij} - \min(x_{ij})}{\max(x_{ij}) - \min(x_{ij})},$$

where x_{ij}^* represents the normalized value, x_{ij} is the original measured value, $\min(x_{ij})$ denotes the minimum value of the variable, and $\max(x_{ij})$ denotes the maximum value of the variable.

5.1.2 Feature Selection

As discussed in Chap. 2, high correlations between variables can introduce bias into the model, affecting the estimation of explanatory variables. This bias can result in inconsistent outcomes and may even hinder model convergence. To address this, Pearson correlation coefficients and the recursive feature elimination (RFE) technique were employed to select features from the training cohort data.

First, the Pearson correlation matrix for 67 indicators was calculated. Twenty variables with a correlation coefficient greater than 0.7 were identified and subsequently removed to reduce multicollinearity. Following a specific feature ranking criterion, RFE starts with the complete set of features and iteratively eliminates the least relevant ones, ultimately selecting the top 26 important features (Table 5.1). Figure 5.4 illustrates the Pearson correlation matrix among these remaining 26 indicators.

Table 5.1 The 26 selected features

Feature	Abbreviation
Age	age
Length of icu stay	los
Heart rate_min	heart_rate_min
Systolic blood pressure_max	sbp_max
Diastolic blood pressure_min	dbp_min
Diastolic blood pressure_max	dbp_max
Mean blood pressure_max	mbp_max
Body temperature_min	Temperature_min
Body temperature_max	Temperature_max
Blood oxygen saturation_min	spo2_min
Blood glucose_min	glucose_min
Blood glucose_mean	glucose_mean
Erythrocyte pressure_min	hematocrit_min
Erythrocyte pressure_max	hematocrit_max
Hemoglobin_min	hemoglobin_min
hemoglobin_max	hemoglobin_max
Leukocyte_min	wbc_min
Leukocyte_max	wbc_max
Total Protein_min	Total_protein_min
Total protein_max	Total_protein_max
Eosinophil count_min	Abs_eosinophils_min
Eosinophil count_max	Abs_eosinophils_max
Partial thromboplastin time_max	ptt_max
Urine volume	urine_output
Acute physiological score	sapsii
Sequential organ failure score	sofa

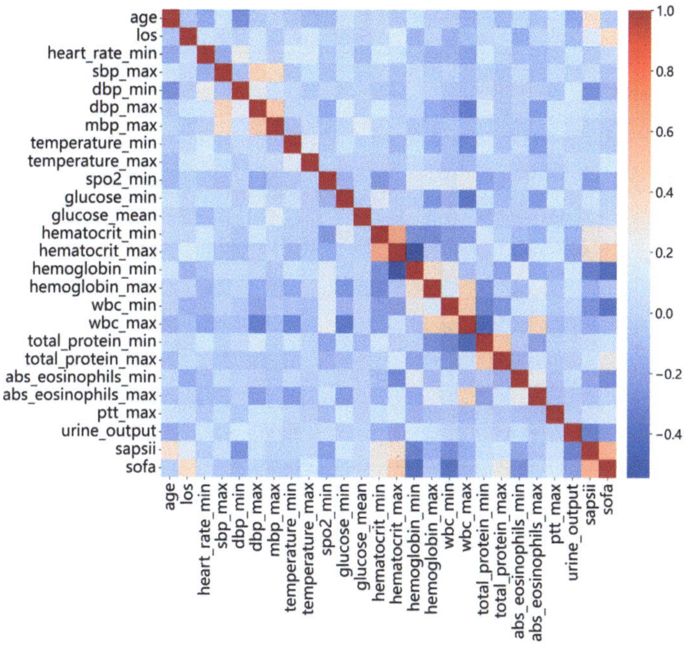

Fig. 5.4 The heatmap of correlation coefficients among variables

5.1.3 Data- and Algorithm-Driven Modeling

The study cohort exhibited a severe data imbalance issue, with the ratio of in-hospital survival to in-hospital mortality at 1150:116. To address this significant class imbalance, diverse methodologies were devised, broadly categorized into two primary approaches: data-driven and algorithm-driven methods. Additionally, hyperparameter optimization and cross-validation through GridSearchCV were applied to prevent overfitting and increase model accuracy. Figure 5.5 shows the simplified schematic workflow for the predictive model.

5.1.3.1 Resampling Method

To address data imbalance, four resampling methods were implemented, the concept of which has been introduced in Chap. 2. Below, we focus on the four specific resampling methods used.

Random oversampling (ROS) balances the dataset by duplicating instances from the minority class. While this method is simple and straightforward, it can lead to overfitting due to the replication of existing data points. The new data point can be generated as follows:

$$N_{\text{new}} = N_{\text{minority}} + (N_{\text{majority}} - N_{\text{minority}}),$$

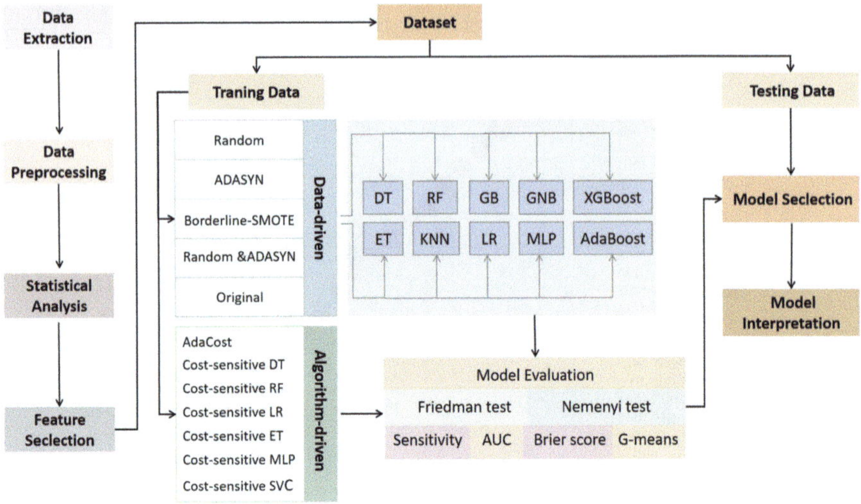

Fig. 5.5 The simplified schematic workflow for the predictive model

where N_{new} is the number of minority class instances after resampling, N_{minority} is the original number of minority class instances, and N_{majority} is the number of majority class instances.

Borderline-SMOTE (BS) oversampling generates synthetic samples only for those minority class instances that are near the decision boundary between classes. This approach helps in better defining class boundaries, thus improving classifier performance on imbalanced datasets. The new data point can be generated as follows:

$$x_{\text{new}} = x_i + \lambda \cdot (x_j - x_i),$$

where x_i is a minority class instance, x_j is one of its k-nearest neighbors, and λ is a random number between 0 and 1.

ADASYN oversampling focuses on generating synthetic data for minority class instances that are harder to learn. It adaptively decides the number of synthetic samples based on the density distribution of minority class examples, thereby reducing the bias introduced by imbalance. The new data point can be generated as follows:

$$x_{\text{new}} = x_i + \lambda \cdot (x_j - x_i).$$

The number of generated samples is determined by

$$G = \sum_{i=1}^{N_{\text{minority}}} \delta_i \cdot \frac{N_{\text{majority}} - N_{\text{minority}}}{N_{\text{minority}}}.$$

where δ_i measures the difficulty of learning minority class instances.

The combination of random undersampling and ADASYN oversampling (ADASYN & ROS) leverages the strengths of both undersampling and oversampling. Initially, random undersampling reduces the size of the majority class. The random undersampling of the majority class can be performed as follows:

$$N_{\text{majority, new}} = N_{\text{majority}} \times r,$$

where r is the undersampling rate. Next, ADASYN oversampling is applied to increase the minority class instances. This combination aims to balance the dataset while mitigating the risk of overfitting and preserving the essential characteristics of the data.

Figure 5.6a–e depict the class distribution of "Survive" and "Death" in the original dataset as well as in resampled datasets, specifically focusing on two key features, namely, wbc_max and sofa. In the original dataset, the class ratio is observed at 805:81. Subsequently, the data was subjected to diverse four resampling methods (ROS, BS, ADASYN, ROS-ADASYN), resulting in class ratios of 805:805, 805:805, 922:919, and 805:805, respectively (Fig. 5.6f).

5.1.3.2 Machine Learning Algorithms

Following resampling, ML algorithms were trained on the resampled training sets, with model validation performed on the corresponding test sets. The ML algorithms utilized in this study include random forest (RF), decision tree (DT), extra tree (ET), logistic regression (LR), gradient boosting (GB), eXtreme gradient boosting (XGBoost), Gaussian Naive Bayes (GNB), multilayer perceptron (MLP), adaptive boosting (Adaboost), and K nearest neighbors (KNN). RF, DT, GNB, MLP, KNN, and ET have been thoroughly discussed in the previous chapters. Below, we will focus on the remaining ML algorithms.

Logistic regression (LR) is one of the most commonly used algorithm models in machine learning for binary classification tasks [80]. Despite its name, which includes regression, it is actually a classification model. The main idea is to establish a regression formula for the classification boundary based on existing data and use it for classification. The conditional probability distribution of the binomial logistic regression model is as follows:

$$P(Y = 1|X) = \frac{1}{1 + e^{-(\omega \cdot X + b)}}, \tag{5.1}$$

$$P(Y = 0|X) = 1 - P(Y = 1|X), \tag{5.2}$$

where X is the input and Y is the output, ω and b are parameters, referred to as the weight vector and bias, respectively. $\omega \cdot X$ represents the dot product of ω and X.

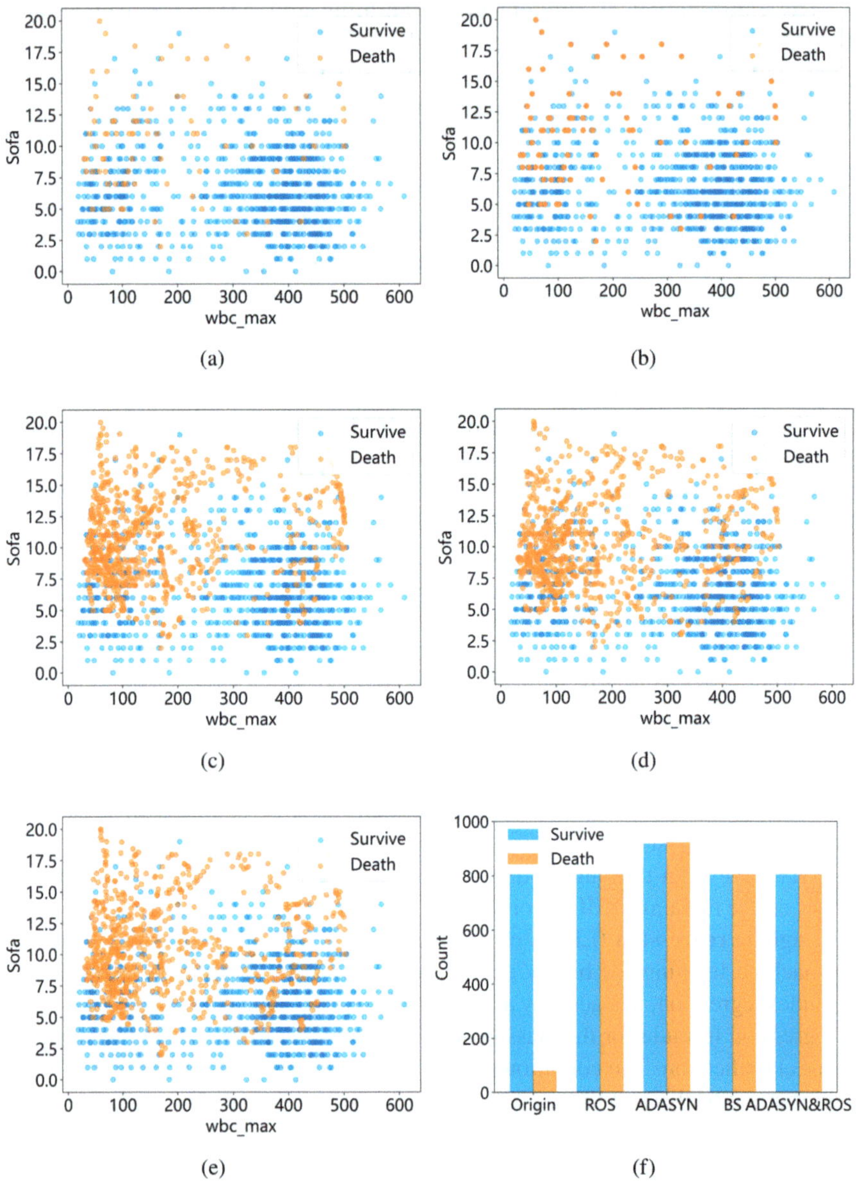

Fig. 5.6 **(a)**–**(e)** The class distribution of "Survive" and "Death" in the original dataset, as well as in datasets resampled using ROS, BS, ADASYN, and ROS-ADASYN techniques. **(f)** The counts of "Survive" and "Death" for each of the five datasets

For a given input instance, we can calculate $P(Y = 1|X)$ and $P(Y = 0|X)$ using Eqs. (5.1) and (5.2). By comparing these conditional probability values, we classify the instance into the class with the higher probability.

Deep forest (gcForest) is a type of ensemble learning in nature, which is also endorsed with the strength of deep neural network [81]. It is designed to handle real-world challenges, such as high-dimensional data and class imbalance. High-dimensional data often includes numerous irrelevant features, making the training process time-consuming and resource-intensive. The cascade structure of gcForest addresses this by progressively refining feature representation. Additionally, gcForest employs a cost-based strategy to handle class imbalance, assigning different weights to different classes to mitigate the impact of imbalanced data distributions.

The central part of gcForest is its cascade forest structure. In this structure, each level of the cascade receives feature information processed by its preceding level and outputs its processing result to the next level. This structure ensures that the information flow and feature transformation are progressively refined through the cascade.

Let X denote the original input feature vector. The input to the first layer of the cascade is X. For the t-th layer, the input feature vector $X^{(t)}$ is a concatenation of the original input X and the output of the previous layer $P^{(t-1)}$,

$$X^{(t)} = [X, P^{(t-1)}],$$

with $P^{(0)} = X$.

A critical idea underpinning the cascade forest structure is the class distribution produced by every tree for each input example. This distribution is computed by counting the percentage of different classes of examples at the leaf node where the input example falls. The resulting class vector is obtained by averaging the class distributions across all trees in the same forest. For a classification task with k classes, if m base learners (e.g., random forests or completely random tree forests) are trained in each layer, each base learner outputs a k-dimensional class vector. The combined output of all base learners in the t-th layer is a $(m \times k)$-dimensional vector for each sample. This vector is then concatenated with the original feature vector for the next layer's input.

Formally, let $f_j^{(t)}(X^{(t)})$ denote the k-dimensional class vector output by the j-th base learner in the t-th layer. The combined class vector $P^{(t)}$ for the t-th layer is

$$P^{(t)} = \frac{1}{m} \sum_{j=1}^{m} f_j^{(t)}(X^{(t)}).$$

gcForest's ability to automatically adjust its depth and enhance diversity is crucial for its performance. By adapting the number of cascade layers based on the validation set performance, gcForest ensures that it does not overfit or underfit the data. The cascading process terminates when the accuracy on the validation set stops improving, thereby determining the optimal number of layers automatically.

Gradient boosting (GB) is an ensemble learning technique used for regression and classification tasks, which builds a model in a stage-wise fashion from weak learners (usually decision trees). The main idea behind gradient boosting is to combine the predictions of several base models to improve robustness and accuracy. The algorithm minimizes a loss function by adding models that correct the errors of the combined ensemble. Below are the key steps of GB.

1. Initialization: Initialize the model with a constant value, typically the mean of the target variable y,

$$G_0(x) = \frac{1}{n}\sum_{i=1}^{n} y_i,$$

   where $G_0(x)$ is the initial prediction for all x, and y_i are the target values for the i-th instance.
2. Compute residuals: For each iteration t, the residuals (pseudo-residuals) are computed as the negative gradient of the loss function with respect to the current model predictions. That is,

$$r_i^{(t)} = -\left.\frac{\partial L(y_i, G(x_i))}{\partial G(x_i)}\right|_{G(x_i)=G_{t-1}(x_i)},$$

   where $r_i^{(t)}$ is the residual for the i-th instance at iteration t and $L(y_i, G(x_i))$ is the loss function.
3. Fit a base learner: A new base learner $h_t(x)$ is trained to predict the residuals computed in the previous step, capturing the patterns in the residuals. Here,

$$h_t(x) \approx r_i^{(t)}.$$

4. Model update: The current model is updated by adding the new base learner, scaled by a learning rate η. This learning rate controls the contribution of each base learner to the final model:

$$G_t(x) = G_{t-1}(x) + \eta h_t(x).$$

5. Repeat: Steps 2–4 are repeated for a predefined number of iterations or until the model performance on a validation set stops improving.

Extreme gradient boosting (XGBoost) is a highly efficient and flexible implementation of gradient boosting, designed to optimize both speed and performance [82]. It incorporates several advanced features to handle large-scale data and improve the accuracy of predictions. Gradient boosting is an ensemble learning method that iteratively trains multiple weak learners and combines their predictions through weighted summation to construct a strong learner. The objective of each training round is to reduce the residual errors from the previous round's predictions, thereby continuously enhancing the overall model performance.

Suppose we have a training dataset $(x_i, y_i)_{i=1}^{n}$, x_i represents the input features, and y_i represents the target values. Our goal is to learn a model $F(x)$ such that the predicted values $\hat{y}_i = F(x_i)$ are as close as possible to the true values y_i. In the gradient boosting framework, the model $F_t(x)$ at iteration t is built upon the model from the previous $t-1$ iterations plus a new weak learner $h_t(x)$. That is,

$$F_t(x) = F_{t-1}(x) + h_t(x).$$

To optimize the model, we minimize a loss function, which can be written as

$$\mathcal{L}^{(t)} = \sum_{i=1}^{n} L(y_i, \hat{y}_i^{(t)}) + \sum_{k=1}^{t} \Omega(f_k), \tag{5.3}$$

where $L(y_i, \hat{y}_i^{(t)})$ is the loss function that measures the difference between the actual and predicted values. $\Omega(f_k)$ is the regularization term for the k-th tree, typically defined as

$$\Omega(f) = \gamma T + \frac{1}{2}\lambda \sum_{j=1}^{T} w_j^2,$$

where T is the number of leaves in the tree, w_j are the leaf weights, γ is the regularization parameter for the number of leaves, and λ is the regularization parameter for the leaf weights.

Using the Taylor expansion, (5.3) can be expanded as

$$L(y_i, \hat{y}_i^{(t)}) \approx L(y_i, \hat{y}_i^{(t-1)}) + g_i^{(t)} h_t(x_i) + \frac{1}{2} h_i^{(t)} h_t(x_i)^2,$$

where $g_i^{(t)}$ and $h_i^{(t)}$ are the first- and second-order gradients, with

$$g_i^{(t)} = \frac{\partial L(y_i, \hat{y}_i^{(t-1)})}{\partial \hat{y}_i^{(t-1)}}, \quad h_i^{(t)} = \frac{\partial^2 L(y_i, \hat{y}_i^{(t-1)})}{\partial \hat{y}_i^{(t-1)2}}.$$

The objective function becomes

$$\mathcal{L}^{(t)} \approx \sum_{i=1}^{n} \left[L(y_i, \hat{y}_i^{(t-1)}) + g_i^{(t)} h_t(x_i) + \frac{1}{2} h_i^{(t)} h_t(x_i)^2 \right] + \Omega(f_t).$$

For each tree, the leaf node predictions are constants, and the tree structure is denoted by $q(x_i)$, representing the leaf node to which sample x_i belongs. Thus, the model can be expressed as

$$h_t(x_i) = w_{q(x_i)},$$

where w_j is the weight of the j-th leaf.

Substituting this into the objective function and summing over all leaf nodes,

$$\mathcal{L}^{(t)} = \sum_{j=1}^{T} \left[\sum_{i \in I_j} \left(g_i w_j + \frac{1}{2} h_i w_j^2 \right) \right] + \Omega(f_t),$$

where I_j is the set of instances in leaf j.

By taking the derivative of the objective function with respect to the leaf weights and setting it to zero, we obtain the optimal weight for each leaf

$$w_j^* = -\frac{\sum_{i \in I_j} g_i}{\sum_{i \in I_j} h_i + \lambda}.$$

Substituting the optimal weights back into the objective function gives the optimized objective function

$$\mathcal{L}^{(t)} = -\frac{1}{2} \sum_{j=1}^{T} \frac{\left(\sum_{i \in I_j} g_i \right)^2}{\sum_{i \in I_j} h_i + \lambda} + \gamma T.$$

The steps involved in the XGBoost algorithm are as follows:

1. Initialization: Initialize the model with a constant value, typically the mean of the target variable y:

$$\hat{y}_i^{(0)} = \frac{1}{n} \sum_{i=1}^{n} y_i.$$

2. Gradient and Hessian calculation: For each iteration t, compute the gradients $g_i^{(t)}$ and Hessians $h_i^{(t)}$.
3. Tree construction: Build a new tree f_t to fit the residuals $g_i^{(t)}$ and $h_i^{(t)}$. Calculate the optimal leaf weights w_j and the gain from splitting nodes.
4. Model update: Update the model by adding the contributions of the new tree:

$$\hat{y}_i^{(t)} = \hat{y}_i^{(t-1)} + \eta \cdot f_t(x_i),$$

where η is the learning rate.
5. Iteration: Repeat steps 2–4 for a predefined number of iterations or until convergence.

Light gradient boosting machine (LightGBM) is a machine learning algorithm based on the gradient boosting framework, designed specifically for handling large-scale data and efficient training. Its core principle involves building an ensemble of weak models, where each step attempts to reduce the gradient of the loss function. This iterative process constructs tree models using weighted samples, thereby progressively improving prediction accuracy. LightGBM is applicable to various types of loss functions, including mean squared error for regression tasks and cross-entropy loss for classification tasks. The general form of the loss function can be expressed as

$$L(y, \hat{y}) = \sum_{i=1}^{n} l(y_i, \hat{y}_i),$$

where y_i represents the true labels and $\hat{y}_i$ represents the predicted values by the model.

The objective in the gradient boosting framework is to minimize the loss function. For the i-th iteration, the goal is to find a new base model $h_i(x)$ that, when added to the existing model, minimizes the loss function

$$h_i(x) = \arg\min_h L(y, F_{i-1}(x) + h(x)),$$

where $F_{i-1}(x)$ is the ensemble of the first $i - 1$ models and $h(x)$ is the prediction of the i-th model.

In each iteration, the residuals, which represent the difference between the predicted values and the true labels, need to be computed to fit the next base model. The residual r_i is calculated as

$$r_i = y - F_{i-1}(x),$$

where $F_{i-1}(x)$ is the prediction from the previous iteration.

To fit the residuals, a new base model $h_i(x)$ is trained. Typically, decision trees are used as the base models. This process can be formulated as

$$h_i(x) = \arg\min_h \sum_{j=1}^{n} l(y_j, F_{i-1}(x_j) + h(x_j)).$$

When constructing decision trees, LightGBM employs a Leaf-wise growth strategy. At each leaf node, it calculates the reduction in the loss function when samples are split into left and right child nodes, selecting the split with the maximum gain. This process can be expressed as

$$\text{Gain} = \frac{1}{2}\left(\frac{G_L^2}{H_L + \lambda} + \frac{G_R^2}{H_R + \lambda} - \frac{(G_L + G_R)^2}{H_L + H_R + \lambda}\right) - \gamma,$$

where G_L and G_R are the sums of the first-order gradients for the left and right child nodes, respectively; H_L and H_R are the sums of the second-order gradients for the left and right child nodes, respectively; λ is the regularization parameter; and γ is the minimum gain threshold for the split.

Categorical Boosting (CatBoost) is an advanced gradient boosting algorithm developed by Yandex, specifically designed to handle categorical features and improve the accuracy and efficiency of the model. Its ability to directly process categorical data and optimize for real-world scenarios makes it a powerful tool for various machine learning tasks.

Unlike traditional gradient boosting algorithms that require extensive preprocessing for categorical features, CatBoost incorporates techniques to process these features directly. For a categorical feature x_i in the dataset, CatBoost employs a method known as Target Statistics to transform it into numerical values. The transformation is done as follows:

$$\mathrm{TS}(x_i) = \frac{\sum_{j=1}^{n} y_j \cdot I(x_j = x_i)}{\sum_{j=1}^{n} I(x_j = x_i) + \alpha}, \tag{5.4}$$

where y_j is the target value of the j-th instance; $I(\cdot)$ is an indicator function that equals 1 if the condition is true, otherwise 0; and α is a smoothing parameter to avoid overfitting.

CatBoost uses an ordered boosting scheme to prevent overfitting and reduce prediction bias. This scheme ensures that the model does not use future data to predict past data during training. For a dataset with n instances, ordered boosting constructs k subsets of the data, where each subset S_k contains instances available up to the k-th iteration. It also builds symmetric decision trees, where the structure of the tree is predefined and identical for both left and right branches at each split. This leads to efficient computation and reduces the complexity of the model.

The steps involved in the CatBoost algorithm are as follows:

1. Initialization: Initialize the model with a constant value, typically the mean of the target variable y

$$F_0(x) = \frac{1}{n} \sum_{i=1}^{n} y_i,$$

   where y_i is the target value for the i-th instance.
2. Gradient calculation: Based on the current model $F_{t-1}(x)$, for each iteration t, compute the residuals $r_i^{(t)}$,

$$r_i^{(t)} = - \left. \frac{\partial L(y_i, F(x_i))}{\partial F(x_i)} \right|_{F(x_i)=F_{t-1}(x_i)},$$

   where $L(y_i, F(x_i))$ is the loss function.
3. Handling categorical features: Transform categorical features into numerical values using (5.4).

4. Tree construction: Build a symmetric decision tree $h_t(x)$ to fit the residuals $r_i^{(t)}$. Use ordered boosting to construct the tree, ensuring no future data influences the current prediction.
5. Model update: Update the model $F_t(x)$ by adding the newly constructed tree scaled by a learning rate η,

$$F_t(x) = F_{t-1}(x) + \eta \cdot h_t(x).$$

6. Iteration: Repeat the two to five steps for a predefined number of iterations or until convergence.

5.1.3.3 Algorithm-Driven Modeling

The algorithm-driven approach focuses on fine-tuning the learning algorithm or classifier without requiring modifications to the original training dataset [83]. Traditional classification algorithms generally assume that the cost of misclassification is uniform across all response variable values. In this section, we employed cost-sensitive learning methods by deliberately increasing the cost of misclassified samples and adjusting their weights during the training process [84].

Specifically, we implemented the adaptive cost-sensitive boosting (AdaCost) algorithm [28], which dynamically adjusts sample weights based on their classification errors to mitigate the impact of high-cost classification errors. AdaCost achieves this by refining the weights of incorrectly classified samples according to predefined cost values in the cost matrix. This approach imposes a more substantial penalty for high-cost classification errors, thereby increasing the model's sensitivity to such errors during training. By focusing on high-cost classification errors, the model's overall cost sensitivity is enhanced.

Additionally, our research extended to a variety of cost-sensitive learning techniques, including cost-sensitive logistic regression (LR), cost-sensitive decision trees (DT), cost-sensitive support vector classifiers (SVC), cost-sensitive extra trees (ET), cost-sensitive multilayer perceptrons (MLP), and cost-sensitive random forests (RF). These diverse methodologies contribute to a comprehensive strategy aimed at addressing data imbalance issues with a robust theoretical foundation. The detailed implementation and evaluation of these cost-sensitive learning methods are presented in Chap. 2.

5.1.4 Model Evaluation and Selection

5.1.4.1 Model Evaluation

Due to the issue of data imbalance, the assessment of candidate models' performance encompasses various metrics, including the area under the receiver operating

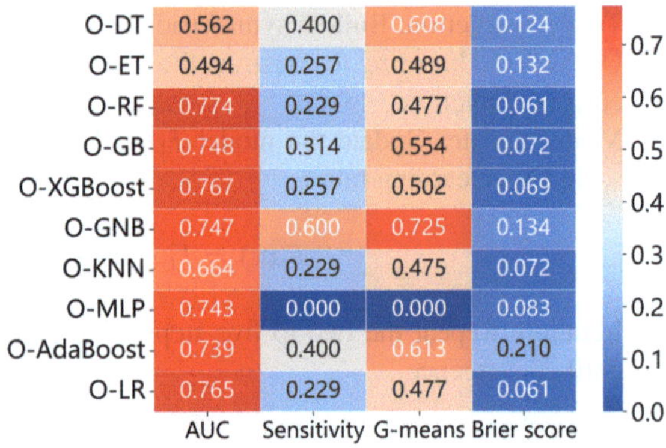

Fig. 5.7 AUC, sensitivity, G-means, and Brier score for ten ML models on the original dataset

characteristic curve (AUC), sensitivity, Brier score, geometric mean score (G-means), and precision-recall area under the curve (PR-AUC). Aside from G-means and PR-AUC, the other metrics were introduced in Chap. 2. Here, the definition of the G-means and PR-AUC will be discussed.

The G-means score provides a comprehensive approach to performance evaluation, particularly useful for identifying and addressing imbalances in datasets. It can be calculated as follows:

$$\text{G-mean} = \sqrt{\text{Sensitivity} \times \text{Specificity}}.$$

PR-AUC evaluates a model's performance under varying trade-offs between precision and recall. The PR-AUC is then computed by integrating the precision-recall curve

$$\text{PR-AUC} = \int \text{Precision}(\text{Recall})\, d(\text{Recall}).$$

The higher the PR-AUC and G-mean, the better the model's performance in handling imbalanced datasets.

Figure 5.7 presents the predictive performance for ten ML models without resampling techniques. It is evident that, prior to implementing data balancing techniques, the original models exhibited lower sensitivity and G-means values across the remaining nine models. This issue arises from analyzing severely imbalanced datasets directly, resulting in a bias toward the majority class, which, in this context, pertains to the survival outcome. Although the original models showed commendable AUC and Brier scores, they suffered from a significantly low true positive rate, making them ineffective for distinguishing deceased patients, thus reducing their practical reliability.

Figure 5.8 provides a comprehensive visual representation of the predictive performance of various combinations of four resampling algorithms paired with ten ML models. After applying data balancing strategies, these resampling-combined models exhibited consistently favorable AUC and retained promising Brier scores. Notably, almost all resampling-combined models showed significant improvements

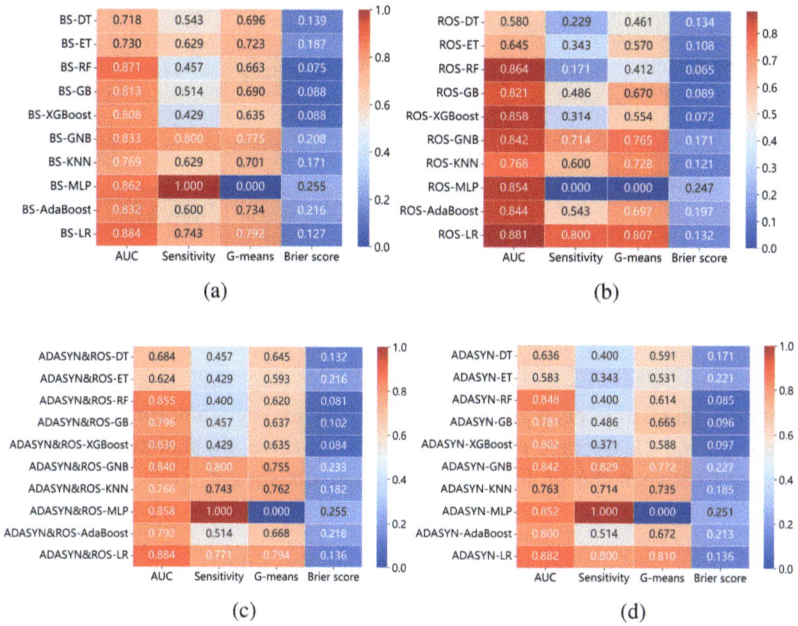

Fig. 5.8 AUC, sensitivity, G-means, and Brier score of 10 ML models with resampling methods. (**a**) BS oversampling. (**b**) ROS oversampling. (**c**) ADASYN and ROS rsampling. (**d**) ADASYN oversampling

in both sensitivity and G-means. This enhancement was most pronounced in models such as ROS-LR, BS-LR, ADASYN-LR, and ADASYN-ROS-LR, indicating a notable increase in the ability of the trained models to discriminate between positive and negative samples.

Figure 5.9 presents a comprehensive evaluation of seven cost-sensitive ML models. The results underscore the efficacy of cost-sensitive learning techniques in improving model performance by assigning higher rewards to minority samples, thus increasing the accuracy of classifying these instances. Notably, the cost-sensitive SVC stood out, exhibiting the highest discrimination levels with an AUC of 0.888 and a Brier score of 0.061. Additionally, this model demonstrated commendable sensitivity of 0.800 and an impressive G-mean of 0.806. Given its exceptional performance, the cost-sensitive SVC appears to be a suitable base model for interpretable analysis.

5.1.4.2 Model Selection

A Friedman test was conducted to further assess the performance of these ML models. The Friedman test is a nonparametric statistical test used to analyze data in which multiple related groups or conditions are compared. The test works by

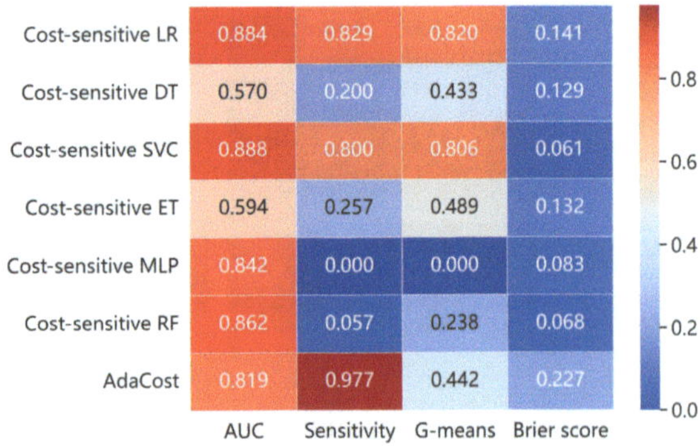

Fig. 5.9 AUC, sensitivity, G-means, and Brier score for cost-sensitive models

ranking the data within each group, calculating the average rank for each group, and then comparing the average ranks to determine if there are statistically significant differences between the groups. The Friedman statistics τ_{χ^2} can be computed as follows:

$$\tau_{\chi^2} = \frac{12N}{k(k+1)}(\sum_j R_j^2 - \frac{k(k+1)^2}{4}),$$

where k and N are the number of algorithms and datasets, R_j is the sum of ranks for the j-th group.

The Friedman test can be followed by post hoc tests (e.g., Nemenyi test) to identify which specific groups differ from each other when significant differences are found. The performance disparity between the two clustering methods is deemed significant when the average rank difference between them surpasses the critical threshold. The critical difference can be computed as follows:

$$CD = q_a\sqrt{\frac{k(k+1)}{6N}}, \tag{5.5}$$

where q_a is the critical value for the Nemenyi test at a specific significance level (alpha), which can be found in Nemenyi's critical values table.

We will illustrate this evaluation process with an example of cost-sensitive learning models. The average rank for each algorithm across the entire dataset was computed. Using a significance level of $\alpha = 0.05$, the test statistics for AUC, sensitivity, G-means, and Brier score yielded values of 19.689, 202.067, 78.918, and 138.860, respectively. All values surpass the critical threshold of 2.324, indicating statistically significant differences among these methods. A post hoc analysis using the Nemenyi test was then performed to discern the nature of these differences. The critical difference (CD) was determined to be 3.185. Figure 5.10 illustrates the CD diagram derived from pairwise Nemenyi tests.

To provide a holistic insight into the predictive performance of the 57 ML models, their AUC and PR-AUC curves are presented in Figs. 5.11 and 5.12. One notable finding is the exceptionally high PR-AUC of 0.977 achieved by the

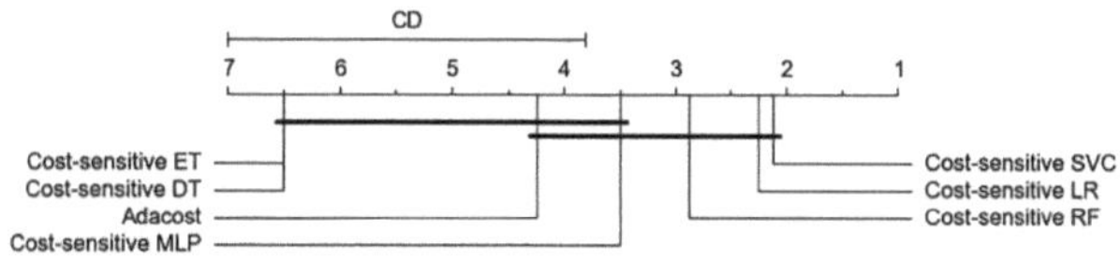

(a) AUC

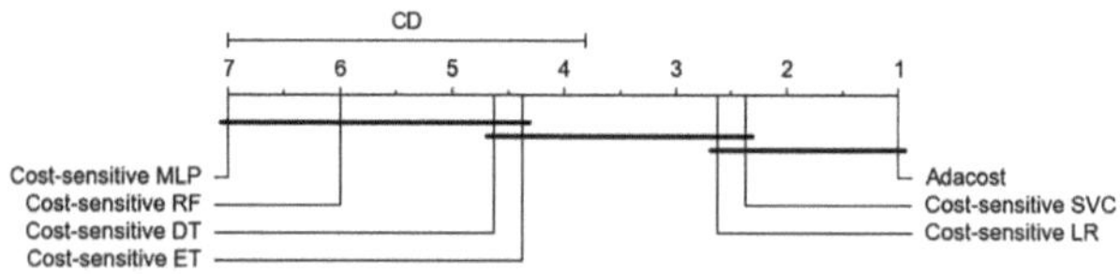

(b) Sensitivity

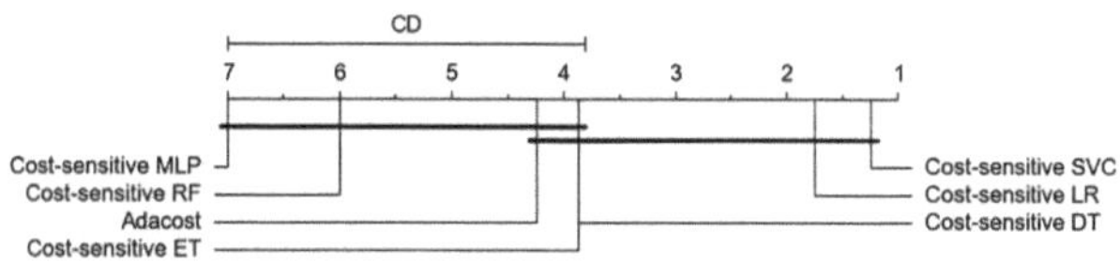

(c) G-means

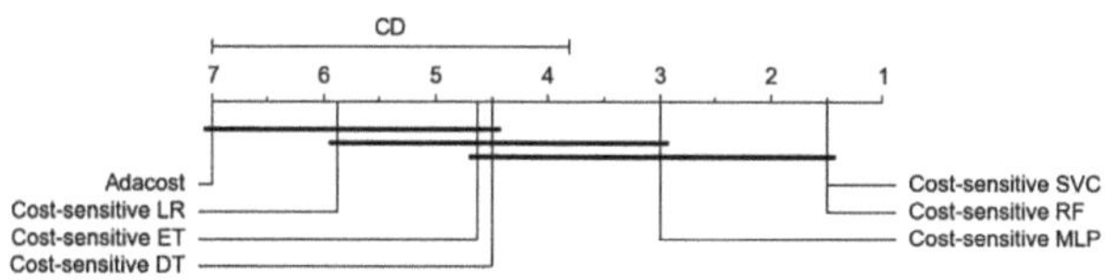

(d) Brier score

Fig. 5.10 The CD diagram of AUC, sensitivity, G-means, and Brier score derived from pairwise Nemenyi tests for the cost-sensitive learning models. Black lines connecting certain groups indicate that the differences between these groups are not significant in multiple comparisons

Adacost model, making it a favorable choice in practical applications focused on identifying mortality. Given the overall performance across four key metrics, the cost-sensitive SVC should be the preferred method for practical clinical diagnosis and applications.

5.1.5 Model Interpretation

The global importance of each feature, estimated using SHAP, was leveraged to understand the overall impact of various features across the study cohort in Fig. 5.13. The SHAP summary plot illustrated the distribution of each feature's

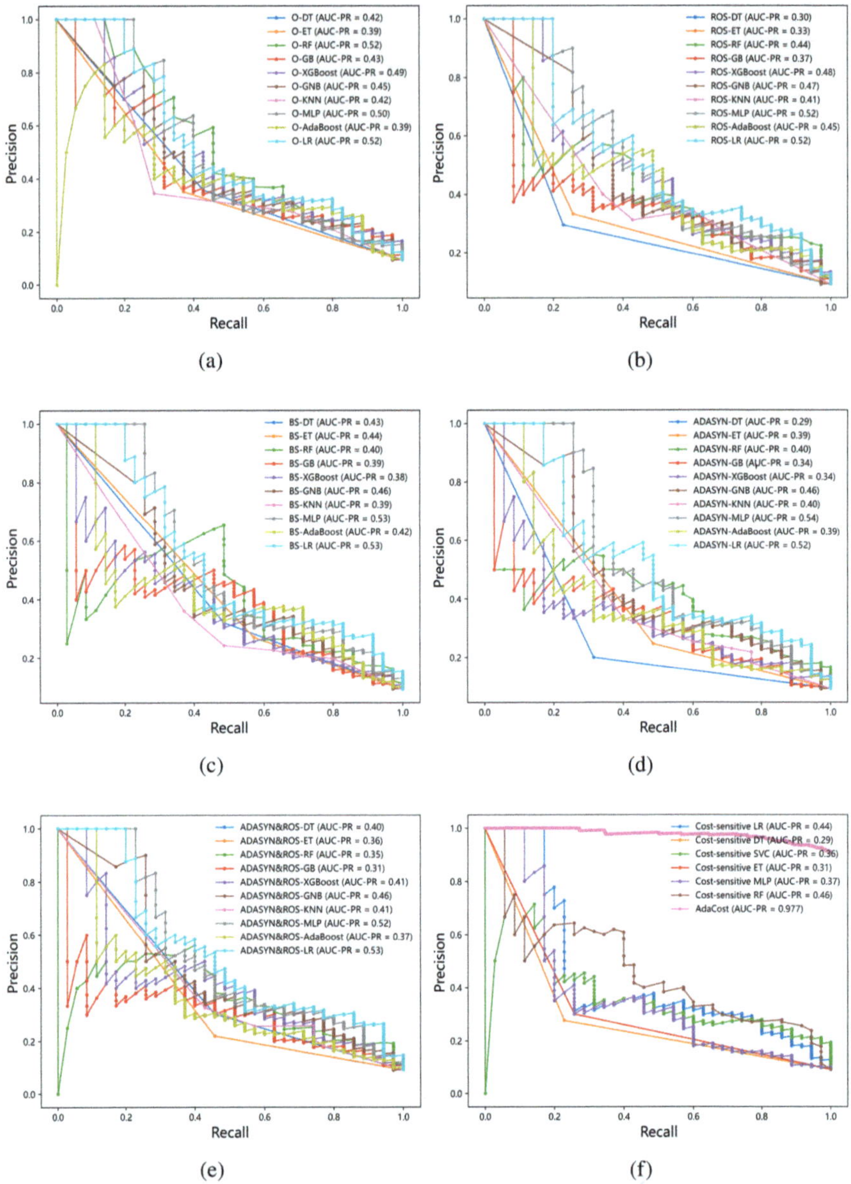

Fig. 5.11 The PR-AUC curves of data- and algorithm-driven approaches for the test cohort

impact on the model output. The color coding facilitated the understanding of how changes in the value of a feature influenced the outcome: red indicates a high feature value, whereas blue indicates a low feature value. The further a point is from the baseline SHAP value of zero, the stronger its effect on the output.

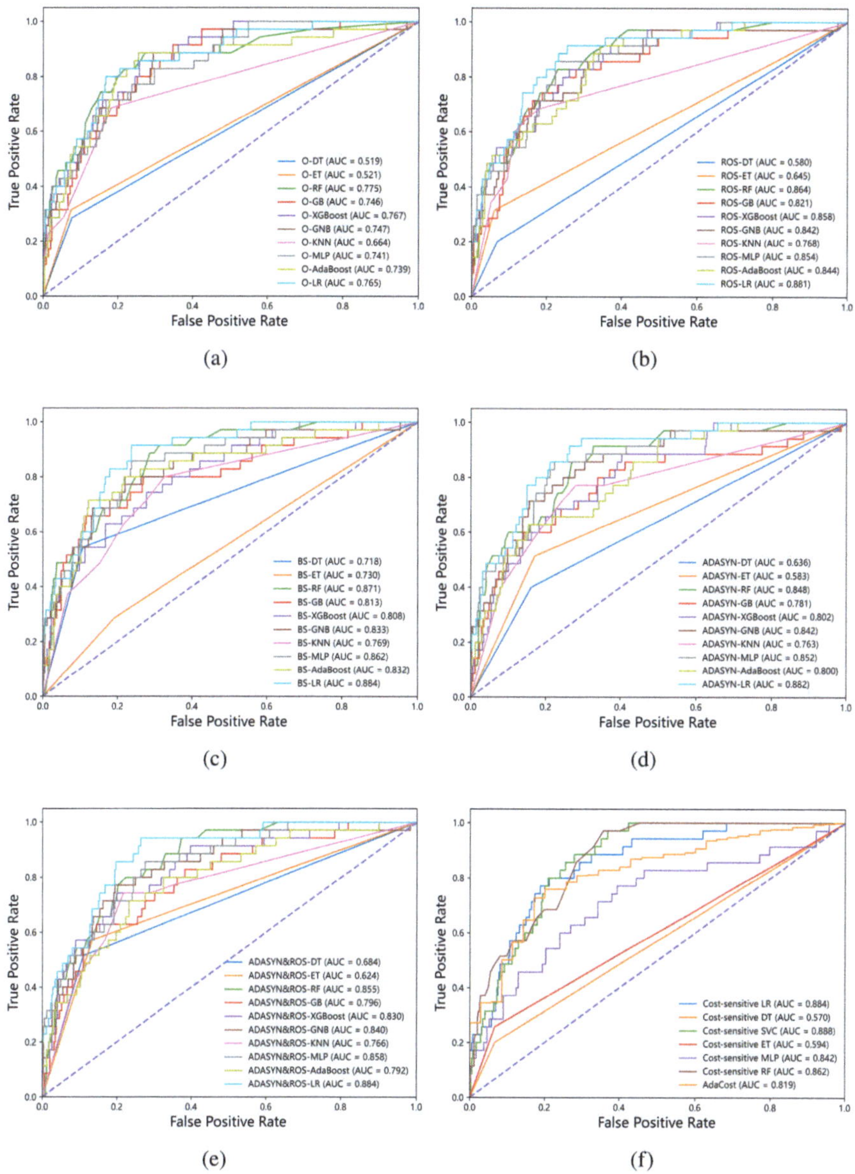

Fig. 5.12 The AUC curves of data- and algorithm-driven approaches for the test cohort

The analysis identifies physiological parameters (SOFA, age, total protein_max), immune function indicators (wbc_max, wbc_min), markers of circulatory dysfunction (dbp_max, temperature_max, urine output), and coagulation function (ptt_max) as determinants in assessing mortality risk for ICU patients. These factors impact

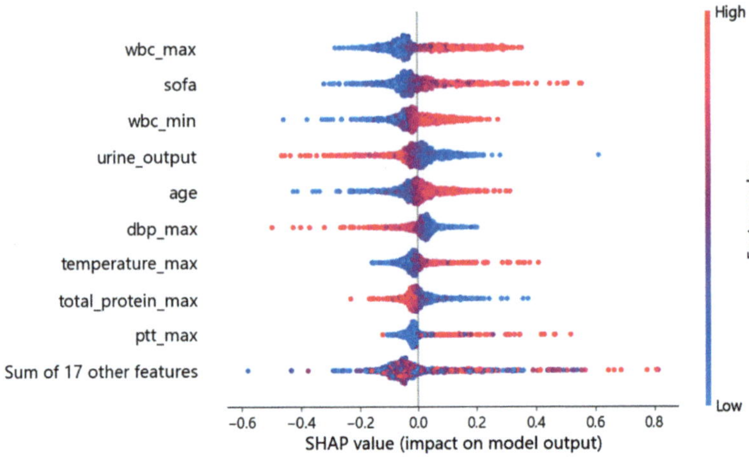

Fig. 5.13 SHAP summary plot of the cost-sensitive SVC model. Each dot represents a patient's feature attribution value. Dots are color-coded based on patient feature values, with red indicating higher values (e.g., higher risk of death) and blue indicating lower values. The distance from the baseline SHAP value of zero reflects the strength of each feature's impact on the model output

the evaluation of patient condition and prognosis, surpassing other indicators such as blood glucose and hematocrit levels. Additionally, the directional impact of SHAP values indicates that elevated levels of SOFA, age, body temperature, white blood cell count, and partial thromboplastin time, represented by a right-tailed distribution, correlate with an increased mortality risk. Conversely, low levels of urine output, total protein, and maximum diastolic blood pressure, represented by a left-tailed distribution, also exhibit significant associations and positively correlate with predicted mortality.

In Fig. 5.14, the SHAP interactive dependency plot is showcased. This visualization, depicting the concurrent variation of two features and its impact on SHAP values and the ultimate model output, serves as a valuable tool for discerning the sensitivity of predictions to various features. In Fig. 5.14a, a noticeable trend emerges: as SOFA values rise, there is a concurrent increase in the corresponding SHAP values. This relationship is interpretable, as an escalation in the SOFA score typically aligns with a deterioration in a patient's condition and organ functionality, ultimately heightening the risk of mortality. When age surpasses 72.5 years, the associated SHAP values consistently remain positive, exhibiting a gradual increase with advancing age. This pattern signifies a persistent escalation in the risk of mortality, aligning with findings from earlier studies [85, 86]. A significant contributing factor to this trend is the increased incidence of two high-fatality complications of RHD—atrial fibrillation and stroke—which rise with age at diagnosis by 5% and 4% per year, respectively [85]. Notably, an interaction between the SOFA score and age emerges during this period. Among patients of identical age, those with elevated SOFA scores exhibit a heightened susceptibility to

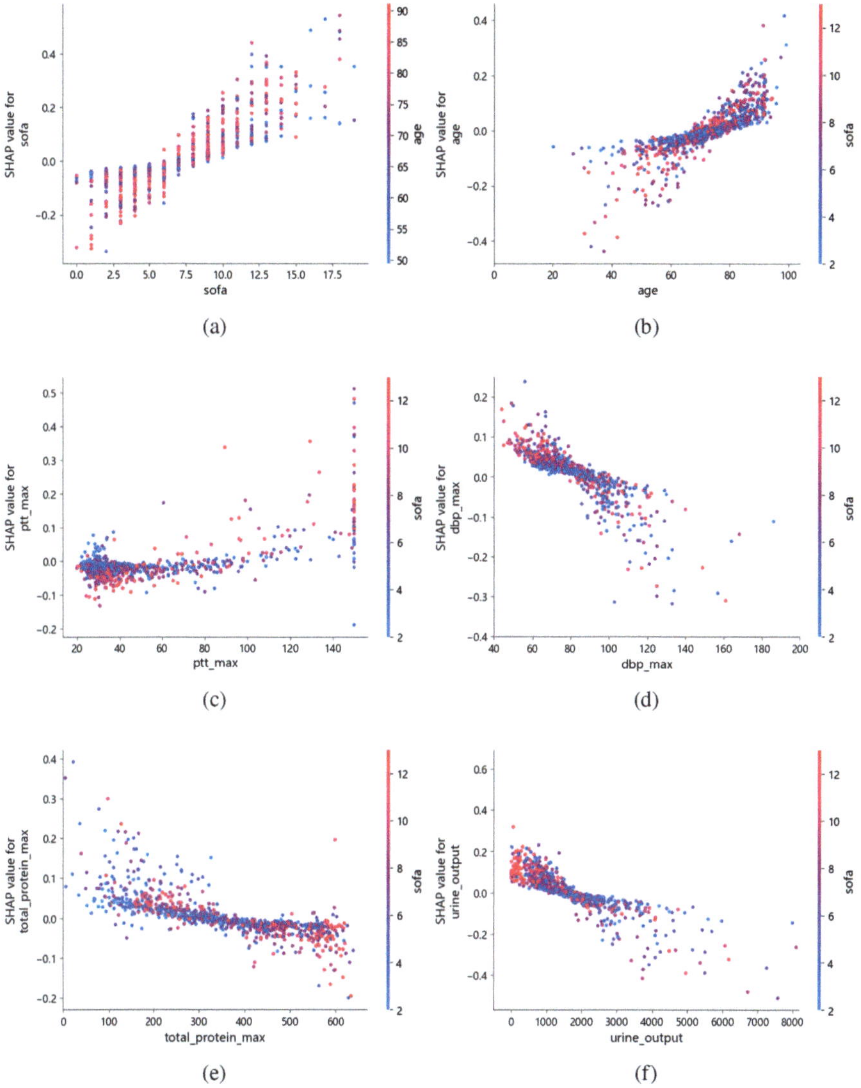

Fig. 5.14 SHAP dependence plot of the cost-sensitive SVC model. The x-axis denotes the value of the primary driver, while the y-axis represents the corresponding SHAP value. The coloration of each point signifies the value associated with the secondary driver

mortality. In contrast, below the age of 72.5 years, this interaction is not discernible (Fig. 5.14b). This is because the SOFA score serves as a straightforward yet potent index, quantifying organ damage by measuring the burden of organ dysfunction in critically ill individuals. It encompasses assessments of cardiovascular, hemostatic, and renal functions [87].

Anticoagulation is crucial for mitigating the progression of RHD. As maximum partial thromboplastin time (ptt_max) extends, the corresponding SHAP values shift from negative to positive, indicating an increased risk of mortality in RHD patients. This suggests potential abnormalities in coagulation function associated with prolonged ptt_max, elevating the risk of thrombotic events or other coagulation-related complications. When ptt_max exceeds 120 seconds, an interaction between ptt_max and SOFA becomes evident: cases with the same ptt_max value but higher SOFA scores exhibit higher SHAP values, indicating an increased risk of mortality (Fig. 5.14c). This phenomenon may partly be due to heightened fibrinogen levels correlating with the presence of cerebral microbleeds in ischemic stroke patients with RHD [88]. Therefore, for patients with RHD, it is essential to consider coagulation function indices like PT, especially those with atrial fibrillation, a history of thromboembolism, or left atrial thrombosis.

For urine_output, dbp_max, and total_protein_max, as their values increase, the corresponding SHAP values also turn positive, indicating a decrease in the associated risk of mortality (Fig. 5.14d–f). This phenomenon can be attributed to improvements in the corresponding physiological conditions. Specifically, elevated urine_output suggests enhanced physiological circulation and renal function. The case report and experimental analysis [89–91] confirmed that RHD adversely affects cardiac function, potentially resulting in a decreased cardiac pumping capacity and reduced tissue perfusion throughout the body. This prompts the kidneys to sense decreased blood flow, leading to a reduction in urine production. However, a significant improvement in the condition is observed after administering cardiac and diuretic medications. This improvement can be attributed to the alleviation of pulmonary vein congestion through treatment, prolonging diastolic time and enhancing cardiac output [71]. Besides, total protein is a reliable indicator of nutritional status [92], and there is report linking malnutrition to elevated mortality in individuals undergoing cardiac surgery [93]. Insufficient total protein levels may lead to inadequate blood volume and fluid retention, potentially exacerbating symptoms of heart failure. Previous investigations have indicated that hypoalbuminemia was common in patients with stable chronic or acute heart failure [94, 95]. It can impact the transport capacity of blood, influencing organ perfusion and oxygenation. Reduced total protein levels may also compromise immune function, increasing the risk of infections, in which the catabolism will increase, including reduced protein synthesis rates and increased protein degradation rates [96]. Further, in patients with RHD, rheumatic inflammation leads to valve damage and structural changes, impacting heart contraction and relaxation and influencing diastolic pressure. Low diastolic pressure decreases tissue perfusion, impairing organ function and coronary artery filling. This can cause pulmonary congestion and trigger acute heart failure by increasing left ventricular (LV) filling pressure. Severe ischemia and atrial fibrillation further compromise LV filling, especially in existing diastolic dysfunction. Severe mitral valve stenosis, common in RHD, exacerbates these issues. Thus, maintaining appropriate diastolic pressure is crucial for proper organ perfusion and assessing the risk of complications like AHF and atrial fibrillation [97].

5.1.6 Conclusion

In comparison to other prevalent infectious diseases such as malaria, HIV/AIDS, and tuberculosis, rheumatic heart disease (RHD) receives relatively less attention despite its significant cardiovascular morbidity and mortality burden. This relative neglect and lack of funding may have contributed to limited progress in basic medical research in this field over the past five decades [66]. Specifically, the current research landscape regarding accurately predicting the mortality risk in RHD remains underexplored. Therefore, this section integrates machine learning, resampling techniques, feature selection, and cost-sensitive learning methods to develop a workflow on a large, imbalanced public dataset, aiming to provide interpretable predictions for mortality risk in RHD patients.

By systematically combining four resampling methods with ten machine learning techniques and considering seven cost-sensitive learning models, a total of 57 machine learning regression prediction models were developed in this study. The utilization of the cost-sensitive SVC yielded the most favorable outcomes, with an AUC of 0.888, sensitivity of 0.800, G-means of 0.806, and a Brier score of 0.061, making it well-suited for the research cohort. Given that the negative class (deaths) is often the minority class, classifiers without balancing practices may exhibit a bias toward predicting the majority class. The application of resampling and cost-sensitive learning methods significantly enhanced sensitivity and G-means for the minority class. Although the improvement in AUC was not statistically significant, these techniques contributed to balancing the weights between high-risk and low-risk patients. Integrating resampling and cost-sensitive learning techniques into mortality risk prediction models should be more widely applied in clinical practice to address the challenges posed by imbalanced data.

Global interpretability enables clinical physicians to understand the response trends of the model across the entire feature space. It is shown that physiological parameters (SOFA, age, total protein, and temperature) and immune function indicators (white blood cell count), along with markers reflecting circulatory dysfunction (diastolic blood pressure, urine output, and partial thromboplastin time), emerged as pivotal determinants for assessing mortality risk. Additionally, this approach offers intrinsic insights into RHD-related mortality, encompassing the prioritization of key features associated with mortality risk, determination of thresholds, and comprehension of how variations in these features influence patient mortality rates.

Finally, it is important to note that this workflow still has the following limitations: (1) Clinical feature-related information was collected retrospectively, which may introduce information bias; (2) due to database constraints or severe missing variables (e.g., body mass index, brain function performance grading scores), the improvement of the predictive model's performance may be affected; (3) some patients died within 24–48 hours of admission, and their initial examination/test results within the first 24 hours may be extremely poor. For this group, the predictive model may have a certain degree of label leakage risk; (4) the study data were sourced from MIMIC-IV 4.0, and the model's generalizability remains to be validated.

5.2 Survival Analysis of Breast Cancer Patients

Breast cancer presents an escalating global health challenge. In the year 2020 alone, approximately 2.3 million women were confronted with breast cancer diagnoses, leading to a staggering 685,000 fatalities across the globe [98]. This not only positions breast cancer as the most prevalent incident cancer on a global scale but also categorizes it as the chief contributor to disability-adjusted life-years lost, surpassing all other forms of malignancies. Confronting breast cancer emerges as a complex endeavor, demanding concerted efforts from healthcare systems, policy formulators, and various stakeholders [99]. As breast cancer is prone to metastasis, with the rising incidence of breast cancer, the prediction of breast cancer survival has become a paramount issue in current breast cancer research.

The significance of utilizing ML to construct a breast cancer survival model is profound. Firstly, ML can analyze vast amounts of clinical data and biological features to identify the most impactful factors on breast cancer survival, thereby achieving more precise survival predictions [100]. Secondly, such models can provide personalized treatment recommendations, tailoring optimal treatment strategies based on individual patient profiles, thus enhancing treatment effectiveness and minimizing side effects. Furthermore, ML models can assist doctors in identifying patients who may experience complex survival dynamics, offering vital insights for clinical decision-making [101].

This section will present a coherent workflow involving ML in the construction of a breast cancer survival model. Initially, the original dataset was normalized and coded. Subsequently, factors exerting significant impacts on patient survival probability were identified through single-factor and multifactor Cox regression analyses. Moving forward, survival analysis models for breast cancer patients were formulated using various techniques, including survival tree, random survival forest, gradient boosted survival tree, and the DeepSurv model within the ML survival framework. The predictive capabilities of these models were evaluated using two key metrics, C-index and IBS. Ultimately, the most effective predictive model's outcomes were subjected to interpretability analysis using the SHAP framework, enabling the derivation of actionable treatment recommendations. For visual clarity, the workflow diagram is presented in Fig. 5.15.

5.2.1 *Data Source and Extraction*

This section utilizes data from the American Institute for Cancer Research SEER (Surveillance, Epidemiology, and End Results) database (https://seer.cancer.gov/)

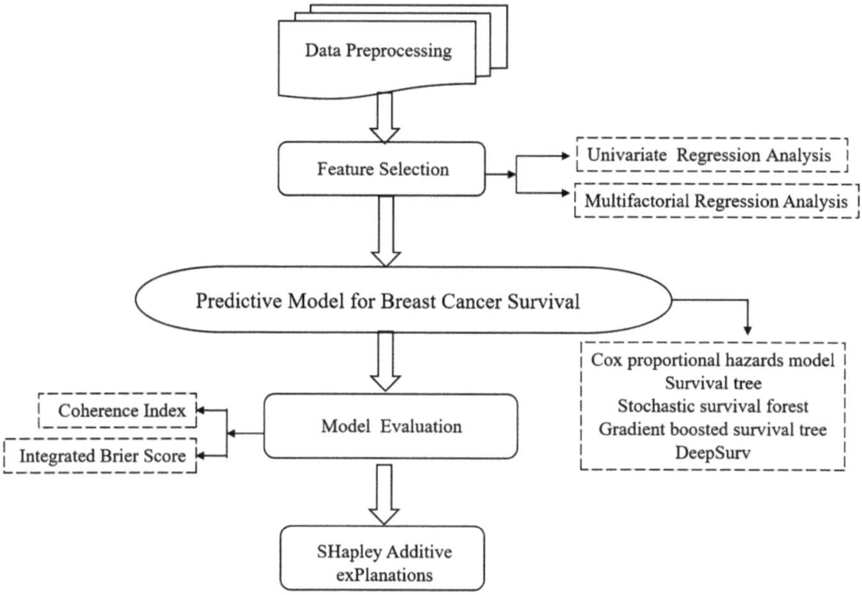

Fig. 5.15 The workflow diagram for prediction of breast cancer survival

to analyze the probability of survival for breast cancer patients within 5 years. The SEER database provides comprehensive information on patients with various tumor types from selected regions in the United States, including patient registration number, sociodemographic details, primary lesion site, tumor size, tumor location code, treatment plans, survival time, and survival status. The study data were coded according to the International Classification of Diseases for Oncology Third Edition (ICD-O-3). To ensure the validity and relevance of the analysis, the study focused on breast cancer patients from the 2014 database, employing specific screening criteria. These criteria included (i) clear pathological diagnosis of breast cancer as the first primary tumor; (ii) diagnosis made in 2014; (iii) complete clinical information, survival time, and survival status for each patient are available; and (iv) exclusion of patients with reporting sources limited to autopsy only or death certificate only as well as patients with missing clinical information.

Ultimately, 41,250 breast cancer patients were included in the study, meeting these rigorous selection criteria. 26 variables are selected to capture crucial factors influencing survival outcomes. These variables encompass key aspects such as age, race, tumor stage, estrogen receptor status, and progesterone receptor status. The independent variables are chosen based on their potential impact on patient survival, while the time variable utilized is survival time. Furthermore, the dependent variable in the analysis model is the patient's survival status.

Table 5.2 Significance analysis for continuous variables

Variable	Risk ratio	P value
Age	1.04	<0.001
Months from diagnosis to treatment	1	0.848
Regional nodes positive ratio	4.05	<0.001
Tumor Size	1.01	<0.001
Number of malignant tumors	1.49	<0.001
Number of borderline tumors	1.1	0.54

5.2.2 *Predictive Model of Breast Cancer Survival*

To optimize the performance of survival prediction models, it is crucial to identify and select the most influential features for model training. This can be achieved through single-factor and multifactor Cox regression analyses on the initial set of features. The goal is to pinpoint the factors that significantly influence the dependent variable, enabling the creation of a streamlined subset of features. This selected subset will serve as the foundation for developing survival prediction models specifically tailored to breast cancer patients.

5.2.2.1 Feature Selection

Tables 5.1 and 5.2 present the results of univariate Cox regression analysis for continuous and categorical variables, respectively, at a significance level of 0.05. Variables with a P-value less than 0.05 significantly influence the outcome variable and should be retained; otherwise, they should be excluded. From the tables, it is evident that the variables Primary site, Laterality, Rare tumors site, Systemic/Sur Seq, T Stage, Surg/Rad Seq, Month from Diagnosis to Treatment, HER2, and Number of Borderline Tumors all have P-values greater than 0.05 and do not exhibit statistical significance. Therefore, these variables should be excluded from further analysis (Table 5.3).

A multifactor Cox regression model is developed to examine the significance of the interaction effects of various variables on patient survival outcomes. The results are illustrated in Fig. 5.16, which presents the hazard ratios (HR) and their 95% confidence intervals within the regression analysis on a coordinate axis, centered around a null line. For categorical variables, the bar typically represents the reference category of the variable, with other categories spread on both sides of the line. If a factor or variable appears on the left side of the null line, it implies that the factor is protective for patient survival events. Conversely, if it appears on the right, it is deemed a risk factor. For example, take the variable "Radiation." If "No/Unknown" is used as the reference category, "Yes" appears on the left side of the null line, suggesting that receiving radiation treatment promotes survival. Similarly, when "stage I" of the "stage" variable is used as the reference, all other

Table 5.3 Significant analysis for categorical variables

Variable	Factor	Risk ratio	P value
Sex	Female	Reference	
	Male	2.0	<0.001
Race	Black	Reference	
	Other	0.47	<0.001
	White	0.68	<0.001
Marital status	Divorced	Reference	
	Married	0.68	<0.001
	Single	1	0.983
	Widowed	1.87	<0.001
Primary site	Nipple	Reference	
	Central portion of breast	2.02	0.037
	Upper-inner quadrant of breast	1.18	0.615
	Lower-inner quadrant of breast	1.47	0.259
	Upper-outer quadrant of breast	1.41	0.3
	Lower-outer quadrant of breast	1.35	0.378
	Axillary tail of breast	1.34	0.461
	Overlapping lesion of breast	1.43	0.286
	Breast, NOS	1.65	0.134
Laterality	Right	Reference	
	Left	0.99	0.742
Grade	I	Reference	
	II	1.41	<0.001
	III	2.62	<0.001
	IV	2.9	0.001
Rare tumors site	Inv carcinoma of no special type-NST	Reference	
	Invasive lobular carcinoma of breast	0.94	0.254
	Mammary pagets disease of breast	1.93	0.014
	Special types of adenocarcinoma of breast	0.8	0.009
	Metaplastic carcinoma of breast	3.29	<0.001
	Salivary gland type tumor of breast	1.57	0.27
	Other epithelial tumors of breast	1.85	<0.001
	Neuroendocrine carcinoma of other sites	1.57	0.525
Stage	I	Reference	
	II	1.98	<0.001
	III	5.1	<0.001
T stage	T0	Reference	
	T1	0.38	0.09
	T2	0.82	0.739
	T3	1.33	0.624
	T4	2.56	0.106

(continued)

Table 5.3 (continued)

Variable	Factor	Risk ratio	P value
N stage	N0	Reference	
	N1	1.8	<0.001
	N2	3.63	<0.001
	N3	5.98	<0.001
Surg prim site	No surgery	Reference	
	Surgery	0.34	<0.001
Surg/Rad seq	Intraoperative radiation	Reference	
	No radiation	2.39	<0.001
	Radiation after surgery	1.55	0.03
	Radiation before and after surgery	4.24	<0.001
	Radiation prior to surgery	2.37	0.021
	Sequence unknown, but both were given	3.48	0.011
	Surgery both before and after radiation	0	0.962
Systemic/Sur seq	Systemic therapy after surgery	Reference	
	No systemic therapy	34,884.56	0.972
	Sequence unknown	15,566.75	0.974
	Surgery both before and after systemic therapy	41,565.18	0.971
	Systemic therapy after surgery	18,388.51	0.974
	Systemic therapy before surgery	43,184	0.971
	Systemic therapy both before and after surgery	36,343.4	0.972
Radiation	None/Unknown	Reference	
	Yes	0.65	<0.001
Chemotherapy	None/Unknown	Reference	
	Yes	1.16	<0.001
ER status	Negative	Reference	
	Positive	0.45	<0.001
PR status	Negative	Reference	
	Positive	0.48	<0.001
HER2	Negative	Reference	
	Positive	1.02	0.719
First malignant primary	No	Reference	
	Yes	0.57	<0.001
Summary stage	Distant	Reference	
	Localized	0.12	<0.001

stages appear on the right side of the null line, indicating that higher stages are associated with a lower probability of survival.

Figure 5.16 reveals that 14 variables, including age, gender, race, marital status, grade, stage, radiotherapy, chemotherapy, regional nodal positivity, ER status, PR status, number of malignancies, tumor size, and summary stage, significantly

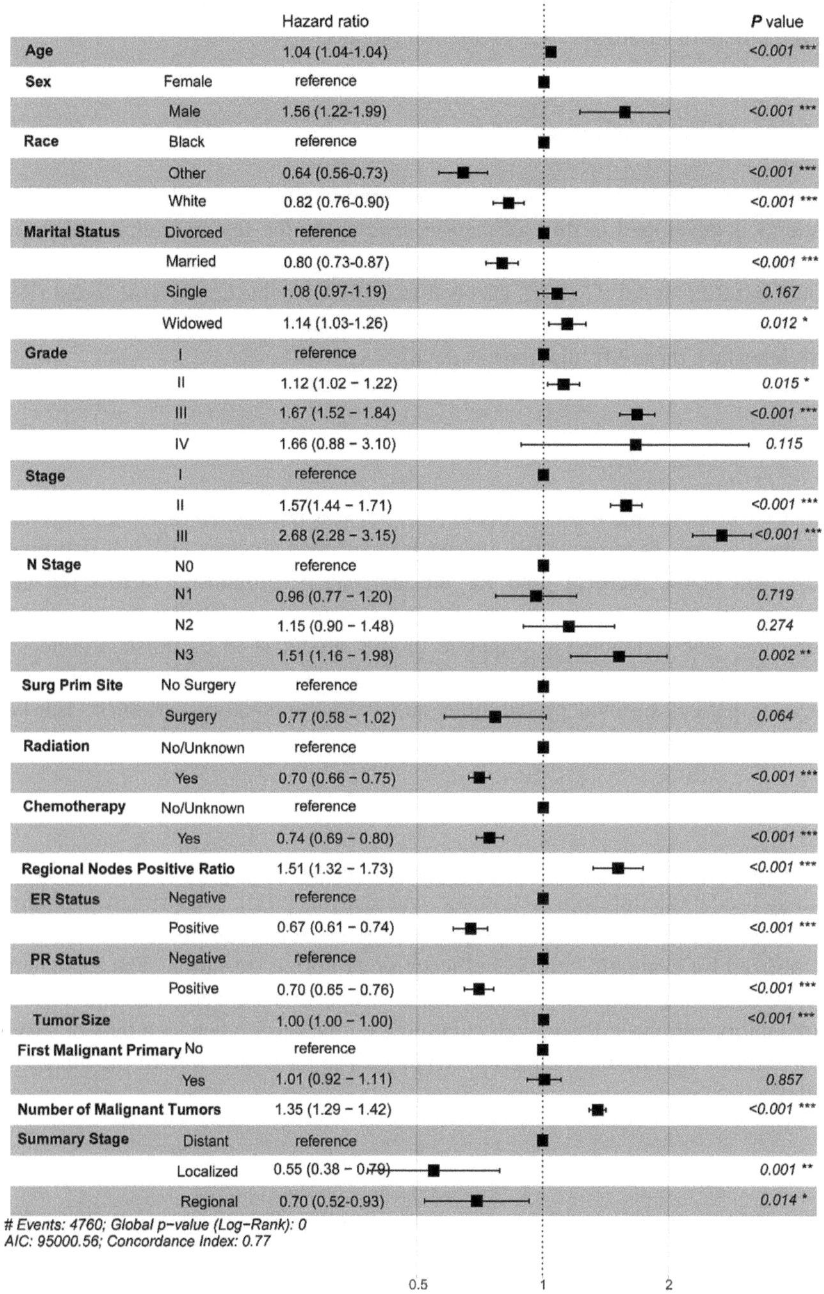

Fig. 5.16 Multifactor Cox regression nomogram for breast cancer

influence the dependent variable. Consequently, these variables are integral to the development of future survival prediction models.

5.2.2.2 Model Framework

A survival probability prediction model specifically designed for breast cancer patients is developed in this subsection, leveraging the features identified through the aforementioned screening process. The models utilized include the Cox proportional hazards model (Coxph), survival tree (ST), stochastic survival forest (RSF), gradient-boosted survival tree (GBST), and DeepSurv model (DeepSurv). Below we will introduce these ML methods in detail.

Cox Proportional Hazards Model

The Cox proportional hazards model (Coxph) was introduced in 1972 and remains the most widely used modeling approach in survival analysis [102]. It is primarily employed in the medical field for the analysis of prognostic factors related to diseases. It can also be utilized for evaluating clinical efficacy or exploring causation. The fundamental principle of this model is to establish a connection between survival time and explanatory variables, thus establishing the relationship between patient survival probabilities and these explanatory variables. The basic structure of the model is as follows:

$$h(t|x) = h_0(t) \cdot e^{(\beta_1 x_1 + \beta_2 x_2 + \ldots + \beta_p x_p)}, \tag{5.6}$$

The Cox proportional hazards model (Coxph), introduced in 1972, remains the most widely used modeling approach in survival analysis [102]. It is primarily employed in the medical field to analyze prognostic factors related to diseases and can also be utilized for evaluating clinical efficacy or exploring causation. The fundamental principle of this model is to establish a connection between survival time and explanatory variables, thereby elucidating the relationship between patient survival probabilities and these explanatory variables. The basic structure of the model is as follows:

$$h(t|X) = h_0(t) \exp(\beta_1 X_1 + \beta_2 X_2 + \cdots + \beta_p X_p)$$

where $h(t|X)$ is the hazard function at time t given the covariates X; $h_0(t)$ is the baseline hazard function, representing the hazard when all covariates are zero; and $\beta_1, \beta_2, \ldots, \beta_p$ are the coefficients for the covariates $X_1, X_2, \ldots, X_p$.

The hazard function describes the instantaneous risk rate of death for individuals who have survived up to time t. The Coxph reveals that the hazard ratio (HR) signifies the risk ratio associated with a unit increase in feature x_m. The calculation formula for HR is given by

$$HR = e^{\beta m}, \tag{5.7}$$

where βm is the coefficient for feature x_m. Interpreting the HR:

- If $HR > 1$, it indicates that an increase in the variable's value is associated with a higher risk, indicating it is a risk factor.
- If $HR < 1$, it suggests that an increase in the variable's value is linked to a lower risk, making it a protective factor.
- If $HR = 1$, it suggests that the variable has no effect on the risk.

Survival Tree

The survival tree method (ST) [103] is an extension of the traditional classification and regression tree techniques, specifically designed to handle survival data. This method is particularly useful in medical research where data is often right-censored, meaning that for some subjects, the event of interest (such as death) has not occurred by the end of the study period.

The key feature of the survival tree method is its ability to recursively partition the data into more homogeneous subgroups based on survival times. This partitioning is performed without making any assumptions about the underlying distribution of the data or the nature of the independent variables. Instead, the method uses a nonparametric approach to determine the best splits.

The process begins with the entire dataset at the root node and involves the following steps.

1. Recursive partitioning: The dataset is divided into two groups at each node based on a chosen splitting criterion.
2. Splitting criterion: The optimal split at each node is determined by maximizing a statistical measure, which can be formally defined as

$$\max_{s} \{statistic(node_h, s)\}, \tag{5.8}$$

   where $statistic(node_h, s)$ represents the chosen test statistic (e.g., log-rank test) for node h at split point s and s denotes all possible split points for node h.
3. Stopping rule: The recursive splitting continues until a prespecified stopping criterion is met, such as a minimum number of subjects in a node or a maximum tree depth.

The resulting survival tree structure allows researchers to identify subgroups of patients with distinct survival experiences, providing insights into the factors that influence survival and facilitating more personalized treatment strategies.

Random Survival Forest

Random survival forest (RSF) [104], introduced by Ishwaran in 2008, is a non-parametric model tailored for the analysis of survival data. RSF begins by creating multiple bootstrap samples from the original dataset. For each bootstrap sample, a survival tree is grown. Unlike traditional decision trees, survival trees are designed to handle censored data. The splitting criterion is based on survival differences, typically using the log-rank test to choose the best split at each node. This process is repeated recursively until a stopping criterion is met, such as a minimum number of observations in terminal nodes.

At each node, the algorithm selects a subset of features randomly and determines the optimal split based on the chosen criterion. The goal is to maximize the difference in survival times between the resulting child nodes. The optimal split s is determined by (5.8).

Each survival tree generates a cumulative hazard function (CHF) for the terminal nodes. The CHF estimates the hazard at any given time point for the observations in that node. For a given survival tree, the CHF at terminal node h is calculated as

$$H(t|h) = \frac{\sum_{i \in \text{node } h} (T_i \leq t) \Delta D_i}{\sum_{i \in \text{node } h} (T_i \geq t) R_i}, \tag{5.9}$$

where T_i represents the survival time of individual i, ΔD_i is the indicator variable for event occurrence for individual i, and R_i indicates the number at risk for individual i at time t.

Once all trees in the forest are built, the ensemble method combines the CHFs from each tree to produce an aggregate CHF. This is typically done by averaging the CHFs across all trees

$$\hat{H}(t|x) = \frac{1}{M} \sum_{m=1}^{M} H_m(t|x),$$

where $H_m(t|x)$ is the CHF from the m-th tree and M is the total number of trees in the forest.

In summary, RSF leverages the power of randomization and ensemble learning techniques to provide robust insights into survival data patterns, enhancing the understanding of complex survival phenomena in medical research.

Gradient Boosting Survival Tree

The gradient boosting survival tree (GBST) [105] is an extension of the gradient boosting framework specifically designed for survival analysis. It integrates the principles of gradient boosting with survival trees to effectively manage censored survival data. By iteratively fitting survival trees to the negative gradients of a suitable loss function, GBST can capture complex relationships within the data, providing robust and accurate survival predictions. Here is an overview of the GBST methodology.

Gradient boosting is an ensemble learning method that builds a model in a stage-wise fashion by sequentially adding predictors to correct the errors of existing models. For survival analysis, the goal is to predict the time to event while accounting for censored observations. In survival analysis, a suitable loss function is required to handle censored data. A commonly used loss function is the negative partial log-likelihood from the Cox proportional hazards model. The negative partial log-likelihood for the i-th instance is given by

$$L_i = -\left(\delta_i\left(\beta^\top x_i - \log \sum_{j\in R(T_i)} e^{\beta^\top x_j}\right)\right),$$

where δ_i is the event indicator (1 if the event occurred, 0 if censored), x_i is the feature vector for the i-th instance, and $R(T_i)$ is the set of individuals at risk at time T_i.

In the gradient boosting framework, the model is built iteratively. At each iteration t, a new tree $h_t(x)$ is added to the model to correct the errors of the existing model $F_{t-1}(x)$. The new tree is fitted to the negative gradients (pseudo-residuals) of the loss function with respect to the current model. The model at iteration t is updated as

$$F_t(x) = F_{t-1}(x) + \eta h_t(x),$$

where η is the learning rate and $h_t(x)$ is the new tree fitted to the negative gradients.

The survival trees in GBST are constructed similarly to other survival trees but are fitted to the pseudo-residuals from the gradient boosting framework. The splitting criterion in survival trees typically aims to maximize the difference in survival times between the child nodes, often using the log-rank test. For each survival tree, the CHF is calculated for the terminal nodes using (5.9).

The final model combines the predictions from all the trees. The cumulative hazard function for a new instance is obtained by averaging the CHFs from all trees. The overall CHF $\bar{H}(t|x)$ for a new instance x is

$$\bar{H}(t|x) = \frac{1}{M}\sum_{m=1}^{M} H_m(t|x),$$

where $H_b(t|x)$ is the CHF from the b-th tree and B is the total number of trees in the model.

DeepSurv Model

The DeepSurv algorithm [106] is a neural network model that integrates deep learning techniques into the nonlinear Cox Proportional Hazards framework. It is designed to model the nonlinear relationship between risk factors and survival time. This model employs the learned weights of the neural network to predict the impact of patient survival data on their hazard rates. The stability and predictive performance of DeepSurv depend on the configuration of its network parameters.

Similar to other neural network models, the usage of DeepSurv involves a few key steps. First, the structure of the neural network needs to be determined, including the number of layers and nodes in each layer. The input layer should have the same number of nodes as the covariates, while the output layer contains only one node representing the log-risk function. During usage, crucial model parameters also need to be set and optimized, such as the learning rate, choice of activation functions, and selection of regularization methods. By harnessing the power of deep learning, DeepSurv allows for the modeling of intricate nonlinear relationships in survival analysis. It offers a flexible framework for capturing complex interactions between risk factors and survival outcomes, enhancing the predictive capabilities of the model.

5.2.3 *Model Optimization and Evaluation*

The modeling process incorporated the optimal parameters of the models, which were trained via grid search and cross-validation. The optimal parameter combinations for each model are displayed in Table 5.4.

Using the optimal parameter combinations for each model, prediction models were constructed on the training set and subsequently evaluated using the C-index and IBS on the test set. The model performances are presented in Table 5.5.

Table 5.4 Model optimal parameters

Model	Parameter
Coxph	alpha = 0.0001
ST	max_depth = 7, max_features = 9, min_samples_leaf = 7, min_samples_split = 5
RSF	n_estimators = 100, max_depth = 9, min_samples_leaf= 4
GBST	n_estimators = 100, min_samples_leaf= 4, learning_rate = 0.01
DeepSurv	n_layers = 3, n_nodes = 10, learning_rate = 0.01, activation = relu, decays = 0.005667, momentum = 0.887, l2_reg = 6.551, dropout = 0.661

Table 5.5 Models' performance

Model	C-index	IBS
Coxph	0.745	0.071
ST	0.747	0.079
RSF	0.817	0.075
GBST	0.775	0.077
DeepSurv	0.778	0.076

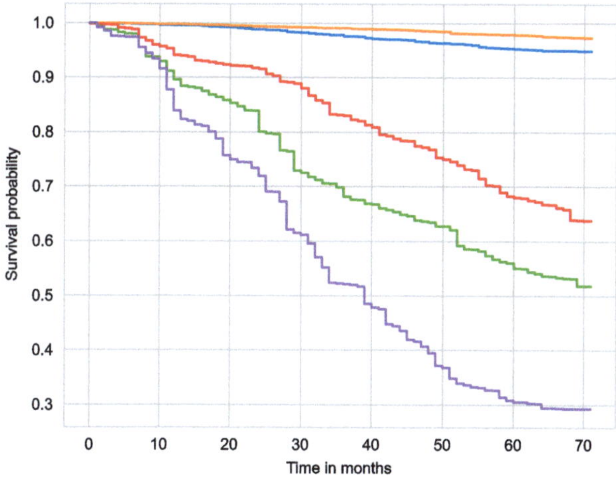

Fig. 5.17 Predictive survival probabilities of five breast cancer patients

As observed from Table 5.5, the RSF model showcased the best prediction performance, improving the C-index metric by 0.072 and lowering the IBS by 0.002 compared to Coxph. Consequently, the RSF model was employed to predict the survival probabilities of five breast cancer patients in the test set (Fig. 5.17). The survival rates of these five patients displayed a downward trend over time and exhibited significant differences at specific moments. This suggests that the model possesses strong discriminative ability in predicting patient survival probabilities and can accurately and efficiently anticipate changes in these probabilities. Therefore, healthcare professionals can leverage the survival probability predictions from this model to offer personalized treatment recommendations to patients.

5.2.4 Global and Individual Survival Analysis Interpretations

While certain tree models, such as the RSF and GBST, can illustrate the importance of each feature within the model through feature importance ranking, they fall short in elucidating how individual sample features contribute to the prediction results. To bridge this gap, the SHAP framework is introduced. This framework not only highlights the importance of each feature in predicting outcomes but also

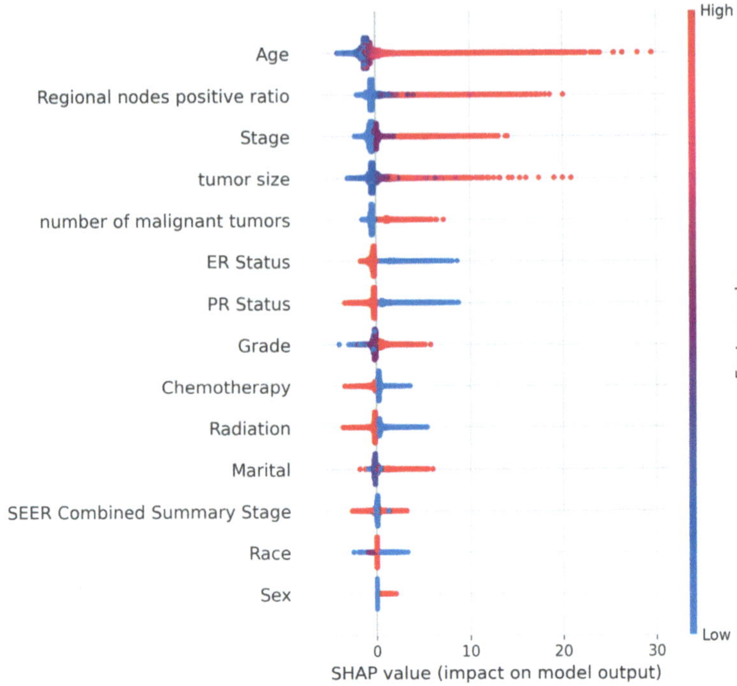

Fig. 5.18 SHAP summary plot for survival probability. Each point represents a Shapley value for a feature and a specific sample. The color of the points indicates the value of the feature (red for high, blue for low). Features are sorted by their importance, as determined by the mean absolute value of the Shapley values

clearly delineates the degree of influence and the positive or negative effects of each feature on individual sample predictions. The SHAP framework facilitates the explanation of survival probability predictions for breast cancer patients from two distinct angles: global and local. The global perspective enables an analysis of the factors that significantly influence the prognostic outcomes of breast cancer patients, while the local perspective evaluates the specific impact of features on the prediction results for each patient.

Figure 5.18 presents the feature importance ranking and the positive and negative impact effects of the RSF model within the SHAP framework. The top five variables influencing mortality risk for breast cancer patients are age, stage, regional nodes positive ratio, tumor size, and ER status. The first four variables are positively correlated with mortality risk, meaning that higher values for these variables correspond to an elevated risk of death. In contrast, ER status, a categorical variable, decreases the risk of death when its value is 1 (indicating a negative ER status).

Figure 5.19 provides a detailed SHAP breakdown for two breast cancer patients with contrasting predicted outcomes. This analysis elucidates how individual features contribute to the model's predictions for each patient. In the figures, $f(x)$

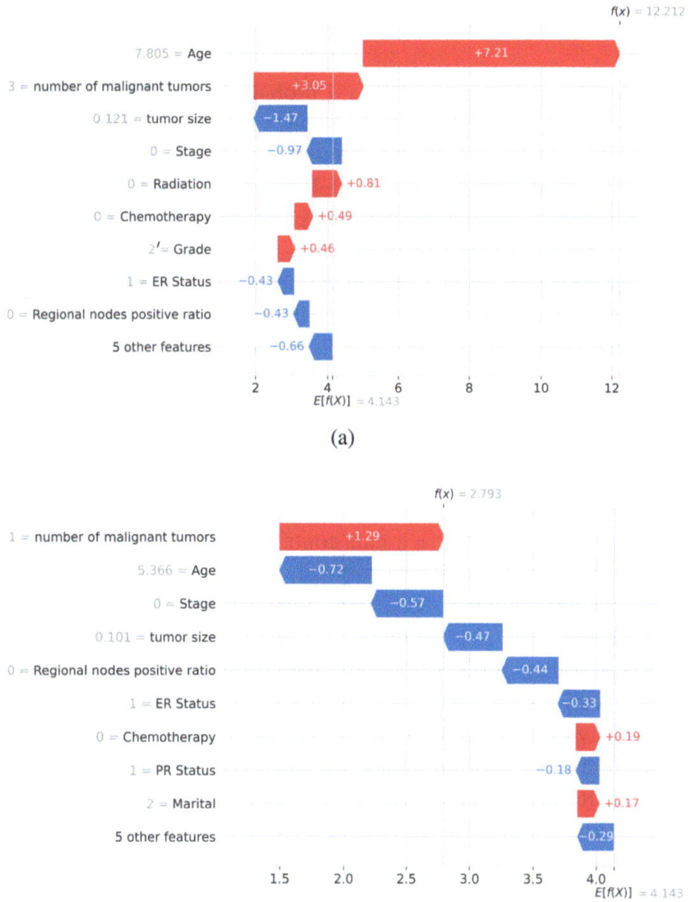

Fig. 5.19 SHAP breakdown for survival probability by two breast cancer patients. The x-axis represents the SHAP value, which quantifies the impact of each feature on the prediction. The y-axis lists the features, ordered by their importance for the specific prediction. The length of the bar indicates the magnitude of the SHAP value. Red bars indicate positive contributions (increasing the risk of death), while blue bars indicate negative contributions (decreasing the risk of death)

represents the predicted value for the sample, and $E[f(x)] = 4.143$ is the baseline value, corresponding to the mean predicted value. In Fig. 5.19a, the sample has a predicted value of $f(x) = 12.212$, which is higher than the baseline, indicating a higher survival probability and thus a lower risk of mortality. Key beneficial factors for this case include age and the number of malignancies. In contrast, Fig. 5.19b shows a sample with a predicted value of $f(x) = 2.793$, which is lower than the baseline, suggesting a lower survival probability and consequently a higher risk of mortality. For this case, age, stage, and tumor size are significant adverse factors.

5.2.5 Conclusion

This section developed a bespoke survival probability prediction model for breast cancer patients. Initially, feature selection was performed using single-factor and multifactor Cox regression analyses, identifying 14 variables that significantly impact patient survival probabilities. Models were constructed using the ST, RSF, GBST, and DeepSurv, and these models were compared with the traditional Coxph. All ML algorithms outperformed Coxph.

Based on the best-performing RSF model, the SHAP method was employed to elucidate the influence dynamics of the variables for both overall and individual cases in the survival analysis. The results indicated that while the importance of attributes within the prediction process varied across patients, four pivotal variables—age, tumor stage, lymph node positivity rate, and tumor size—exerted paramount influence over prediction outcomes. The local variable contribution decomposition diagram can assist physicians in understanding the specific survival scenarios for each patient, subsequently facilitating the provision of personalized treatment strategies. This approach can contribute to mitigating patient risk and augmenting the probability of patient survival.

However, it is important to note that this ML prediction model, based on the SEER database, has the following limitations: (1) Clinical feature-related information was collected retrospectively, which may introduce information bias. (2) SEER primarily focuses on cancer incidence and survival data, often lacking detailed clinical variables such as comorbid conditions, specific treatments, and patient health metrics. This limitation can affect the performance of predictive models. (3) SEER data may lack long-term follow-up information for all patients, potentially impacting survival analysis and long-term outcome predictions. (4) SEER covers specific geographic regions, so the generalizability of the models remains to be validated.

Chapter 6
Machine Learning in Drug Discovery

New drug development is a time-consuming, labor-intensive, and high-risk endeavor. The current state of international scientific research and development for new drugs is a major societal concern. However, this field faces significant obstacles: The rate of development has remained constant, and technical challenges continue to intensify. The frequent failures in drug discovery can be attributed to a lack of understanding of disease onset and progression mechanisms, the difficulty of identifying high-quality drug targets, and a shortage of biomarkers that can reflect disease progression and drug efficacy.

The application of artificial intelligence (AI) in the pharmaceutical industry presents an opportunity to address these challenges. In recent years, machine learning (ML) models have been effectively implemented in various areas, including drug target discovery, drug molecule discovery, information aggregation and integration, and the repurposing of existing drugs for new diseases [107–109]. This technology has significantly decreased the time and cost associated with research and development while increasing the success rate.

6.1 Prediction of Drugs-HPV Target Proteins Interaction

Human papillomavirus (HPV) infection can lead to various gynecological diseases. Although HPV vaccines are available, they are ineffective for individuals already infected with the virus. Based on virulence and infection risk, HPV can be categorized into low-risk and high-risk types, with the latter potentially causing fatal diseases through persistent infection. Current treatments for HPV include interferons, imiquimod, cidofovir, and traditional Chinese medicine. Additionally, innovative therapies such as HPV-specific targeted drugs, immunotherapies, and podophyllotoxin have emerged in recent years. However, no fully effective pharmaceutical interventions have been identified to date [110].

J. Ren, Y. Tao, *Modeling in Life Sciences and Ecology*, Springer Asia Pacific Mathematics Series 4, https://doi.org/10.1007/978-981-95-1038-2_6

Drug-protein interaction, the dynamic interplay between drug and protein molecules, is a fundamental determinant of drug activity. Drugs interact with proteins to alter their functions and processes, resulting in therapeutic effects. These interactions can occur through hydrogen bonds, van der Waals forces, hydrophobic interactions, and ionic pair interactions, significantly impacting pharmacokinetics. Thus, studying drug-protein interactions is crucial in pharmaceutical research for drug discovery, design, and optimization. In drug repositioning, drugs modulate the host by interacting with target proteins, making the identification of drug-target protein interactions essential for drug screening. By leveraging bioinformatics features of pharmaceuticals and target proteins, ML-based methods can identify potential drug-target protein interactions, facilitating the screening of candidate drugs against HPV.

This section aims to introduce ML models that utilize bioinformatics features of drugs and target proteins to predict potential anti-HPV drugs among antiviral medications.

6.1.1 Data Collection and Configuration

The data on drug-protein interactions were extracted from the DrugBank database. DrugBank is a comprehensive and unified online resource and knowledge base for drug information, provided free of charge by the University of Alberta. It contains chemical, biological, and clinical data for thousands of available medications. Detailed information on drug uses, ingredients, dosages, side effects, and interactions can be queried in the DrugBank database. To gain access to DrugBank, users must establish an account on the official website (https://go.drugbank.com) and then submit an email application detailing the purpose of their research.

Version 5.1.10 of DrugBank was utilized in this section: A total of 21,626 drug-protein interactions, including 7,627 drugs and 4,884 target proteins, were extracted from the DrugBank database. These data were subjected to filtering procedures: For drugs, those without SMILES chemical structure notation and drugs that the RDKit package in Python could not recognize were excluded. For target proteins, those without amino acid sequences and proteins that the PyPro module in Python could not recognize were excluded. These operations yielded a final set of 6,141 pharmaceuticals and 4,405 target proteins for 17,380 drug-protein interaction pairs.

An interaction between a drug and its target protein is considered a positive instance. Conversely, drug-target protein pairs that lack an established interaction are deemed negative instances. Therefore, the total sample space equals $6141 \times 4405 = 27{,}051{,}105$. Among these tens of millions of samples, only 17,380 drug-target protein pairs are known to interact. Within the remaining over 20 million drug-target protein pairs, there are undoubtedly interactions that humans have not yet discovered.

Four experimental configurations were conducted to evaluate the model performance.

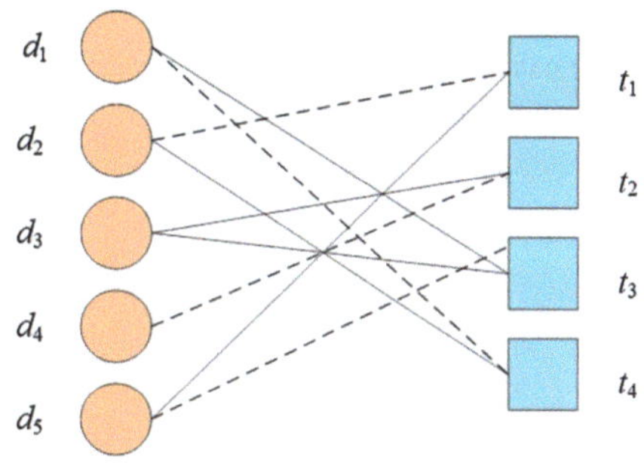

Fig. 6.1 The training and testing dataset for E1 configuration

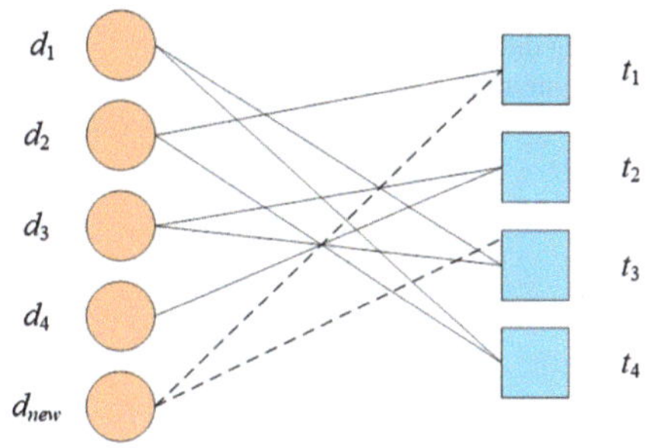

Fig. 6.2 The training and testing dataset for E2 configuration

1. Prediction of drug-plate protein interaction (E1)
   The test set's medications and plate proteins also appear in the training set in this configuration. This procedure primarily demonstrates the model's ability to predict drug-plate protein interactions. In this scenario, 20% of the drug-plate protein pairs in the dataset are randomly selected as the test set, while the remaining 80% serves as the training set. Figure 6.1 depicts d_i as pharmaceuticals and t_i as plate proteins. The dashed lines represent drug-protein pairs from the test set, while the solid lines represent those from the training set.
2. Novel drug prediction (E2)
   In this configuration, the test set medications do not appear in the training set. This configuration evaluates the ability of the model to forecast the interaction between novel pharmaceuticals and known targets. To facilitate this, 20% of the drugs are randomly selected, and the drug-protein pairs involving these drugs are allocated to the test set, with the remaining samples constituting the training set. As shown in Fig. 2.3, d_{new} indicates that the substance is unique to the test set (Fig. 6.2).
3. New target prediction (E3)
   In this configuration, the test set's target proteins aren't present in the training set. This experimental design assesses the model's ability to predict interactions between novel targets and established drugs. To accomplish this, 20% of the target proteins are selected at random, and drug-target protein pairs involving these target proteins are assigned to the test set, with the remaining samples constituting the training set. As shown in Fig. 6.3, new indicates that the target protein is present only in the test set.

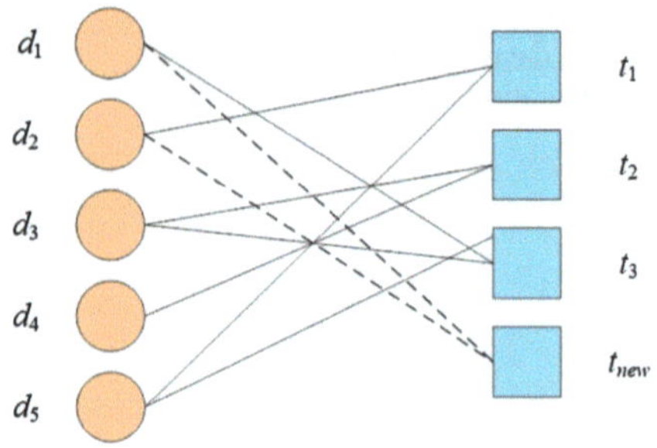

Fig. 6.3 The training and testing dataset for E3 configuration

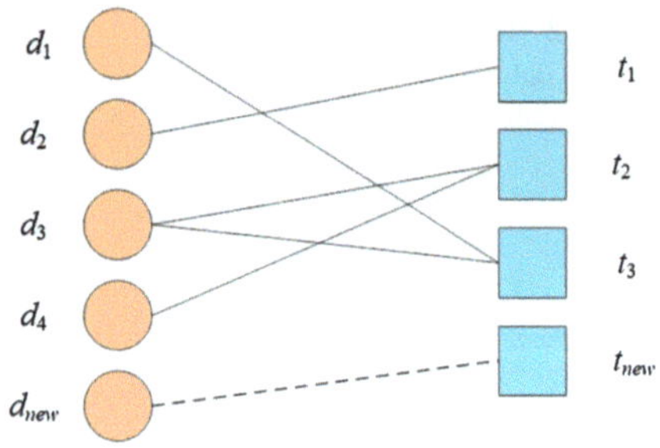

Fig. 6.4 The training and testing dataset for E4 configuration

4. Prediction of novel drugs and targets (E4)
   In this configuration, neither the pharmaceuticals nor the target proteins in the test set are present in the training set. This configuration is intended to evaluate the ability of the model to predict the interaction between novel pharmaceuticals and novel targets. To achieve this, 20% of the drugs and 20% of the target proteins are selected arbitrarily, and the drug-target protein pairs involving these drugs and targets are allocated to the test set. In the meantime, we eliminate all combinations involving these drugs or targets from the dataset, with the remaining pairs comprising the training set. As shown in Fig. 6.4, d_{new} and t_{new} suggest the drug and target protein are unique to the test set.

6.1.2 *Construction of Negative Drug-Target Pairs*

During the process of training a model, a balanced distribution of positive and negative samples is required in the training dataset to ensure the effectiveness. However, if we were to directly extract the data and classify the known interacting drug-target protein pairs as positive samples and the pairs with unknown interactions as negative samples, we would have 17380 positive samples and potentially 27033725 negative samples. In this instance, the ratio of positive to negative samples would be approximately 0.064%, indicating a substantial imbalance within the dataset.

Most studies seek to achieve a balanced dataset by randomly selecting as negative samples those drug-target protein pairs with unknown interactions. Yet, these negative instances do not necessarily indicate that these drug-target protein pairs lack an interaction; thus, they are not genuinely negative. To address this issue, it would be prudent to create more dependable negative samples. Next, we introduce the foundational knowledge required to construct negative samples.

6.1.2.1 Drug/Target Similarity

Drug/target similarity, within the realm of pharmacology and drug discovery, pertains to the extent of similarity or correspondence between a given drug and its target protein or molecule. This concept serves as a cornerstone in comprehending how drugs engage with their designated targets, and it assumes a pivotal role in the advancement of novel drugs and therapeutic methodologies. In this context, we introduce three frequently employed types of interactions: interaction profile similarity, term frequency-inverse document frequency similarity, and molecular fingerprint similarity.

Interaction profile similarity typically refers to a measure of how similar or related the patterns of interactions are between two entities, such as proteins, genes, or users in a network. This concept is commonly used in various fields, including bioinformatics, social network analysis, and recommendation systems. In bioinformatics, it may refer to how similar the sets of proteins that interact with two different proteins are. If two proteins have highly similar interaction profiles, it suggests that they may have similar functions or play roles in similar biological processes. To compute the interaction similarity of drugs, we define an interaction matrix of size, where represents the number of drugs and denotes the number of target proteins in Fig. 6.5.

Each element in this matrix indicates whether there is an interaction between the drug and the target protein: 0 represents an unknown interaction, while 1 indicates

Fig. 6.5 Drug-target protein interaction matrix

	Protein 1	Protein 2	Protein 3	
Drug 1	0	1	0	...
Drug 2	1	0	0	...
Drug 3	1	0	1	...
	...	...	...	...

an interaction. The ith row of the interaction matrix (I_{d_i}) represents the interaction vector of drug d_i. Using this notation, the similarity calculation formula between the two drugs d_i and d_j is as follows:

$$K_{\text{GIP}}(d_i, d_j) = e^{-\text{DIS}(I_{d_i}, I_{d_j})},$$

where DIS$(\cdot, \cdot)$ is the Jaccard distance [111].

Term frequency-inverse document frequency (TF-IDF) similarity is a measure commonly used in natural language processing and information retrieval to determine the similarity between two documents or pieces of text. It evaluates the importance of individual terms within a document in relation to a larger collection of documents (a corpus). Drugs can be viewed as texts, allowing the computation of TF-IDF semantic similarity between them. Here's a breakdown of how TF-IDF similarity works:

1. Term frequency (TF): It measures how frequently a term (word) appears within a document. The more often a term appears in a document, the more important it is likely to be in that document.
2. Inverse document frequency (IDF): It measures the importance of a term across a collection of documents. If a term appears frequently across many documents, it is considered less important (common) compared to terms that appear in fewer documents (rare). IDF is calculated as the logarithm of the total number of documents divided by the number of documents containing the term.
3. TF-IDF score: The TF-IDF score for a term in a document is calculated by multiplying its TF within that document by its IDF across the corpus. This score reflects the importance of the term in the specific document relative to its importance in the entire corpus.

To calculate the TF-IDF similarity between two documents, one can typically follow these steps:

1. Conduct word extraction. The SMILES structure of each drug is treated as a text, and every five characters are taken as a word, denoted as LINGO. The method involves setting five characters as a sliding window and sequentially selecting five characters as LINGO, as illustrated in Fig. 6.6.
2. Conduct a corpus. After word extraction, each drug's SMILES structure is represented as a text containing several LINGOs. Define a set $D = \{d_1, d_2, \ldots, d_n\}$, where each element represents a text; a set $W = \{w_1, w_2, \ldots, w_n\}$, which comprises m LINGOs that have appeared in all texts; a word-text matrix X, where

$$X = \begin{bmatrix} x_{11} & x_{12} & \ldots & x_{1n} \\ x_{21} & x_{22} & \ldots & x_{2n} \\ & & \ldots & \\ x_{m1} & x_{m2} & \ldots & x_{mn} \end{bmatrix},$$

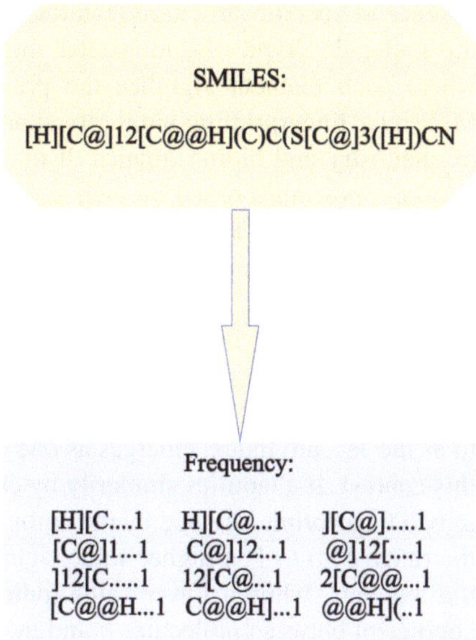

Fig. 6.6 The process of term extraction

which represent the frequency of LINGOs appearing in the texts. x_{ij} represents the frequency of LINGO w_i appearing in text d_j. A higher $x_i j$ suggests a higher importance of the LINGO in the text. The jth column vector x_j of the word-text matrix represents text d_j and is referred to as the word vector. In this case, the matrix X can also be represented as $X = \{x_1, x_2, \ldots, x_n\}$.

3. Calculate word vectors. The term frequency-inverse document frequency (TF-IDF) is used to represent the weight, defined as

$$\mathrm{TF} - \mathrm{IDF}_{ij} = \frac{tf_{ij}}{tf_{.j}} \log \frac{df}{df_i},$$

where tf_{ij} represents the number of times LINGOw_i appears in text d_j, $tf_{.j}$ represents the sum of the number of times all LINGOs appear in text d_j, dfi represents the number of all texts containing LINGOw_i, and d_f represents the number of texts.

4. Calculate cosine similarity. The semantic similarity between two texts is represented by the normalized inner product (cosine similarity) of the two word vectors. Therefore, the similarity between text d_i and text d_j can be computed as

$$\text{Cosine similarity} = \frac{x_i \cdot x_j}{\|x_i\| \|x_j\|},$$

where $\cdot$ denotes the inner product and $\| \cdot \|$ denotes the norm of the vector.

A molecular fingerprint serves as a simplified representation of a molecule's structure and properties, encoding vital information regarding the presence or absence of specific structural features, substructures, or chemical properties within the molecule. Typically, molecular fingerprints take the form of binary vectors, where each element signifies the presence or absence of a particular feature. Molecular fingerprinting similarity stands as a fundamental concept in the realms of chemistry and bioinformatics. It plays a pivotal role in evaluating the likeness between molecules based on their structural and chemical attributes. This concept finds indispensable application in drug discovery, chemoinformatics, and the analysis of chemical compounds. The process involves the identification of specific substructures within the molecular composition, effectively converting the intricate molecular structure into a series of binary fingerprint sequences. Consequently, the structural similarity of small molecules can be succinctly conveyed through the resemblance of their molecular fingerprints. The Tanimoto coefficient, often referred to as the Jaccard index, emerges as one of the most frequently employed metrics in this context. It quantifies similarity by calculating the intersection of bits set to one in two fingerprints, divided by the union of those bits. This coefficient's value spans the range of 0 to 1; a higher value signifies a greater degree of similarity between the two sets, while a lower value indicates a diminished level of similarity. The coefficient between molecular p and molecular q can be calculated as

$$T(p,q) = \frac{p \cdot q}{\|p\|^2 + \|q\|^2 - p \cdot q}, \tag{6.1}$$

where $\cdot$ denotes the inner product and $\| \cdot \|$ denotes the norm of the vector.

Similarity integration, in the context of data analysis and information retrieval, refers to the process of combining or aggregating multiple similarity measures or sources of similarity into a single, comprehensive similarity score. This integration aims to capture a more holistic and accurate measure of similarity between objects, whether they are documents, images, products, or other data points. Moving forward, let us delve into a widely adopted technique for amalgamating similarity scores. Similarity integration entails the fusion of similarity scores originating from various sources or metrics to formulate a comprehensive, unified similarity score. An exemplary algorithm emblematic of this process is for drugs D_i and D_j, their comprehensive similarity (CS) is calculated as follows:

$$CS_{ij} = 1 - \prod^{n}(1 - cs_{ij}^{n}).$$

Here, cs_{ij}^{n}, $(n = 1, 2, 3)$, respectively, represent the IP similarity, TF-IDF similarity, and molecular fingerprint similarity between the drugs D_i and D_j.

6.1.2.2 Framework for Negative Sample Selection

Based on the assumption that dissimilar drugs may not target the same molecular targets, we introduce a negative sample selection process designed to construct a

Fig. 6.7 Diagram of the framework for negative sample selection

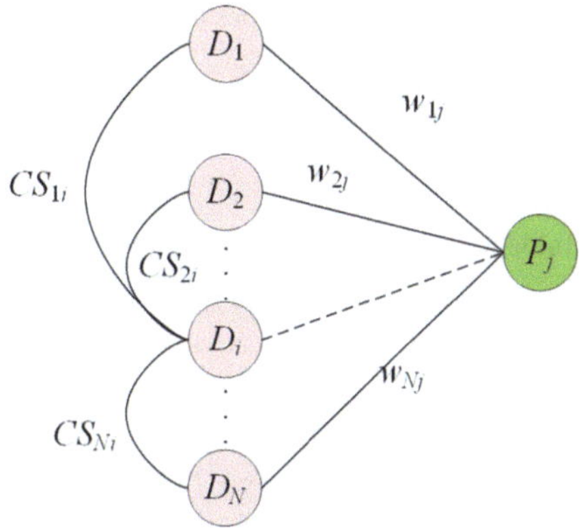

more dependable set of negative instances. Assume that when an interaction is observed between a specific drug D_l and a target protein P_j, the score is recorded as $w_{lj} = 1$; otherwise, the score is recorded as $w_{lj} = 0$. The weighted score (spc_{ijl}) of the interaction of drug D_l with the target protein P_j is calculated as

$$spc_{ijl} = w_{lj} \times CS_{li},$$

where w_{lj} is a weighting coefficient and CS_{li} represents the comprehensive similarity between drugs D_l and D_i. Then the composite score of drug D_i targeting protein P_j is

$$SPC_{ij} = \frac{\sum^{l} spc_{ijl}}{\sum^{l} w_{lj}}.$$

A low value of SPC_{ij} indicates that the drug D_i is dissimilar to the drug D_l that targets the protein P_j, suggesting D_i is less likely to target the protein P_j and therefore be considered a negative case (Fig. 6.7).

Based on the framework above, drug-target combinations with low scores $(SPC_{ij} < 0.2)$ are added to the pool of prospective negative samples. To preserve a balanced ratio of positive to negative samples, 17380 negative candidate samples are randomly selected. Therefore, the dataset contains 34,760 entries, including 17,380 negative and 17,380 positive samples.

To illustrate the superiority of the dataset constructed based on the negative sampling framework, we use a RF as the benchmark model. In this scenario, we employ 1928-dimensional drug PubChem molecular fingerprints and protein molecular descriptors as features. We randomly allocate 20% of the samples for testing, with the remainder designated for training. We compare the performance of random under-sampling and similarity-based sampling methods on the test set. The results of the two sampling methods are presented in Table 6.1. Analysis of the results from the table reveals that the performance of the similarity-based sampling

Table 6.1 Evaluation metrics for two sampling methods

Sampling method	Accuracy	AUC	F1 score	Precision	Recall
Similarity-based sampling	0.829	0.908	0.824	0.831	0.817
Random under-sampling	0.787	0.867	0.788	0.807	0.770

Table 6.2 Partial explanations of PubChem molecular

Bit position	Bit substructure	Bit position	Bit substructure
1	>= 4 H	421	C=O
2	>= 8 H	422	C=S
3	>= 16 H	423	N=N
4	>= 32 H	424	N=O
5	>= 1 Li	425	N=P

method surpasses that of the random under-sampling method across all evaluation metrics. Consequently, the subsequent work selects the dataset constructed using the similarity-based negative sampling framework as the foundation.

6.1.3 *Predictive Model for Drug-Target Protein Interactions*

6.1.3.1 Feature Construction

PubChem molecular fingerprints are selected as the feature representation for pharmaceuticals [112]. PubChem molecular fingerprints are utilized predominantly for describing and comparing the structures of compounds. These signatures are generated by calculating molecular descriptors and employing a hash algorithm to convert compounds, allowing for an efficient similarity comparison. They are used frequently in virtual screening of compounds, drug design, chemoinformatics, and other related fields. In addition, they are crucial instruments for large-scale virtual screening and analysis utilizing ML techniques. In the following, the procedure entails converting the drug's SMILES structure into a Padel-recognizable format using RDKit. The PubChem molecular signatures are then calculated using the Padel-descriptor software [113].

The PubChem molecular fingerprints are partially explained in Table 6.2. For instance, if the number of hydrogen atoms (H) is 4, the first position is represented as 1; otherwise, it is represented as 0. If the molecule contains a C=O structure, the 421st position is represented as 1; otherwise, it is represented as 0. The PubChem molecular fingerprints consist of 881-dimensional features, with each element being either 0 or 1.

For the target proteins, this section utilizes the PyPro module in Python to compute structural and physicochemical features based on the protein's amino acid sequence. Five protein descriptors are employed [114], including the following:

- Amino acid composition (ACC): This descriptor comprises 20-dimensional features, representing the probabilities of occurrence for standard amino acids in the protein sequence.
- Dipeptide composition (DPC): This descriptor consists of 400-dimensional features. It considers both amino acid frequencies and the sequential information of amino acids in the protein sequence.
- Composition, transition, and distribution (CTD) descriptors: This descriptor encompasses 147-dimensional features. CTD descriptors capture specific patterns of amino acid distribution that correspond to unique structural or physicochemical properties on protein or peptide chains.
- Normalized Moreau-Broto autocorrelation (MBauto): This descriptor contains 240-dimensional features.
- Moran autocorrelation (Moranauto): This descriptor includes 240-dimensional features. In total, 1,047 protein descriptors are obtained in this study.

Having drug and protein molecular descriptors alone is inadequate; subsequent algorithms require drug-protein pair features as input in the ML model. To construct the drug-protein pair features, the drug and protein features are concatenated. Let the drug feature vector be denoted as $(d_1, d_2, \ldots, d_n)$ and the protein feature vector as $(p_1, p_2, \ldots, p_j)$. The feature representation of a drug-protein pair can then be expressed as $(d_1, d_2, \ldots, d_n, p_1, p_2, \ldots, p_j)$. In this section, the drug features are 881-dimensional and protein features are 1,047-dimensional, resulting in a feature dimension of 1,928 for each drug-protein pair.

During data collection, it is common to gather various features with different properties, dimensions, and scales. When combining multiple features in a model, discrepancies in their properties, dimensions, and scales prevent direct use and integration of the features. If the magnitudes and scales of different features significantly differ, direct analysis using the raw features can lead to biased model results, as features with larger values will dominate the model's influence compared to features with smaller values. To establish a unified standard and enhance result reliability, this study performs data standardization, ensuring that all features are on a consistent scale.

6.1.3.2 Feature Selection

The variance filtering method is utilized to eliminate features with zero variance, resulting in 1758 remaining features. Following this, a random forest (RF) model is utilized to rank feature importance by computing the mean Gini index for each feature. If a data set contains samples from K classes, Gini index is defined as

$$GI_m = 1 - \sum_{k=1}^{K} p_{mk}^2,$$

where p_{mk} is the relative frequency of class mk.

The importance of feature X_j at node m is the change in the Gini index before and after the split of node m, which is

$$VIM_{jm} = GI_m - GI_l - GI_r,$$

Then we can sum up all the nodes to get the overall importance of feature X_j in the tree i, which is

$$VIM_{ij} = \sum_{m \in M} VIM_{jm}.$$

Subsequently, this calculation can be extended to all trees within the RF, and the average can be taken to ascertain the importance of feature X_j across the entire ensemble. Herein,

$$VIM_j = \sum_{i=1}^{n} VIM_{ij},$$

where n is the total number of the forest.

Ultimately, the importance scores of all features can be normalized to ensure a total sum of importance scores equating to 1. The conventional normalization approach involves dividing the importance score of each feature by the total sum of importance scores across all features. This process can be formulated as follows:

$$VIM_j = \frac{VIM_j}{\sum_{i=1}^{K} VIM_{ij}}.$$

Due to the variation in training results caused by different sampling and the resulting different importance scores, a threshold needs to be set for feature selection after ranking feature importance using RF. Fivefold cross-validation determines the selected features' accuracy under different thresholds. Figure 6.8a presents the importance scores for each feature, and Fig. 6.8b illustrates the accuracy variation concerning the threshold. It is found that the model performs best when the threshold is set to 0.001. Thus, features with an importance greater than 0.001 are chosen, leaving 167 remaining features.

The features selected from the previous methods are further subjected to feature screening using genetic algorithm (GA). In the context of optimization problems, it refers to the evolution of a certain number of candidate solutions into improved solutions via iterative optimization, ultimately leading to the optimal solution. The first step in solving a problem with the GA is to define the objective function and problem variables. The solutions to a problem are represented by a string of numbers in genetic algorithms, and the genetic operators operate based on these numerical strings. Consequently, the second stage is encoding the variables. For instance, binary encoding is utilized for feature selection: each bit signifies a feature, with

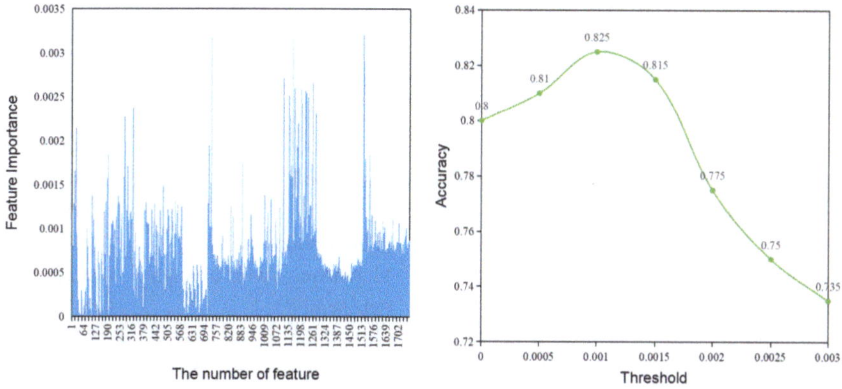

(a) (b)

Fig. 6.8 (**a**) The importance scores of the i-th feature. (**b**) The accuracy of RF model under varied threshold

"1" representing selection and "0" representing non-selection. When using GA, it is necessary to define an evaluation function to distinguish between good and poor individuals.

In genetic operations, selection, crossover, and mutation are utilized. Selection and crossover are the essential operations in genetic algorithms, whereas mutation is an auxiliary operation that improves the algorithm's global search performance. Together, these three operations expedite the genetic algorithm's search for the optimal solution. The primary activities are as follows:

- Selection is performed based on the fitness of individuals, giving higher-fitness individuals a greater chance of being selected, thus passing useful genetic information to the next generation.
- Crossover involves exchanging certain positions between two individuals to generate new individuals. This allows for better combinations of genes within the population and helps avoid getting trapped in local optima.
- Mutation refers to randomly changing certain genes in the genetic string to generate new individuals.

After the GA feature selection operation, the remaining 104-dimensional features are used as input features. Table 6.3 displays the 20 most significant features out of 104-dimensional features.

6.1.4 Model Construction and Evaluation

In this subsection, we construct drug-target protein interaction prediction models using the 20 important features identified in the previous analysis. The methods

Table 6.3 The top 20 most important features

Feature	FI	Feature	FI
W	0.003176479	_ChargeC3	0.002139333
MoranAuto_Hydrophobicity5	0.002913068	_NormalizedVDWVD3025	0.002042075
_ChargeT23	0.002598142	_SolventAccessibilityT13	0.002033629
_SolventAccessibilityD3001	0.002584034	PubchemFP411	0.001999215
_SolventAccessibilityC3	0.002512885	E	0.001944392
_PolarityD1001	0.002497724	_PolarizabilityD2001	0.001927659
_SecondaryStrD1001	0.002418639	MoranAuto_AvFlexibility5	0.001844002
_HydrophobicityD3001	0.002314624	PubchemFP314	0.001836896
PubchemFP407	0.002276388	GY	0.001748457
_PolarizabilityD3001	0.002201267	_NormalizedVDWVD2001	0.001737794

employed include rotation forest (RoF), artificial neural network (ANN), extreme gradient boosting (XGBoost), logistic regression (LR), decision tree (DT), adaptive boosting (AdaBoost), and RF. In previous chapters, we have already introduced the latter five algorithms. In the following, we will focus on RoF and ANN.

6.1.4.1 Artificial Neural Network

Artificial neural networks (ANNs) are data processing models inspired by biological neural networks [115]. They consist of numerous interconnected artificial neurons that process information through weighted connections. The core concept of ANNs is to model input data by continuously adjusting these weights, ultimately solving complex problems.

ANNs mimic two key characteristics of the human brain: acquiring knowledge through learning processes and storing knowledge in the connection weights between neurons. This characteristic allows ANNs to learn relationships between inputs and outputs from data without relying on precise mathematical models. This "black box" feature enables ANNs to achieve various nonlinear mappings, making them powerful tools for engineering analysis and prediction.

A typical ANN comprises an input layer, one or more hidden layers, and an output layer. The input layer receives external data and transmits it to the hidden layers. The hidden layers are the core of the ANN, processing input signals through weighted sums and activation functions before passing the results to the output layer. The output layer generates the model's predictions. During training, ANNs use forward propagation and backpropagation to continuously adjust weights and biases, minimizing prediction errors.

Forward propagation refers to the process where data flows from the input layer through the hidden layers to the output layer. In this process, each neuron computes a weighted sum of the inputs and applies an activation function to produce an output signal. Given the input $\mathbf{x}$, weights $\mathbf{W}$, biases $\mathbf{b}$, and activation function f, the output

a of a hidden layer neuron is given by

$$\mathbf{a} = f(\mathbf{W}\mathbf{x} + \mathbf{b}).$$

The output of the final layer can be expressed as

$$\mathbf{y} = g(\mathbf{W}'\mathbf{a} + \mathbf{b}'),$$

where $\mathbf{W}'$ and $\mathbf{b}'$ are the weights and biases from the hidden layer to the output layer and g is the activation function for the output layer.

The result of forward propagation is the model's prediction output, with the difference between this output and the actual value representing the error. A loss function L is used to measure this error, with common choices including mean squared error and cross-entropy.

Backpropagation is the process of adjusting weights using the error. It involves computing the gradient of the loss function with respect to each neuron's output and propagating these gradients backward through the network. For the output layer neurons, the error gradient $\delta^{(2)}$ is

$$\delta^{(2)} = \frac{\partial L}{\partial \mathbf{y}} \cdot g'(\mathbf{z}'),$$

where $\frac{\partial L}{\partial \mathbf{y}}$ is the derivative of the loss function L with respect to the output $\mathbf{y}$ of the network and $g'(\mathbf{z}')$ is the derivative of the activation function used in the output layer with respect to its input $\mathbf{z}'$.

For the hidden layer neurons, the error gradient $\delta^{(1)}$ is

$$\delta^{(1)} = (\mathbf{W}')^T \delta^{(2)} \cdot f'(\mathbf{z})$$

where $(\mathbf{W}')^T$ is the transpose of the weight matrix connecting the hidden layer to the output layer and $f'(\mathbf{z})$ is the derivative of the activation function used in the hidden layer with respect to its input $\mathbf{z}$.

Weights $\mathbf{W}$ and biases $\mathbf{b}$ are updated using gradient descent,

$$\mathbf{W} \leftarrow \mathbf{W} - \eta \frac{\partial L}{\partial \mathbf{W}},$$

and

$$\mathbf{b} \leftarrow \mathbf{b} - \eta \frac{\partial L}{\partial \mathbf{b}},$$

where η is the learning rate.

By repeating forward propagation and backpropagation, the model iteratively optimizes its parameters to minimize prediction errors. ANNs have been widely applied in image recognition, speech recognition, natural language processing, and predictive analysis. Their powerful learning capabilities and flexibility make them highly effective in handling complex nonlinear problems.

6.1.4.2 Rotation Forest

Rotation forest (RoF) is an innovative ensemble learning method designed to enhance the accuracy and diversity of individual classifiers by creating multiple rotated feature sets [116]. This technique leverages the power of principal component analysis (PCA) to transform the feature space, thereby fostering classifier diversity and improving overall prediction performance.

The core idea of RoF lies in its ability to extract and rotate features in such a way that each classifier in the ensemble is exposed to a unique perspective of the data. The process begins with the division of the original feature set into several nonoverlapping subsets. For each subset, PCA is applied independently to generate new rotated versions of the data. This rotation is crucial as it introduces variations in the feature space, leading to a diverse set of classifiers within the ensemble.

Once the rotated feature sets are created, each classifier in the ensemble is trained on a different rotated version of the data. This diversity among classifiers is instrumental in reducing the risk of overfitting and enhancing the model's generalization capabilities. The final prediction of the RoF is obtained by aggregating the predictions of all individual classifiers, typically through majority voting or averaging, which combines the strengths of each classifier to produce a robust and accurate prediction.

The rotation process involves applying PCA to each subset of features. Given a feature subset, PCA is used to obtain the principal components, which are then used to rotate the original features. This transformation ensures that each classifier is trained on a distinct rotated feature set, contributing to the overall diversity and performance of the ensemble. For instance, if we denote the original feature set as X, and a subset of features as X_i, PCA is applied to X_i to obtain the principal components P_i. The rotated features $\tilde{X}_i$ are then derived by multiplying X with P_i. Each base classifier C_i is trained on these rotated features, and the ensemble prediction is an aggregation of the predictions from all classifiers.

One of the notable advantages of RoF is its flexibility. It can incorporate any base classifier, such as DTs, support vector machines, or neural networks, making it adaptable to various types of data and problem domains. Additionally, the method's inherent ability to increase classifier diversity significantly reduces the likelihood of overfitting, making it a powerful tool for complex prediction tasks.

RoF has demonstrated remarkable success across a range of applications, including pattern recognition, medical diagnosis, financial forecasting, and image classification. Its unique approach to feature rotation and ensemble learning not only enhances prediction accuracy but also provides a robust framework for handling diverse and complex datasets.

Table 6.4 Model performance under setting E1

Model	Accuracy	AUC	F1 score	Precision	Recall
LR	0.663	0.714	0.650	0.665	0.635
AdaBoost	0.715	0.784	0.702	0.724	0.681
RF	0.835	0.914	0.831	0.838	0.824
DT	0.758	0.806	0.746	0.772	0.721
XGBoost	0.842	0.915	0.839	0.844	0.836
ANN	0.785	0.834	0.789	0.763	0.816
RoF	0.844	0.921	0.837	0.863	0.813

Table 6.5 Model performance under setting E2

Model	Accuracy	AUC	F1 score	Precision	Recall
LR	0.660	0.719	0.650	0.672	0.629
AdaBoost	0.717	0.790	0.709	0.734	0.685
RF	0.820	0.898	0.816	0.839	0.794
DT	0.731	0.783	0.722	0.749	0.697
XGBoost	0.822	0.890	0.817	0.846	0.790
ANN	0.745	0.803	0.748	0.755	0.741
RoF	0.824	0.900	0.814	0.865	0.770

6.1.5 Model Evaluation

This subsection demonstrates the performance of seven candidate ML models under experimental settings E1–E4. Table 6.4 presents the performance metrics for various models under the E1 setting, focusing on Accuracy, AUC, F1 Score, Precision, and Recall. Among the models, RoF and XGBoost demonstrate the highest accuracy, with scores of 0.844 and 0.842, respectively, indicating their superior performance in correctly classifying the test set. The RF and ANN models also show strong performance, with accuracy scores of 0.835 and 0.785, respectively. LR and AdaBoost exhibit lower accuracy, highlighting their relatively weaker performance in this setting.

Table 6.5 presents the performance metrics for various models under the E2 setting. Among the models, RoF exhibits the highest accuracy (0.824) and AUC (0.900), indicating its superior performance in classifying the test set. XGBoost also performs well with an accuracy of 0.822 and an AUC of 0.890, showing a balanced performance across all metrics. RF, while slightly behind XGBoost and RoF, still shows strong results with an accuracy of 0.820 and an AUC of 0.898. LR and DT have lower accuracy and AUC scores, highlighting their relatively weaker performance in this setting. ANN shows moderate performance, with an accuracy of 0.745 and an AUC of 0.803. Overall, RoF and XGBoost demonstrate the most robust performance in the E2 setting.

Table 6.6 presents the performance metrics for various models under the E3 setting. In this scenario, XGBoost stands out with the highest accuracy (0.743) and

Table 6.6 Model performance under setting E3

Model	Accuracy	AUC	F1 score	Precision	Recall
LR	0.650	0.700	0.641	0.644	0.638
AdaBoost	0.639	0.682	0.615	0.644	0.588
RF	0.715	0.793	0.648	0.820	0.536
DT	0.657	0.688	0.631	0.668	0.597
XGBoost	0.743	0.812	0.720	0.772	0.675
ANN	0.680	0.725	0.674	0.673	0.674
RoF	0.713	0.797	0.635	0.841	0.510

Table 6.7 Model performance under setting E4

Model	Accuracy	AUC	F1 score	Precision	Recall
LR	0.650	0.700	0.641	0.644	0.638
AdaBoost	0.639	0.682	0.615	0.644	0.588
RF	0.715	0.793	0.648	0.820	0.536
DT	0.657	0.688	0.631	0.668	0.597
XGBoost	0.743	0.812	0.720	0.772	0.675
ANN	0.680	0.725	0.674	0.673	0.674
RoF	0.713	0.797	0.635	0.841	0.510

AUC (0.812), as well as the highest F1 Score (0.720) and strong performance in Precision (0.772) and Recall (0.675). The RF and RoF models also show solid performance, with RoF exhibiting the highest precision (0.841), despite having a lower recall (0.510). LR and DT models have moderate accuracy and AUC values, with LR achieving 0.650 and 0.700, respectively. ANN shows a balanced but moderate performance across all metrics, with an accuracy of 0.680 and an AUC of 0.725. AdaBoost has the lowest performance across most metrics, indicating its relative inefficacy in this setting. Overall, XGBoost demonstrates the most balanced and robust performance among the models evaluated in the E3 setting.

Table 6.7 presents the performance metrics for various models under the E4 setting. XGBoost exhibits the highest performance with an accuracy of 0.743 and an AUC of 0.812, alongside a strong F1 Score of 0.720, precision of 0.772, and recall of 0.675. RF and RoF also perform well, with RoF achieving the highest precision of 0.841, although it has a lower recall of 0.510. LR, and DT models show moderate performance, with LR attaining an accuracy of 0.650 and an AUC of 0.700. The ANN model presents balanced but moderate metrics, with an accuracy of 0.680 and an AUC of 0.725. AdaBoost shows the lowest performance across most metrics. Overall, XGBoost demonstrates the most balanced and robust performance among the models evaluated in the E4 setting.

According to the experimental results, the RoF performs better in the E1 and E2 experimental settings, whereas XGBoost performs better in the E3 and E4 experimental settings. Overall, ensemble learning models demonstrate satisfactory performance across all evaluation indicators in all experimental settings. However, it is important to note that the predictive efficacy decreases progressively from E1 to E4. This decline is due to the fact that in the E2 to E4 settings, the drugs or target proteins in the test set were not included in the training set, which prevented the model from effectively learning their patterns.

6.1.6 Conclusion

This section outlines a ML workflow for predicting how medications would interact with HPV target proteins using molecular descriptors. An initial assortment of descriptors is calculated for both drugs and proteins. To mitigate overfitting concerns stemming from high-dimensional features, feature selection techniques, including variance filtering, importance ranking based on RF, and genetic algorithm, are employed to enhance the generalization capacity of the model. As a result, a modified feature set with 104 variables is chosen as the ML model's foundation. Subsequently, several prominent ML methods are examined and discussed, namely, LR, DT, AdaBoost, RF, XGBoost, artificial neural network, and RoF. The performances of these models are rigorously compared across four experimental settings using five evaluation metrics. This analysis shows that both XGBoost and RoF demonstrate superior overall performance, rendering them suitable candidates for predicting anti-HPV drugs in future research endeavors.

6.2 Omics-Informed Prediction of Anticancer Drug Response

Cancer constitutes a profound public health challenge, as evidenced by the escalating global trends in its incidence and mortality rates. Conventional cancer therapies, including chemotherapy, radiation therapy, and surgical interventions, engender substantial physical detriments to afflicted individuals, demanding protracted periods for recuperation [117]. Furthermore, they may damage healthy tissues, decrease the antitumor response, and increase tumor metastasis [118]. These conventional treatments are time-consuming and expensive.

In recent decades, biomedical researchers have made significant progress in understanding diseases, leading to the development of new anticancer drugs and innovative treatment approaches like targeted therapy and immunotherapy [119, 120]. These approaches not only provide clinical benefits and improve patients' quality of life but also enhance survival rates and offer potential cures. However, current methods for testing drug response rely heavily on time-consuming and resource-intensive clinical trials. Therefore, establishing predictive models is necessary to overcome these limitations.

Cancer is caused by mutations in critical cancer genes and is classified as a genomic disease. Existing literature has shown that cancer cell lines can to some extent characterize the characteristics of the original cancer genes [121]. Hence, an imperative avenue of research in precision oncology revolves around the meticulous examination of varying cancer cell line sensitivities to distinct pharmaceutical agents. Concurrently, the swift advancement of high-throughput sequencing technology has ushered in a new era of comprehending cancer at the molecular stratum, thereby furnishing copious volumes of multi-omics data for the realm of precision oncology. The capacity for conducting high-throughput drug screenings across a plethora of cancer cell lines further unveils an unparalleled prospect for delving into the prognostication of drug reactions [122]. Therefore, based on emerging multi-omics data, including transcriptomics, proteomics, genomics, and epigenomics, establishing predictive models to decode the relationship between genetic features and drug response is a crucial research direction in precision oncology.

6.2.1 Data Collection and Preprocessing

6.2.1.1 Data Source

The omics and drug sensitivity data used in this section come from the genomics of drug sensitivity in cancer (GDSC) database (https://www.cancerrxgene.org/) [123], while the drug chemical information is sourced from the organic small molecule bioactivity (PubChem) database (https://pubchem.ncbi.nlm.nih.gov/) [124]. A brief overview of these two databases is provided below.

The GDSC database contains rich omics data and comprehensive sensitivity response information of various tumor cells to drugs. The database allows retrieval based on three main aspects: drugs, molecules, and cells. The GDSC database stores two types of drug sensitivity data for cell lines: the original data, which includes the half maximal inhibitory concentration (IC50) values of drug-cell line responses, and the AUC values. The IC50 value represents the concentration of a drug or compound required to induce apoptosis in half of the tumor cells. It also reflects the tolerance of various tumor cell lines to anticancer drugs. A higher IC50 value indicates a higher concentration of drug required to induce apoptosis, indicating greater resistance of the cells to the drug. Conversely, a lower IC50 value indicates a lower concentration of drug required, indicating higher sensitivity of the cell line to the drug. The GDSC database also contains a wealth of cancer-related genomic information. Cell lines are characterized by genomic and transcriptomic sequencing, yielding multi-omic profiles that include gene expression, copy number variations, and genetic mutations. Moreover, the drug information module of the GDSC database contains information on anticancer drugs' known protein targets. This not only enhances the utility of the database by providing genomic and multi-omic profiles but also enables an in-depth understanding of the interactions between these profiles and various anticancer drugs.

The PubChem database, developed by the U.S. National Center for Biotechnology Information (NCBI), serves as a public repository for a wealth of information on drug chemistry. It houses extensive data on chemical structures of drugs and results of compound assays. PubChem allows the retrieval of a multitude of chemical structures, biochemical experimental data of molecules or drugs, and information on compound chemical structures. To obtain the chemical structure data of the relevant drugs, structure-data files (SDF) based on chemical structures are downloaded from the PubChem database. These non-digital SDF files necessitate further processing to compute the molecular descriptors of the pertinent drugs, a task achieved through the use of the software tool PaDEL-Descriptor [113]. By inputting the SDF file of a drug, it computes one-dimensional, two-dimensional, and three-dimensional descriptors and molecular fingerprints. These include atom-type electro-topological state descriptors, McGowan volume, molecular linear free energy relationship descriptors, ring counts, Laggner-identified chemical substructure counts, as well as Klekota-Roth binary fingerprints and chemical substructure counts, among others.

6.2.1.2 Drug Feature Data

The drug target data underwent a binary encoding process, wherein a binary code was employed. In this encoding scheme, the presence of a "1" signifies potential interactions between a given drug and its target protein, as documented within the drug information dataset, while a "0" denotes the absence of reported potential interactions. Subsequently, a novel feature matrix was meticulously crafted, wherein each row of this matrix represents an individual anticancer drug and each column corresponds to a distinct target protein. Post removal of duplicated drug entries, we meticulously curated a final dataset encompassing interactions between 346 unique pharmaceutical compounds and 332 specific target proteins.

Concurrently, one-dimensional and two-dimensional structural attributes pertaining to the chemical composition of these pharmaceutical agents were meticulously extracted from the respective structure-data file (SDF) formats of the drugs, utilizing the PaDEL-Descriptor toolset. Within this context, a comprehensive feature matrix was meticulously generated, wherein each row stands as a unique entity representing an individual anticancer drug, while each column encapsulates a specific feature value elucidating the chemical characteristics of the drug. Ultimately, this endeavor yielded a comprehensive dataset encompassing 1444 distinct chemical structural features, all within the ambit of 146 distinct pharmaceutical agents.

6.2.1.3 Genomic Data

In the context of genomic research, three distinct categories of data are sourced from the GDSC (Genomics of Drug Sensitivity in Cancer) database: gene expression, copy number variation, and gene mutations.

The raw gene expression data extracted from the GDSC database is presented in the form of a matrix. Each row within this matrix corresponds to a specific cancer cell line, while each column represents an individual gene. Numerical values within this matrix are integers, conveying the levels of gene expression.

Table 6.8 Number of features

Data type	Number of original features	Number of features after processing
Gene expression	37600	37261
Copy number variation	24502	23772
Gene mutation	23156	23156

This comprehensive dataset encompasses information from 1431 cancer cell lines and encompasses a total of 37602 unique gene features. Similarly, the raw data pertaining to copy number variation is also structured as a matrix. In this case, each row signifies a distinct cancer cell line, and each column correlates to a specific gene. The matrix entries are integers, signifying gene copy numbers. The dataset detailing copy number variation spans 1371 cancer cell lines and incorporates data for 18001 unique gene features.

The dataset concerning gene mutations encompasses all genetic mutations observed in the sequenced models and is also presented in matrix form. In this matrix, each row represents a cancer cell line, while each column characterizes the genotype of a particular gene. The elements of this matrix are binary, with "1" indicating the presence of a mutant allele and "0" denoting the wild-type allele. This dataset encompasses information from 1428 cancer cell lines and 23156 gene features. Across these three distinct sets of original genomic data, instances of missing values are encountered. To address this data gap, two distinct methodologies are employed based on the distribution characteristics of the variables. Variables exhibiting low coverage rates and missing rates surpassing 50 percent are systematically removed from the analysis. For other variables with lower rates of missing data, the K-nearest neighbors (KNN) algorithm is applied for imputation. In this context, the value of K is set to 5, and missing data points are individually estimated for each omics dataset.

Following the comprehensive processing of all omics data to rectify missing values through the aforementioned techniques, there is a notable reduction in the number of features present in both the gene expression and copy number variation datasets. The specific alterations in the feature count for each type of omics data are meticulously documented in Table 6.8, facilitating a clearer understanding of the impact of data processing on the datasets.

6.2.1.4 Integrating Omics Data and Drug Features

The schematic representation of the procedural integration of cancer omics data and drug attributes is elucidated in Fig. 6.9.

In the initial phase, a similarity matrix is meticulously formulated to encapsulate the resemblances inherent in diverse cancer cell lines. This is achieved through the utilization of a kernel function-based similarity measure, with the gene expression profiles of the cell lines serving as the foundational data. Notably, for this purpose,

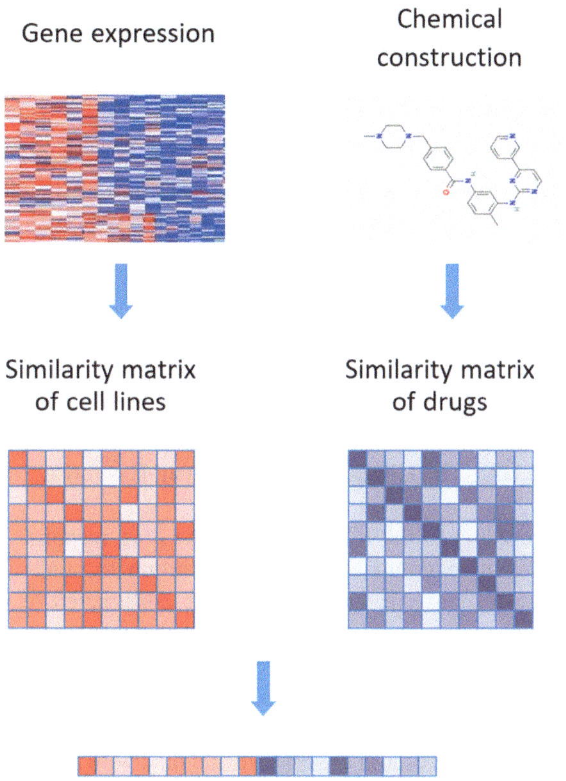

Fig. 6.9 The process of integrating omics data and drug features

the radial basis kernel function has been selectively adopted, owing to its capacity to discern intricate relationships and patterns within the data, which is

$$K(c_a, c_b) = e^{-\gamma \|X_{c_a} - X_{c_b}\|^2},$$

where c_a and c_b represent two different cell lines, X_{c_a} and X_{c_b} are the values corresponding to the feature types of the cell lines, and γ is the parameter of the kernel function, which is set to 0.01 in this case. Each row in the matrix represents a similarity vector of a cell line with other cell lines, capturing the similarity relationships within the cell line dataset. Then, a similarity matrix is constructed to represent the similarity between anticancer drugs using the Tanimoto coefficient similarity (6.1). Each row in the matrix represents a similarity vector of a drug with other drugs, capturing the structural similarities between the drugs. Subsequently, feature vectors are extracted from the aforementioned two similarity matrices and concatenated. The concatenated vector serves as the input feature for subsequent ML tasks. This integration method effectively utilizes the similarity matrices to represent the samples, resulting in a reduced feature space.

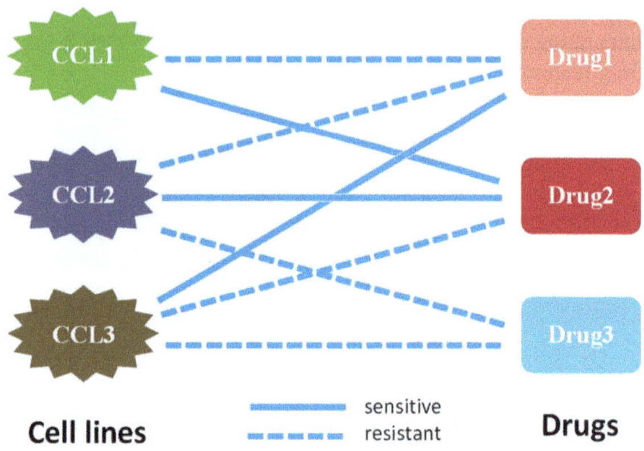

Fig. 6.10 Binary graph of drug responses

6.2.1.5 The Classifications for Drug Response

According to the IC50 values in the drug response data, the data is divided into a sensitive group and a resistant group to study the classification problem of drug response. A bipartite graph is constructed to represent the aforementioned relationships, as shown in Fig. 6.10. The two types of nodes in the bipartite graph represent drugs and cancer cells, respectively, and the edges represent the sensitive (S) or resistant (R) response of the corresponding cancer cells to drug treatment. A drug classification threshold [125] was applied to define data with an IC50 value below the threshold as sensitive and data with an IC50 value above the threshold as resistant. The process generated cell line-drug sensitivity labels for 962 cell lines and 189 drugs, with 25,037 instances categorized as "sensitive" and 25,807 instances categorized as "resistant." Further analysis showed that out of the 962 cell lines, 887 had complete information for all 3 types of omics data, and out of the 189 drugs, 146 had complete information. This resulted in a total of 38,364 data points, which will be used for subsequent research.

6.2.2 *Modeling of Anticancer Drug Response*

6.2.2.1 Model Construction and Evaluation

Predictive models for anticancer drug response are constructed based on the hypothesis that similar drugs exhibit similar responses when treating similar cell lines. The following classification machine learning (ML) models are implemented: logistic regression (LR), support vector machine (SVM), decision tree (DT), random forest (RF), and extreme gradient boosting (XGBoost). The basic processes of these algorithms have already been introduced in previous chapters. To reduce reliance on heuristics and brute force search, all models employed Bayesian optimization to find optimal hyperparameters. The original dataset was randomly divided into an

Table 6.9 Performance comparison of five models

Model	Accuracy	AUC	Sensitivity	Specificity	F1 score
LR	0.762	0.799	0.534	0.895	0.623
SVM	0.746	0.782	0.432	0.928	0.555
DT	0.783	0.767	0.705	0.828	0.704
RF	0.823	0.889	0.779	0.849	0.764
XGBoost	0.853	0.910	0.762	0.905	0.791

80% training set and a 20% testing set. The five models were trained on the same training set. Table 6.9 presents a performance comparison of these five models.

Table 6.9 presents the performance comparison of five machine learning models in terms of accuracy, area under the curve (AUC), sensitivity, specificity, and F1 score. Among the models, XGBoost achieves the highest accuracy (0.853) and AUC (0.910), indicating its superior performance in predicting anticancer drug response. It also demonstrates a high specificity (0.905) and the highest F1 score (0.791), balancing both precision and recall effectively. Random forest follows, with strong performance across all metrics, notably an accuracy of 0.823 and an AUC of 0.889. The decision tree model shows decent performance, particularly in sensitivity (0.705) and specificity (0.828). Logistic regression and SVM show lower sensitivity (0.534 and 0.432, respectively) and overall performance compared to the other models, suggesting they may not capture the complex relationships as effectively as ensemble methods like RF and XGBoost. Overall, XGBoost stands out as the most robust model for this classification task.

A fivefold cross-validation was performed to assess the performance of five models. The results indicate that the ensemble learning models, RF and XGBoost, achieved higher AUC values and accuracies compared to logistic regression (LR), support vector machine (SVM), and decision tree (DT). Notably, XGBoost outperformed RF. The relatively stable trend lines in the figures indicate good model stability, suggesting that the XGBoost-based model consistently exhibited superior predictive performance on the training set compared to the other models (Fig. 6.11).

To test the predictive ability of the model for new drugs, two drugs, Neratinib and AG-014699, were selected from the PubChem database, which were not included in the training dataset. Predictive analysis was conducted on these drugs by calculating their comprehensive similarity with the 146 drugs in the dataset. The resulting similarity vectors were used as drug information features for Neratinib and AG-014699. By combining these drug features with cell line similarity features and employing the trained XGBoost model, sensitivity scores were predicted for each drug-cell line pair across all cell lines. A higher score indicated a higher likelihood of sensitivity between the drug and the cell line. Subsequently, the sensitivity scores were ranked in descending order for each drug-cell line pair. Table 6.10 presents the top ten potential cell lines for each drug based on their rankings.

According to previous studies [126], the head and neck squamous cell carcinoma cell lines CAL27, SCC25, and FaDu have shown sensitivity to the drug Neratinib.

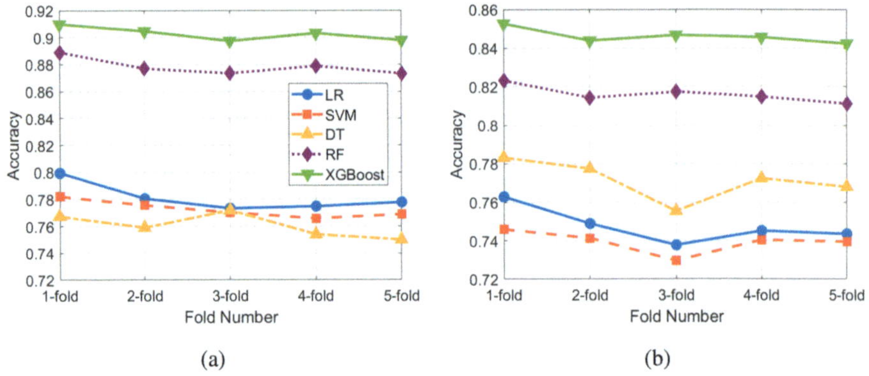

Fig. 6.11 The AUC and accuracy of ML models under various cross-validations

Table 6.10 The top 10 potential cell lines for Neratinib and AG-014699

Drug	Rank	Potential cell line	Drug	Rank	Potential cell line
Neratinib	1	**CAL-27**	AG-014699	1	A673
	2	ME-180		2	**A2780**
	3	GI-ME-N		3	Jurkat
	4	MC-IXC		4	**OVTOKO**
	5	**SCC-25**		5	G-361
	6	CTB-1		6	Farage
	7	**FADU**		7	EJM
	8	NCI-H2141		8	SCC90
	9	KMH-2		9	NB12
	10	NCI-H146		10	Hs-445

Furthermore, Ihnen et al. [127] indicated that the ovarian cancer cell lines A2780 and OVTOKO exhibit sensitivity to Neratinib. In Table 6.10, the bolded cell lines represent those that have been validated in the literature. These findings demonstrate that the XGBoost model can effectively predict the potential efficacy of new drugs based on drug similarity and that the predicted potential cell lines have clinical relevance.

Figure 6.12 presents the top 20 structurally similar compounds to Neratinib and AG-014699 based on their chemical properties. Although these two test drugs were not included in the training dataset, the XGBoost model can infer their potential therapeutic outcomes due to the similarity association rule. This rule suggests that if two drugs have similar chemical properties, they are likely to exhibit similar inhibition effects on cell lines. For instance, neratinib is a tyrosine kinase inhibitor targeting the EGFR, HER2, and HER4 pathways in breast cancer and other solid tumors. Despite not being specifically trained in the previously established model, lapatinib, which was among the 146 anticancer drugs studied during training, shares similar chemical properties with neratinib. The chemical structure similarity

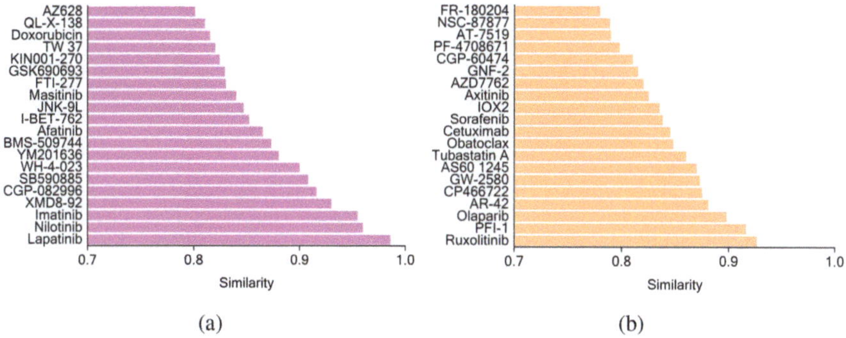

Fig. 6.12 (**a**) The top 20 drugs similar to neratinib in terms of chemical properties. (**b**) The top 20 drugs similar to AG-014699 in terms of chemical properties

between these two compounds is as high as 0.986. Consequently, the algorithm automatically uses lapatinib as a template to infer the therapeutic effect of neratinib, enabling the XGBoost model to predict potential cell lines for neratinib treatment.

Similarly, AG-014699 is an inhibitor targeting the PARP-1, PARP-2, and PARP-3 pathways in ovarian and liver cancer. Although it was not trained in the previously established model, olaparib, another drug studied among the 146 anticancer drugs during training, shares similar chemical properties with AG-014699. The chemical structure similarity between these two compounds is as high as 0.898. Thus, the algorithm uses olaparib as a template to infer the therapeutic effect of AG-014699.

6.2.3 Conclusion

This section is predominantly dedicated to the development of ML models for predicting drug responses using gene expression data and drug chemical structure data. Methods for constructing features from cell line similarity matrices and drug similarity matrices are introduced. A comparison with other ML algorithms highlights the superior predictive performance of the XGBoost-based model, which achieves an accuracy of 0.853 and an AUC of 0.910. Furthermore, the model's stability is demonstrated.

Predictive analysis is also conducted on two drugs, neratinib and AG-014699, which were not included in the training dataset. The top 10 potential cell lines for each drug are identified based on their rankings. Although the model did not directly learn from these drugs, it had previously trained on other drugs with similar chemical properties. These analogous drugs served as reference points for inferring the therapeutic effects of the new drugs.

Chapter 7
Machine Learning in Ecology

The rapid increase in both the quantity and complexity of data being generated daily in the field of ecology demands accompanying advancements in data analytics. Advanced data analysis approaches, such as machine learning (ML), have become indispensable tools for deciphering the complex relationships between system variables and system behaviors, for which conventional analytical methods face limitations or challenges. ML methods offer the ability to process large volumes of data and uncover intricate patterns that traditional techniques may miss.

Meanwhile, there is a growing demand for ML methods that can provide more robust interpretability. Such interpretability is crucial for revealing hidden patterns and formulating effective control measures in ecological studies. Understanding the underlying mechanisms and relationships identified by ML models can lead to more informed decision-making and policy development.

This section introduces the latest research progress in ML, emphasizing its potential to revolutionize data analysis and modeling in ecology. The discussion highlights how ML techniques are being integrated into ecological research to enhance our understanding of complex environmental systems, improve predictions of ecological phenomena, and support sustainable environmental management practices.

7.1 Quantifying Drivers of Air Quality Change

Air pollution has become one of the global major environmental issues due to urbanization and industrialization [128]. According to the World Health Organization (WHO), 95% of the world's population lives in places where air quality exceeds the WHO guideline limits. These higher pollution levels have been linked to downstream effects including ecosystem degradation [129] and more pronounced effects of climate change [130], as well as greater disease burden [131] and decline

J. Ren, Y. Tao, *Modeling in Life Sciences and Ecology*, Springer Asia Pacific Mathematics Series 4, https://doi.org/10.1007/978-981-95-1038-2_7

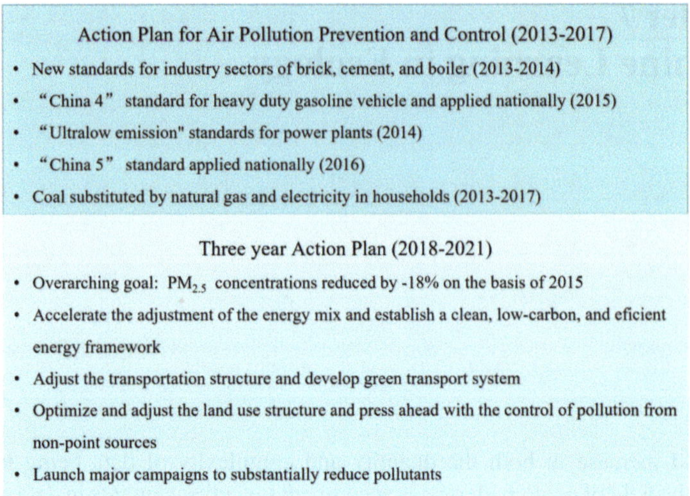

Fig. 7.1 Summary of major air pollution control measures taken from 2013 to 2021

in cognitive function [132] for those exposed. For example, serious air pollution has caused over one-third of deaths from stroke, lung cancer, and chronic respiratory disease, and one-quarter of deaths from ischemic heart disease.

As these represent substantial environmental and health risks, the Chinese government has introduced a series of the strictest emission reduction plans in history since 2013, such as the Action Plan for Air Pollution Prevention and Control in 2013, the 3-year Action Plan for Winning the Battle Against Blue Skies in 2018, and the Tough Battle for Pollution Prevention and Control in 2021 [133, 134]. Included in the ACP are emission control efforts and the deployment of alternative transportation and electricity generation technologies (Fig.7.1). With these rigorous control measures, substantial improvements in air quality have been observed in most major Chinese cities [135].

China's vast geographical expanse leads to diverse meteorological and socioeconomic conditions across different cities, resulting in varying pollution distribution characteristics [136]. Given the spatiotemporal heterogeneity of pollution patterns, it is essential to enhance the analysis of the impact mechanisms of air pollution at the city level to enable precise prevention and control of atmospheric pollution [137]. In this section, we introduce an interpretable ML workflow to quantify the contribution of each feature variation on air quality change at both global (across cities) and local (within a single city) levels. This approach aims to provide actionable recommendations for effective city management.

7.1.1 Data Collection and Processing

A 7-year (2014–2020) annual data set, including air quality, natural, and socioeconomic data, was collected and compiled into training models for 31 major Chinese cities (27 provincial capitals and the 4 municipalities of Beijing, Tianjin, Shanghai, and Chongqing). Air quality index (AQI) is a communication tool to reflect the concentrations of individual pollutants on a common scale, which is constructed based on the levels of six common pollutants, i.e., SO_2, NO_2, O_3, CO, $PM_{2.5}$, and PM_{10}. AQI is derived by taking the maximum of the individual AQI (IAQI) values for each pollutant, where

$$\mathrm{AQI} = \max\{\mathrm{IAQI}_1, \mathrm{IAQI}_2, \mathrm{IAQI}_3, \mathrm{IAQI}_4, \mathrm{IAQI}_5, \mathrm{IAQI}_6\},$$

with IAQI_i is the individual AQI of one of the six pollutants mentioned above (i.e., $1 \leq i \leq 6$). IAQI of the six pollutants that are used in the calculation of AQI were downloaded from the World Air Quality Index project (https://aqicn.org/).

Data were collected on 14 factors considered to potentially affect air quality and are summarized below. The data for average wind speed (WS), precipitation (RF), average temperature (AT), average relative humidity (AH), population (POP), population density (DEN), share of secondary industry in GDP (SEC), number of civilian vehicles, SO_2 emission, NO_x emission, and dust emission of each city was downloaded from the National Statistical Yearbooks of China and the National Meteorological Centre of China. To effectively compare cities across a range of sizes, gross domestic product per km^2 (GDP), number of civilian vehicles per km^2 (CAR), SO_2 emission per km^2 (ESO_2), NO_x emission per km^2 (ENO_x), dust emission per km^2 (EDUS), and electric consumption per km^2 (ELC) were normalized by the urban area of each city in square kilometers. The data of SO_2, NO_x, and dust includes emissions from both industrial and private household sources. Annual normalized difference vegetation index (NDVI) of each city was annually averaged from Spatial Distribution Dataset of Monthly NDVI in China (http://www.resdc.cn/DOI). These predictor variables are summarized in Table 7.1.

Figure 7.2a and b show the annual mean IAQI_i values for six pollutants over 31 major Chinese cities in 2014 and 2020. In both years, high pollutant concentrations were observed in some northern and central cities (Shijiazhuang, Zhengzhou and Jinan). As illustrated in Fig. 7.2c, from 2014 to 2020, significant reductions in levels of SO_2, NO_2, CO, $PM_{2.5}$, and PM_{10} occurred across almost all studied cities, while substantial increases in O_3 levels occurred in 18 of 31 cities. The highest increase in O_3 was observed in Hefei, whereO_3 concentrations rose by 63.60%.

Table 7.1 Summary of predictor variables

Category	Variable (Unit)	Abbreviation
Natural	Average wind speed (km/h)	WS
	Precipitation (mm)	RF
	Average temperature (°C)	AT
	Average relative humidity (%)	AH
	Normalized difference vegetation index	NDVI
Socioeconomic	Gross domestic product per km^2 (100 million RMB/km^2)	GDP
	Population (million)	POP
	Population density (people/km^2)	DEN
	Share of secondary industry in GDP (%)	SEC
	Number of civilian vehicles per km^2 (vehicle/km^2)	CAR
	SO_2 emission per km^2 (ton/km^2)	ESO_2
	ENO_x emission per km^2 (ton/km^2)	ENO_x
	Dust emission per km^2 (ton/km^2)	EDUS
	Electric consumption per km^2 (MkWh/km^2)	ELC
Air quality	Annual mean IAQI of $PM_{2.5}$	$IAQI_{PM_{2.5}}$
	Annual mean IAQI of PM_{10}	$IAQI_{PM_{10}}$
	Annual mean IAQI of NO_2	$IAQI_{NO_2}$
	Annual mean IAQI of O_3	$IAQI_{O_3}$
	Annual mean IAQI of CO	$IAQI_{CO}$
	Annual mean IAQI of SO_2	$IAQI_{SO_2}$
	Change of IAQI for pollutant i	$\Delta IAQI_i$

7.1.2 Modeling of Air Quality Index

7.1.2.1 Feature Selection

Prior to using features as inputs for the machine learning models, a screening process was conducted to check their correlations with one another. This approach helps avoid multiple collinearity problems, thereby making feature selection more reasonable. This subsection employs the partial correlation coefficient (2.8) [138] for screening. In cases where more than two indicators are interrelated, assessing the partial correlation between two indicators after controlling for a third variable or a set of variables proves more practical.

As shown in Fig. 7.3, the partial correlation coefficients for most pairs of variables have absolute values of less than 0.7. Exceptionally, there exist two pairs for which the correlation coefficients are greater than 0.7: POP and GDP, as well as DEN and CAR. These high correlations are easy to understand. In Chinese cities, GDP has been found to increase superlinearly with population due to economies of scale [139], resulting in a positive association between population and GDP per unit area. Furthermore, car ownership in China has been found to be higher in urban

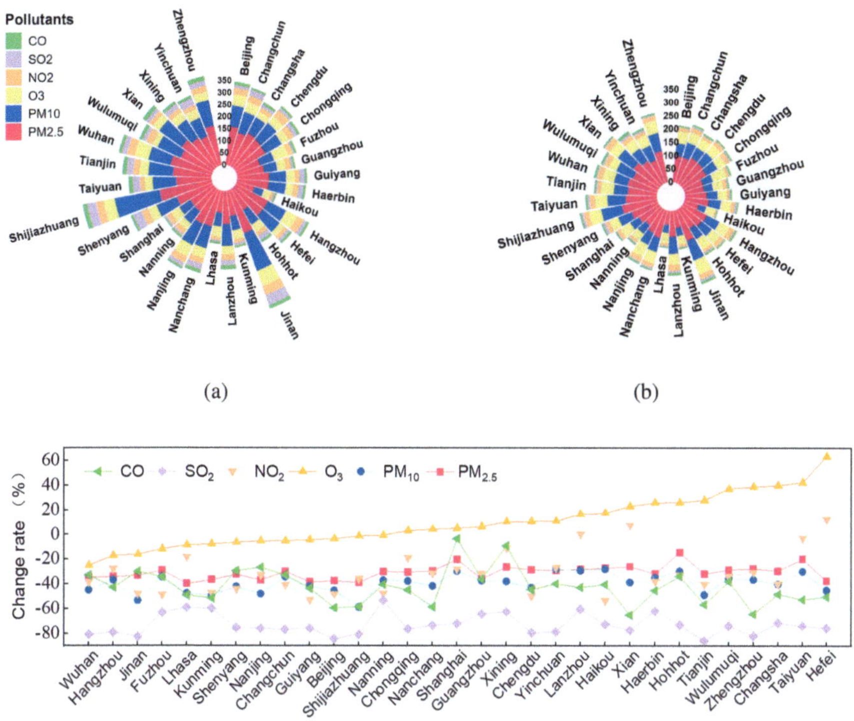

Fig. 7.2 (**a**) and (**b**) annual mean IAQI_i for six pollutants in 2014 and 2020. (**c**) Change rate of annual mean IAQI_i between 2014 and 2020

(i.e., high-density) areas compared to rural ones, even after accounting for factors such as income level [140]. Due to the more important roles of civilian vehicles and GDP than population in pollutant mitigation [141], we consider that these two factors have greater capacity for driving air quality change. Therefore, we eliminate the other two redundant variables, POP and DEN, reducing the final number of variables in use to 12.

7.1.2.2 Model Framework

To predict the IAQI of six pollutants, we used a variety of different machine learning models, namely, categorical boosting (Catboost), random forest (RF), gradient boosting decision tree (GBDT), gradient boosting machine (LightGBM), extreme gradient boosting (XGBoost), support vector regression (SVR), K-nearest neighbor (KNN), deep neural networks (DDN), and BP neural network (BP). The basic processes of the first seven methods have been introduced in the previous chapters. Next, the fundamental concepts of DNN and BP will be discussed in detail.

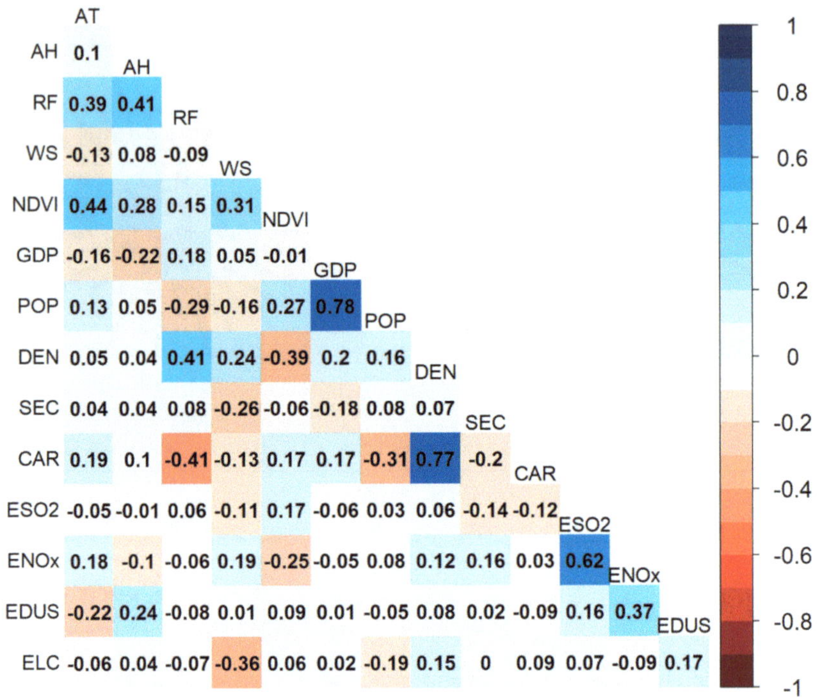

Fig. 7.3 The correlation coefficients of the 14 studied features in 31 major cities from 2014 to 2020, colored from blue (strong positive correlations) to red (strong negative correlations). The less saturated the color, the weaker (i.e., closer to zero) the correlation is

Deep Neural Network

A deep neural network (DNN) is a type of artificial neural network characterized by multiple layers of neurons [142]. It is a crucial model within the field of machine learning. DNNs process and analyze complex data by emulating the way human brain neurons connect and transmit information. The core idea of a DNN is to learn data features through multiple layers of nonlinear transformations, ultimately performing tasks such as classification or prediction.

The fundamental unit in a DNN is the neuron. Each neuron receives input from the previous layer, processes this input through weighted summation and an activation function, and passes the result to the next layer. This process can be viewed as the step-by-step transmission and transformation of information. Activation functions are typically nonlinear, enabling DNNs to handle complex data and express intricate functions.

DNNs can be regarded as an extension and deepening of artificial neural networks (ANNs). The structure of a DNN typically includes an input layer, hidden layers, and an output layer. The input layer is responsible for receiving raw data, while the output layer provides the final prediction results. Hidden layers, situated between the input and output layers, are the critical components of a DNN. The number of hidden layers and the number of neurons in each layer significantly impact the performance of a DNN. More hidden layers enhance the network's ability to represent data but also increase the difficulty and computational complexity of training.

The process of training a DNN involves forward propagation and backward propagation. In forward propagation, input data is processed through each layer of neurons, ultimately producing a prediction in the output layer. Backward propagation, on the other hand, adjusts the weights of each neuron based on the error between the predicted and actual results using the gradient descent algorithm. This iterative process continues until the model's error reaches a satisfactory level.

Backpropagation (BP) neural networks are a type of multilayer feedforward neural network that use the error backpropagation algorithm for training [143]. BP neural networks are widely applied in fields such as classification, regression, and pattern recognition, making them one of the most classical and commonly used models in artificial neural networks.

A BP neural network also consists of an input layer, hidden layers, and an output layer. The input layer is responsible for receiving external input data, the hidden layers perform feature extraction and nonlinear transformation, and the output layer produces the final prediction results. The number of hidden layers can vary depending on the specific application requirements and the complexity of the network.

The core of BP neural networks involves two main steps: forward propagation and backpropagation. In the forward propagation phase, input data passes through the input layer to the hidden layers and the output layer. Each neuron processes the input data by performing a weighted sum and then applying an activation function to produce an output that is passed to the next layer. The specific calculation for the output of each neuron is given by

$$y = f\left(\sum_i w_i x_i + b\right),$$

where y is the output of the neuron, x_i are the inputs, w_i are the weights, b is the bias, and f is the activation function (such as sigmoid, tanh, or ReLU).

In the backpropagation phase, the error between the predicted output and the actual output is computed first. This error is then propagated backward through the network, and the weights and biases are adjusted using the gradient descent algorithm. The specific steps are as follows:

1. Error calculation: The error between the predicted output and the actual output is computed using a loss function, such as mean squared error.
2. Backpropagation of error: Starting from the output layer, the error term δ for each neuron is calculated and propagated backward. For a neuron in the output layer, the error term is given by

$$\delta = (y - t) f'(net).$$

Here t is the target output, and $f'(net)$ is the derivative of the activation function with respect to the input of the neuron.

3. Weight update: The weights and biases are updated using the computed gradients and a learning rate. The weight update rule is

$$\Delta w = -\eta \delta x,$$

where η is the learning rate, x is the input to the neuron, and δ is the error term.

The advantages of BP neural networks include their strong generalization ability and theoretical maturity, making them suitable for various types of data and tasks. However, they also have notable disadvantages, such as high computational complexity, susceptibility to local minima, and a strong dependence on large amounts of training data. These drawbacks can be mitigated in practice through algorithmic improvements and the use of more powerful computational resources.

For performance improvement, the tree-structured Parzen estimator (TPE) was employed to search a set of optimal parameters for ML models [144]. The ML models were developed using a training set, to which 80% of the samples from the dataset were randomly assigned. The remaining 20% of the samples were assigned as the test set. Figure 7.4 shows our workflow for model training, testing, and evaluation.

7.1.2.3 Model Evaluation

To verify the reliability of the models, three indicators are employed: the coefficient of determination (R^2), root mean square error (RMSE), and mean absolute error (MAE). These metrics are used to evaluate the performance of each machine learning model. Additionally, the robustness of all tested models is assessed using tenfold cross-validation. Tables 7.2, 7.3, and 7.4 present the performance of these nine models in predicting the IAQI of six pollutants.

The R^2 values indicate the proportion of the variance in the dependent variable that can be predicted from the independent variables. Higher R^2 values suggest better model performance. CatBoost consistently achieved the highest R^2 values across all pollutants, with particularly high values for $IAQI_{PM_{2.5}}$ at 0.89, $IAQI_{PM_{10}}$ at 0.88, and $IAQI_{CO}$ at 0.82. DNN also performed well, especially for $IAQI_{PM_{2.5}}$ at 0.86 and $IAQI_{PM_{10}}$ at 0.88, but showed weaker performance for $IAQI_{O_3}$ at 0.44. KNN had the lowest R^2 values, indicating poor performance in predicting all pollutants, particularly $IAQI_{O_3}$ at 0.11.

RMSE measures the average magnitude of the prediction error, providing an indication of how well the model predicts the dependent variable. CatBoost again demonstrated superior performance with the lowest RMSE values across most pollutants, such as $IAQI_{PM_{2.5}}$ at 10.2 and $IAQI_{PM_{10}}$ at 8.6. DNN showed competitive RMSE values, such as $IAQI_{PM_{2.5}}$ at 11.1 and $IAQI_{PM_{10}}$ at 10.0, although these

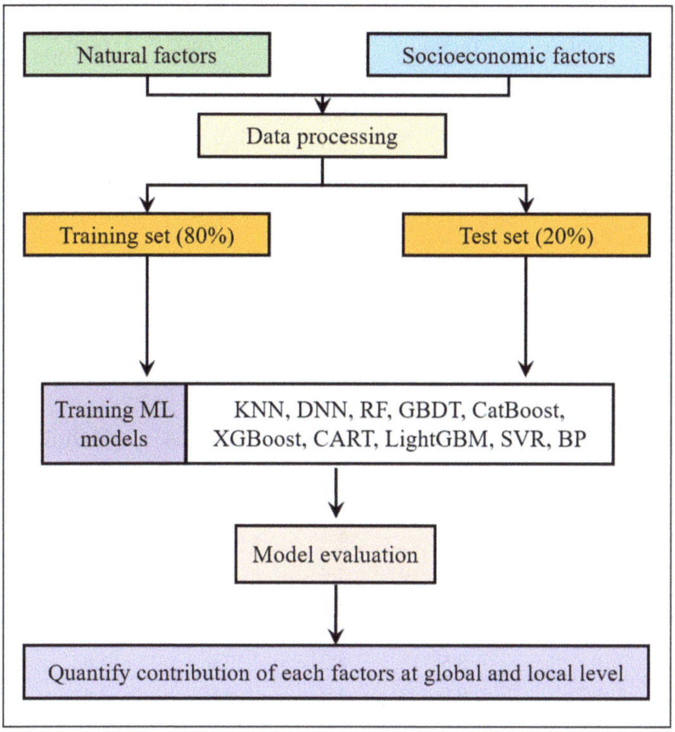

Fig. 7.4 Simplified schematic of ML workflow

Table 7.2 R^2 of ML models

Model	$IAQI_{PM_{2.5}}$	$IAQI_{PM_{10}}$	$IAQI_{O_3}$	$IAQI_{NO_2}$	$IAQI_{SO_2}$	$IAQI_{CO}$
SVR	0.81	0.81	0.65	0.65	0.69	0.63
BP	0.81	0.80	0.47	0.71	0.72	0.75
KNN	0.48	0.48	0.11	0.466	0.44	0.45
DNN	0.86	0.88	0.44	0.81	0.75	0.54
RFR	0.82	0.84	0.50	0.74	0.62	0.74
GBDT	0.72	0.71	0.35	0.64	0.63	0.64
LightGBM	0.81	0.83	0.40	0.78	0.67	0.75
XGBoost	0.83	0.80	0.43	0.73	0.71	0.73
CatBoost	0.89	0.88	0.78	0.86	0.88	0.82

were higher than those of CatBoost. KNN had the highest RMSE values, such as $IAQI_{PM_{2.5}}$ at 20.4 and $IAQI_{PM_{10}}$ at 16.7, reflecting poor prediction accuracy.

MAE measures the average absolute errors between predicted and actual values, providing a straightforward interpretation of prediction accuracy. CatBoost consistently achieved the lowest MAE values across all pollutants, with notable values for $IAQI_{PM_{2.5}}$ at 8.5 and $IAQI_{PM_{10}}$ at 7.0. DNN also performed well with competitive

Table 7.3 RMSE of ML models

Model	$IAQI_{PM_{2.5}}$	$IAQI_{PM_{10}}$	$IAQI_{O_3}$	$IAQI_{NO_2}$	$IAQI_{SO_2}$	$IAQI_{CO}$
SVR	13.1	10.7	4.8	4.2	3.5	2.0
BP	13.4	11.4	6.2	3.7	3.5	1.8
KNN	20.4	16.7	8.0	5.0	4.8	2.5
DNN	11.1	10.0	6.0	3.2	3.3	2.3
RFR	12.6	10.0	5.5	3.8	3.8	1.7
GBDT	15.3	12.3	6.7	4.2	4.0	2.1
LightGBM	12.2	10.5	6.1	3.4	4.0	1.7
XGBoost	11.6	11.1	5.8	3.6	3.4	1.7
CatBoost	10.2	8.6	3.8	3.0	1.4	1.1

Table 7.4 MAE of ML models

Model	$IAQI_{PM_{2.5}}$	$IAQI_{PM_{10}}$	$IAQI_{O_3}$	$IAQI_{NO_2}$	$IAQI_{SO_2}$	$IAQI_{CO}$
SVR	11.0	8.6	3.8	3.4	2.5	1.5
BP	11.2	9.5	4.7	3.2	2.6	1.3
KNN	15.8	12.0	6.1	3.6	3.9	1.8
DNN	8.9	8.5	4.5	2.5	2.5	1.5
RFR	10.6	7.9	3.9	3.1	2.7	1.3
GBDT	12.5	10.0	5.1	3.3	3.1	1.5
LightGBM	9.8	8.5	4.2	2.6	2.8	1.2
XGBoost	9.7	9.0	4.4	3.0	2.4	1.3
CatBoost	8.5	7.0	2.9	1.9	2.3	1.1

MAE values, such as $IAQI_{PM_{2.5}}$ at 8.9 and $IAQI_{PM_{10}}$ at 8.5. KNN again showed the highest MAE values, indicating less accurate predictions, especially for $IAQI_{PM_{2.5}}$ at 15.8 and $IAQI_{PM_{10}}$ at 12.0.

Overall, CatBoost emerged as the best-performing model across all three evaluation metrics (R^2, RMSE, and MAE), demonstrating superior predictive accuracy and reliability for all six pollutants. Thus, we selected CatBoost to assess feature importance. Since we subsequently applied SHAP to the CatBoost output (see below), our overall workflow is therefore an implementation of Tree SHAP [145], which is an algorithm for reducing computation time when SHAP is used in conjunction with a tree-based ML model.

7.1.3 Quantifying Drivers at Global Levels

Figure 7.5 shows the global contributions of our 12 features to levels of each of the 6 pollutants according to Catboost models. Each dot represents a SHAP value that measures the feature contribution to an individual prediction, while an aggregate of dots indicates the magnitude, commonality, and direction of the variable's global

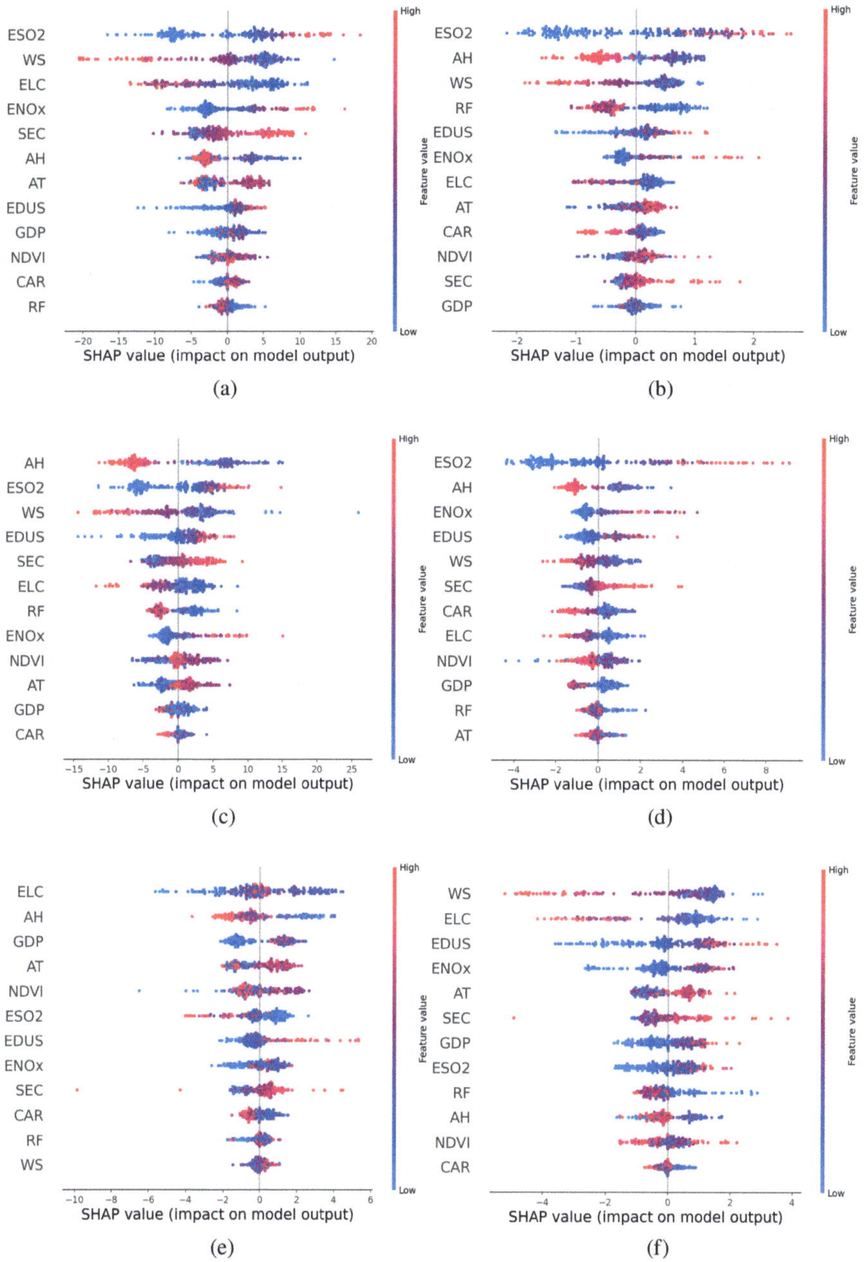

Fig. 7.5 SHAP summary plots for IAQI by pollutants. The x-axis represents the SHAP value. The color from blue to red represents the feature's value from low to high. The y-axis represents feature importance, ranked in descending order. **(a)** SHAP summary plots for $IAQI_{PM_{2.5}}$. **(b)** SHAP summary plots for $IAQI_{CO}$. **(c)** SHAP summary plots for $IAQI_{PM_{10}}$. **(d)** SHAP summary plots for $IAQI_{SO_2}$. **(e)** SHAP summary plots for $IAQI_{O_3}$. **(f)** SHAP summary plots for $IAQI_{NO_2}$

effect. The mean SHAP value across all cities represents a measure of global importance on the entire data set; features are listed from top to bottom in each subplot according to this value. It is clear that the ranks of feature importance are different for different pollutants.

As shown in Fig. 7.5a, for ESO_2, ENO_x, SEC, AT, and EDUS, the larger these factors, the higher their SHAP value, and hence the larger their impacts on aggravating $PM_{2.5}$ levels. On the other hand, WS, ELC, and AH have inhibitory effects: the higher these factors, the lower $PM_{2.5}$ levels are predicted to be. The six pollutants share several common top influencing variables (AH, ENO_x, and EDUS), and the directions of the effects of most variables are consistent across the pollutants. However, ESO_2 has less uniform patterns: It positively contributes to O_3 mitigation due to the chemical interactions between O_3 and SO_2 [146], and negatively impacts the mitigation of other pollutants.

Figure 7.6 shows how each feature's SHAP value changes with the feature's own value for $PM_{2.5}$, capturing the thresholds for air quality mitigation. The dominant socioeconomic factor for determining $PM_{2.5}$ concentration was ESO_2. Its effects toward enhancing $PM_{2.5}$ rapidly increased as emissions did, until a value of approximately 1.534 ton/km^2 was reached. SO_2 emissions above this level caused slower increases in ESO_2's SHAP values up until about 11.706 ton/km^2, at which point $PM_{2.5}$ leveled off. The other socioeconomic features directly related to emissions, ENO_x and EDUS, followed similar saturation-like patterns as ESO_2. When $15.2\% \leq \text{SEC} \leq 35.5\%$, its SHAP value remained unchanged and negative. ELC also had substantial effects on $PM_{2.5}$ concentrations; it was predicted to adversely affect air quality when 0.047 MkWh/km^2 $\leq$ELC ≤ 0.508 MkWh/km^2.

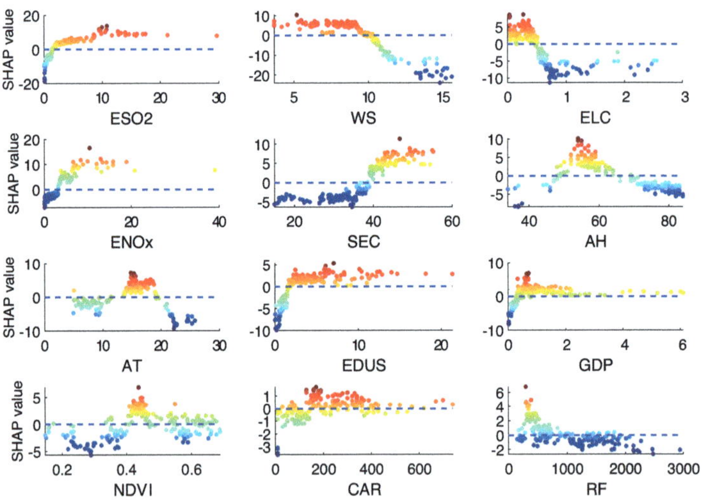

Fig. 7.6 SHAP dependence plot for $IAQI_{PM_{2.5}}$. Each dot corresponds to a single prediction from the dataset. The x-axis represents the value of the variable, and the y-axis is the corresponding SHAP value

When ELC exceeded this range, there was a downward trend in its SHAP values, which might be due to an electricity generation mix increasingly oriented toward clean technologies. SEC's SHAP values for predicting $PM_{2.5}$ followed an increasing logistic pattern, the opposite of those of ELC. They were positive when SEC $\geq$ 38.31%. The relationship between GDP and its SHAP value was a left-skewed inverted V, implying that the highest impact of GDP on $PM_{2.5}$ concentrations occurred at moderately low values of GDP per unit area. Notably, the SHAP values still remained positive as GDP increased. CAR had relatively small SHAP values compared to the other features when predicting $PM_{2.5}$, as well as less variation in SHAP values across its range; hence, its effect on $PM_{2.5}$ is of lesser importance.

WS was the most important natural feature in the determination of $PM_{2.5}$ levels, and the second-most important overall. For low average wind speeds, the effects of WS on $IAQI_{PM_{2.5}}$ were positive, with little variation below an inflexion point at approximately 9.1 km/h. Following this, a downward trend was observed: WS's effect on $PM_{2.5}$ remained positive (i.e., associated with increasing $PM_{2.5}$ levels) until WS $\approx$ 10.3 km/h and became negative thereafter, in agreement with the trends from previous observations [147]. The relationship between AH and its SHAP value followed an inverted-V pattern. We estimated that when 48% $\leq$AH $\leq$ 66%, the SHAP value is larger than 0, implying that AH substantially enhances $PM_{2.5}$ concentrations when the humidity is in this range. Based on its SHAP values, AT presents a positive impact for aggravating pollution when temperature is in the range of 13.6°C–19.2°C. Intuitively, static wind, relatively low temperature, and humidity would easily cause $PM_{2.5}$ pollution. The SHAP values of NDVI and RF were small in absolute value, indicating that these factors' ranges of relative contributions were slight.

An understanding of these environmental thresholds is useful in establishing strategies for air pollution mitigation. SHAP operates in a manner that explicitly predicts changes in AQI values relative to baseline for given feature inputs. Hence, our results include estimates of how much AQI will be raised or lowered by incremental change in a given feature, which values of each feature produce the greatest such difference in AQI, and the aforementioned threshold feature values where nonlinearities in these AQI changes appear. This makes our results highly suited to policy-making at the city level, since local officials will be able to use them to estimate what will be necessary to achieve their desired targets.

7.1.4 Quantifying Drivers at Local Levels

Figure 7.7 shows the contribution of each feature to $\Delta IAQI_i$ in all 31 cities by using bar plots. While ESO_2, ENO_x, and EDUS were all highly associated with $\Delta IAQI_{O_3}$, the directions of their contributions are different at the city level (Fig. 7.7a). An example of this is how O_3 declined with decreasing NO_x emissions in 21 of 31 cities but increased in response to reducing NO_x emissions in the others. Temperature

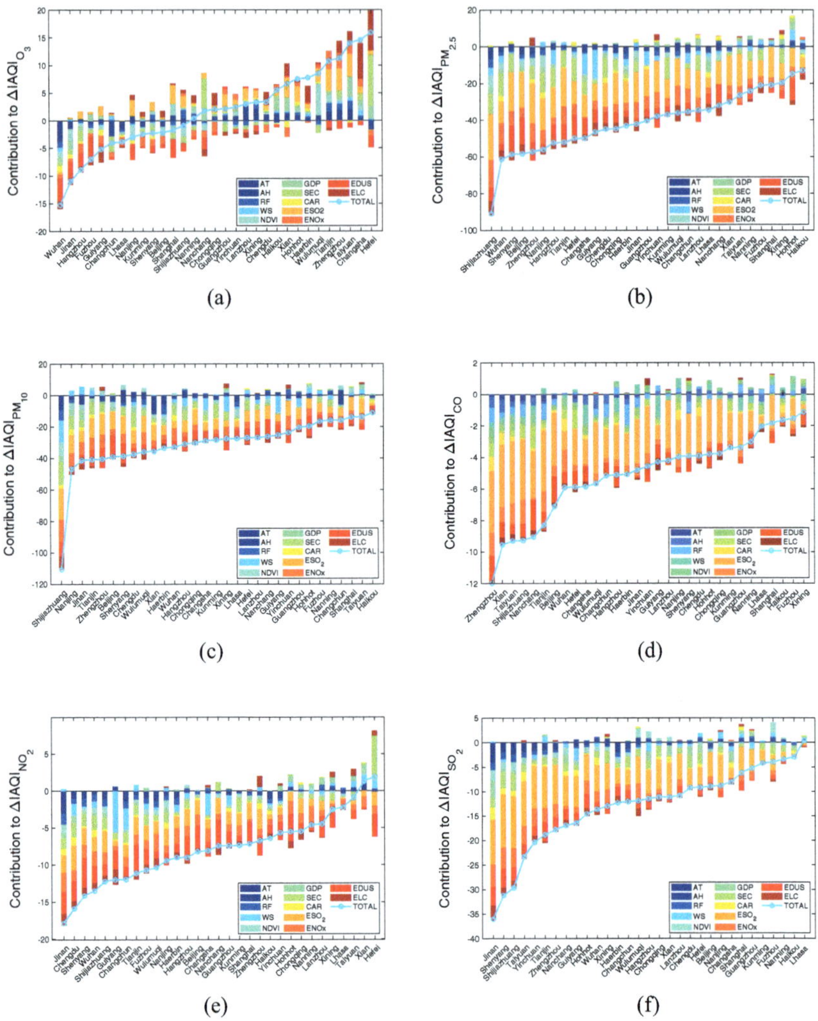

Fig. 7.7 Feature contribution to $\Delta IAQI_i$ at the local city level. (**a**) Feature contribution to $\Delta IAQI_{O_3}$. (**b**) Feature contribution to $\Delta IAQI_{PM_{2.5}}$. (**c**) Feature contribution to $\Delta IAQI_{PM_{10}}$. (**d**) Feature contribution to $\Delta IAQI_{CO}$. (**e**) Feature contribution to $\Delta IAQI_{NO_2}$. (**f**) Feature contribution to $\Delta IAQI_{SO_2}$

warming was also found to play a positive role in elevating $IAQI_{O_3}$ in most cities, though the contribution was relatively weak compared to those of NO_x and SO_2. CAR's contribution was small in absolute value, indicating that the increasing number of motor vehicles does not significantly impact ozone pollution. This may be attributable to the implementation of urban motor vehicle restriction policies. Besides, the relatively consistent, green, and well-established public transportation systems in these provincial capital cities could also somewhat mitigate ozone pollution. The contributions of GDP, SEC, and ELC are heterogeneous at the local level. In several cities (like Taiyuan, Xi'an, and Haikou), increasing GDP has produced greater ozone pollution. This may be due to regional patterns of demand in these cities being, to some extent, focused mainly on economic growth at the expense of environmental quality. Hefei has an industrial structure dominated by secondary industries and high energy use intensity, and it also experienced the most significant increase in ozone concentration caused by increasing SEC and ELC. Thus, although the local deployment of upgraded industrial structure and electricity generation technologies has a mitigating impact on ozone pollution, this effect is limited.

7.1.5 Conclusion

In addition to accurately predicting the changes of AQI, understanding the direct causes of and key drivers behind these changes is of paramount importance to environmental studies. In this regard, efforts to embrace ML models and incorporate the most relevant and influential factors into these models are required to enhance predictive accuracy and discover potential interactions. In the present study, Catboost models coupled with SHAP, an emerging interpretation method, are applied to investigate the roles of driving factors for AQI changes in 31 Chinese major cities from 2014 to 2020. Compared to the variable importance offered by tree-decision methods, the SHAP method is more versatile, as it can provide information on the magnitude and direction of each feature's contribution at both global and local levels. We used the global interpretations offered by SHAP to generate a ranking of the most important features and found that that control measures on the emission of NO_x and dust will lead to optimal conditions for all studied forms of pollution. We found substantial heterogeneity in the effects of the other features, emphasizing the importance of fine-tuned local strategies for pollution control that depend on the observed values of each feature in each city.

We found nonlinear trends in how economic factors affect pollutant concentrations, which can be used to formulate "win-win" local development plans which simultaneously maximize growth while maintaining a healthy environment. For instance, the variation of SEC's dependency on $IAQI_{PM_{2.5}}$ suggests that an economic development mode that relies on secondary industry will aggravate $PM_{2.5}$ pollution, but moderate SEC development could be advocated in those areas that have not yet reached the turning point of about 35% (see Fig. 7.6), in agreement with a previous study [141]. Besides, higher SEC exacerbates $PM_{2.5}$ pollution, while higher GDP mitigates it, concluding that $PM_{2.5}$ pollution is a problem derived from the mode of economic development rather than economic development overall. This is in concordance with previous studies: The initial process of industrialization in China led to a sharp increase in air pollution [148], but as China's GDP has continued to rise in recent years, the country's externalities from environmental damage have decreased [149]. On a similar note, the setting of GDP growth targets is commonplace at all levels of Chinese government, including in cities [150]. The selection of these targets has been shown to have an impact on factors important for environmental protection such as energy efficiency [151] and green technology development [152], so choosing them in order to strike the right balance between growth and low emissions is key. However, few Chinese cities have achieved this balance so far [153]. Cities aiming to do so could use our results and modeling approach to set their growth targets in a way that explicitly takes into account potential environmental effects. As such, we have built a single framework for global and local interpretations by integrating local explanations of each prediction into a global understanding of the model structure.

7.2 Prediction of Algal Growth Tendency

Algae occasionally referred to as phytoplankton are similar to terrestrial plants in that they possess chlorophyll and depend on sunlight to live and develop. The majority of phytoplankton are buoyant and float toward the surface of the ocean's upper layer, where sunlight can reach them. Additionally, algae need inorganic elements such as phosphates, sulfur, and nitrates, which they use to make proteins, lipids, and carbohydrates.

In a healthy ecosystem, phytoplankton is the base of several aquatic food webs, feeding various animals, such as shrimp, snails, and jellyfish. It also can fix carbon at the bottom of the food chain and support fisheries and ecosystems worldwide. However, when too many nutrients and high water temperatures are available, phytoplankton may grow out of control and form harmful algal blooms (HABs). They can occur in canals, lakes, coastal waters, and even swimming pools, having numerous adverse effects on the aquatic ecosystem and public health, including massive fish deaths, contamination of drinking water, and human illness caused by toxins. Satellite data has shown that algal blooms have boomed worldwide during the last 30 years, and this trend is projected to load further stress from the rise in water temperature induced by climate change and the increase in anthropogenic fertilizer intake produced by human activities [154].

Algal growth in a water body is a complex activity characterized by many interior physical, chemical, and biological mechanisms and impacted by a series of external environmental factors [155]. Quantifying the water parameters' independent and synergistic contributions to HABs and their critical thresholds are practical tools for water quality management. In this subsection, we introduced an ML workflow coupled with several explainable AI (XAI) methods to predict the algae density and understand the interactions between water parameters.

Table 7.5 Summary of predictor variables

Variable	Abbreviation	Unit
Algae Density	AD	cell/ml
Water Temperature	TEMP	°C
pH	pH	NA
Dissolved Oxygen	DO	mg/L
Electrical conductivity	CON	μs/cm
Water Turbidity	TUR	NTU
Permanganate Index	PI	mg/L
Ammonia Nitrogen	NH	mg/L
Total Phosphorus	TP	mg/L
Total Nitrogen	TN	mg/L
Chlorophyll a	Chla	mg/L
Ratio of total nitrogen to total phosphorus	N-P	NA

7.2.1 Study Area and Dataset

Lake Liangzi (30°05’-3018’N, 114°21’-114°39’E) boasts an intact ecosystem that is rich in species and natural properties and is one of the best protected freshwater lakes in China. Lake Liangzi’s aquatic vegetation is well developed, which makes it a typical grass-type lake. The lake is located on the south bank of the middle reaches of the Yangtze River in southeastern Hubei Province, spanning Wuhan city’s Jiangxia District (formerly Wuchang County) and Ezhou City. Lake Liangzi covers an area of 304.3 km^2 and is 4.16 meters deep on average. At present, the trophic state of Lake Liangzi has shifted from poor to mesotrophic and is locally close to eutrophic [156]. The phytoplankton community is mainly dominated by Chlorophyta and Cyanophyta, respectively, accounting for 43.5% and 41.8% of the total phytoplankton abundance [157].

Data were downloaded from China’s National Environmental Monitoring Centre (www.cnemc.cn), which automatically monitors water quality. The CNEMC checks water quality every 4 hours and publishes the most recent data. Monitoring indices include water temperature (TEMP), pH, dissolved oxygen (DO), electrical conductivity (CON), permanganate index (PI), water turbidity (TUR), ammonia nitrogen (NH), total nitrogen (TN), and total phosphorus (TP) (Table 7.5). In this section, data were collected from four water quality monitoring stations in Liangzi Lake between November 2020 and March 2023.

7.2.2 Predictive Model of Algae Density

To predict the algae density, this subsection applied several ML models, including SVR, KNN, CART, GBDT, XGBoost, LightGBM, and CatBoost. These models

Table 7.6 Performance of ML models

Model	R^2	RMSE	MAE
SVR	0.74	27560.5	16304.4
KNN	0.836	10208.9	4060.8
CART	0.814	11158.8	3879.7
GBDT	0.764	11110.4	5600.6
CatBoost	0.89	8069.2	3720.8
XGBoost	0.871	8735.5	4061.6
LightGBM	0.885	8176.4	3859.7
CatBoost(CMA-ES)	0.913	7758.3	3421.1
XGBoost(CMA-ES)	0.894	8020.1	3701.9
LightGBM(CMA-ES)	0.903	7982.9	3680.1

have been previously described, so detailed descriptions will not be reiterated here. For training and testing these ML models, input variables observed in November 2020 and March 2023 were utilized. The data was split into training and testing sets with a ratio of 0.8:0.2, ensuring that observations from an entire year were included in both datasets. This approach facilitates robust model evaluation and enhances the predictive accuracy for algae density estimation.

Table 7.6 presents the performance metrics of various machine learning models evaluated using R^2, RMSE, and MAE. The CatBoost model optimized with CMA-ES achieved the highest R^2 value at 0.913, indicating the best fit, along with the lowest RMSE at 7758.3 and MAE at 3421.1, reflecting superior predictive accuracy. Other models like LightGBM and XGBoost, both with and without CMA-ES optimization, also performed well, with R^2 values exceeding 0.85 and relatively low RMSE and MAE values. In contrast, the SVR model had the lowest R^2 at 0.74 and the highest RMSE at 27560.5 and MAE at 16304.4, indicating the least accurate predictions among the models evaluated. Based on its overall high model performance, CatBoost(CMA-ES) was selected to assess feature importance and conduct SHAP analysis.

7.2.3 *Analysis of Algae Growth Mechanism*

Table 7.7 summarizes the values for the three indices: the absolute value of the Shapley value (SHAPABS), FI, and VIF. Each index provides a different perspective on the importance and relevance of the input variables in the context of a predictive model.

SHAPABS and FI were sorted from high to low values, while VIF was sorted from low to high. Higher SHAPABS and FI values indicate that the variable has a greater impact on model performance. A higher VIF indicates that the variable has a highly collinear relationship with the other variables, which is helpful to

Table 7.7 Summary of indices for relative importance of input variables

SHAPABS		FI		VIF	
Feature	Value	Feature	Value	Feature	Value
TEMP	8363.0	PI	19.64	Chla	1.05
PI	6254.5	TEMP	19.52	CON	1.07
CON	2888.9	CON	17.62	TUR	1.13
DO	1632.6	TN	9.68	NH	1.22
TP	1553.3	Chla	8.95	PI	2.21
pH	1539.9	TUR	7.44	TP	2.82
TUR	1509.6	TP	6.12	pH	3.15
Chla	1463.2	pH	3.70	DO	3.27
TN	1149.9	DO	3.45	TEMP	5.88
NP	906.8	NP	3.09	NP	6.25
NH	252.2	NH	0.79	TN	6.30

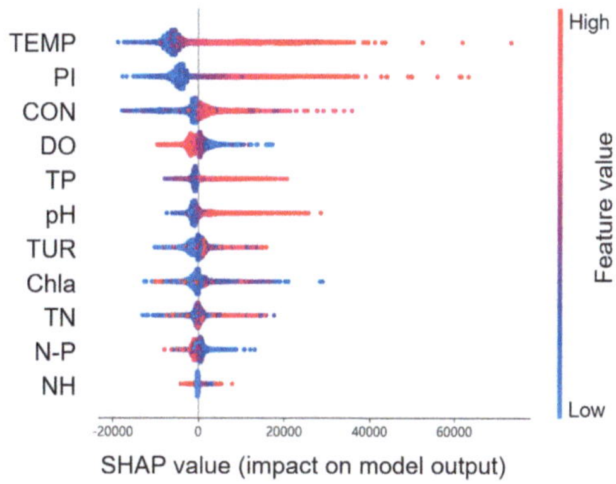

Fig. 7.8 SHAP summary plot. The x-axis of the dual feature dependency plot indicates the value of the main feature; the y-axis shows the SHAP value of this feature; and the color of the point represents the value of the second feature

solve multicollinearity when building the ML model. SHAP and FI suggested a different relative importance of input variables, while both indices showed that TEMP, PI, and CON were the top three important factors. However, there are some differences in the use of VIF, with PI and TEMP ranked as the fifth and ninth most important factors, respectively. NP and NH were the bottom two factors in the ranks of SHAPABS and FI values, indicating that these two factors may not be the dominantly effective factors. The highest VIF was observed for Chla, and the lowest VIF was observed for TN.

The summary scatterplot is shown in Fig. 7.8. The coverage of the points reflects the degree of aggregation of the sample feature value, and the color of the points indicates the feature value itself. It can be inferred that the color transition of TEMP is smooth, which means its influence on algal density is proportional to its value. The larger the TEMP value, the larger its contribution value. Similarly, the values of PI, CON, TP, pH, and TN are proportional to their contribution values, and DO and N-P are inversely proportional to their contribution values.

Figure 7.9 illustrates the impact of various features on algae density using SHAP values. TEMP has a strong influence, with higher values generally having a positive

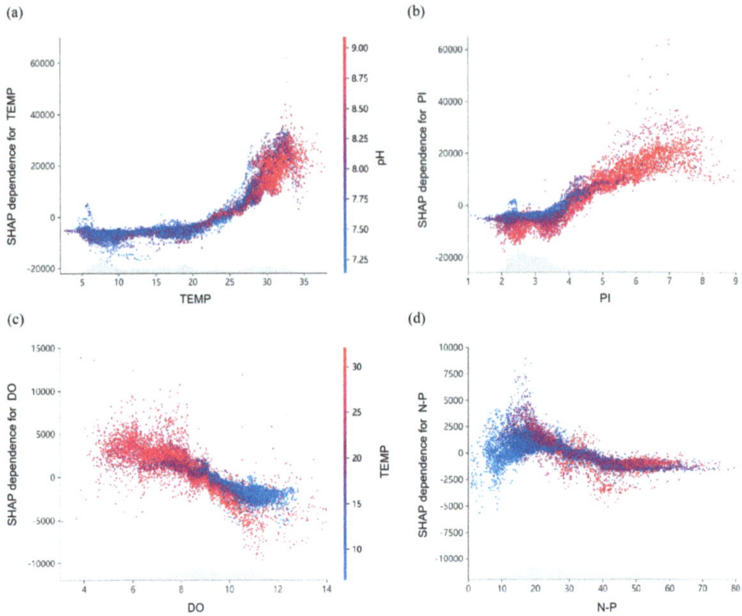

Fig. 7.9 SHAP double feature dependence plots. (**a**) SHAP feature dependence plot (water temperature and pH). (**b**) SHAP feature dependence plot (permanganate index and water temperature). (**c**) SHAP feature dependence plot (dissolved oxygen and water temperature) (**d**) SHAP feature dependence plot (ratio of total nitrogen to total phosphorus and total Nitrogen)

impact and lower values having a negative impact. PI also significantly impacts algae density, with higher values positively correlated, suggesting that an increase in the permanganate index leads to higher predictions. CON shows significant impact as well, with higher values positively influencing algae density and lower values having a negative effect. Both DO and TP have noticeable impacts on algae density. Higher values of DO and TP generally contribute positively, while lower values negatively affect the predictions. pH and TUR also have clear impacts, with higher values positively influencing algae density. Chla and TN significantly affect algae density, with higher values contributing positively. Finally, the N-P ratio and NH have moderate impacts. Higher values of these features are associated with positive effects on algae density, while lower values have a negative influence.

In Fig. 7.9a, when the water temperature is below 22.5°C, most SHAP values are negative, indicating that the water temperature inhibits algae growth. The SHAP values show stability below a water temperature of 20°C. However, when the water temperature rises above 20°C, the SHAP values drastically increase, indicating significant algal growth due to the rise in water temperature. This suggests that the water temperature of 20°C acts as a threshold beyond which the algal growth pattern shifts significantly. When the water temperature is between 22.5°C and 33°C, the SHAP value is positive and increases consistently, indicating that this range is the optimum temperature for algae growth. These results are concordant with published data, yielding mean optimum growth temperatures for cyanobacteria is 27.2°C and for chlorophytes is 26.3°C [158]. When the water temperature exceeds 33°C, the SHAP values show a decreasing trend, indicating that the promotion effect of water temperature on algal growth is gradually weakening. The pH has an apparent interaction with the water temperature range of 25–33°C. When the water temperature in this interval is the same, the smaller the pH value, the more significant the contribution of water temperature to algal growth, indicating that the tepid water in this temperature interval is more suitable for algal growth.

In Fig. 7.9b, most data points show negative SHAP values when the PI concentration is roughly below 4 mg/L, a range of PI concentrations that inhibits algal growth. When the PI concentration exceeds 4 mg/L, the SHAP values become positive, indicating that this PI concentration range has a promotional effect on algal growth. In particular, when PI concentration exceeded 3.5 mg/L, SHAP values show an almost linear increase, indicating a linear growth pattern of algae. Previous work has also shown a similar positive correlation between Chrysophyta and Cryptophyta densities and PI concentrations [159]. The interaction between PI and water temperature occurred mainly in the interval from the PI concentration of 2 to 4.5 mg/L. Specifically, when PI concentrations are the same, low TEMP has higher SHAP values than high TEMP, and it can be inferred that lower water temperature may have a more significant promotion effect on algal growth in this PI concentration range.

In Fig. 7.9c, the value of DO negatively correlates with its SHAP value, suggesting that higher DO concentration is associated with lower contribution to algae growth. This inhibiting effect of algae density on the DO concentration is consistent with a theoretical framework for algal community structure [160]. The dissolved oxygen has an apparent interaction with the water temperature in the range of 9.1–12.9 mg/L. In particular, with the same DO concentration, lower TEMP has higher SHAP values than higher TEMP, implying a more significant promotion of algae growth.

In Fig. 7.9d, when the N-P ratio is less than 29.8, most data points have positive SHAP values; when the N-P ratio is greater than 29.8, the SHAP values are negative. This suggests that algal growth may be accelerated with a low N-P ratio. Once this ratio exceeds 29.8, algae growth will be inhibited. This result is consistent with nutrient effects on the Cyanophyta biomass [161] and the previous experimental observation that blue-green algae tended to be rare when this ratio exceeded 29 [162].

To further investigate the interaction effects between water temperature, water pH, dissolved oxygen, and permanganate index on algal density, we show the two-way PDPs to study the variation of algae density with the changed feature values in Fig. 7.10. The x and y axes represent the values of main effect and interaction effect characteristics. The contours represent the predicted algal density values of the CatBoost (CMA-ES) model for different values of main and interaction effects.

Figure 7.10 illustrates the partial dependence of algae density on TEMP in conjunction with other features: PI, CON, DO, and pH. Each contour plot shows

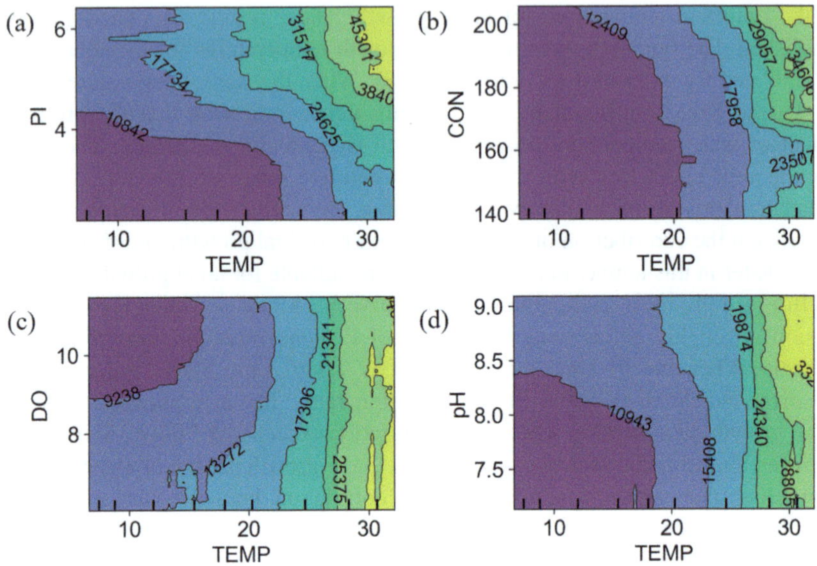

Fig. 7.10 The two-way partial dependence plots of temperature for the CatBoost (CMA-ES) model. The black bars on the axes show the deciles of the feature values in the training data

how algae density varies with changes in TEMP and another feature. Figure 7.10a indicates that algae density increases with higher PI values, especially at higher temperatures. At lower temperatures (below 15°C), algae density remains relatively low regardless of PI. Significant increases in algae density are observed when both TEMP and PI are high, suggesting a synergistic effect of these factors. Figure 7.10b shows the interaction between conductivity and temperature. Algae density increases with higher conductivity values, particularly when TEMP exceeds 22.5°C. At lower temperatures, algae density remains low, even with high conductivity. This suggests that temperature has a dominant influence, with high conductivity enhancing algae density mainly at higher temperatures. Figure 7.10c illustrates the relationship between dissolved oxygen and temperature. Algae density is relatively low at lower temperatures (below 20°C), regardless of DO levels. As TEMP increases beyond 20°C, algae density rises significantly, with higher DO levels contributing to this increase. This indicates that both high temperature and high dissolved oxygen levels are conducive to higher algae density. Figure 7.10d presents the interaction between pH and temperature. Algae density increases with higher pH levels, particularly when TEMP exceeds 20°C. At lower temperatures, algae density remains low across all pH levels, suggesting that temperature is the more critical factor. The highest algae density is observed at high pH values and temperatures above 25°C, indicating that alkaline conditions combined with warm temperatures favor algae growth.

Figure 7.10 demonstrates that while individual features like permanganate index, conductivity, dissolved oxygen, and pH contribute to algae density, their effects are substantially magnified when combined with high temperatures. These insights are crucial for understanding the environmental conditions that promote algae growth and can inform management strategies for aquatic ecosystems.

7.2.4 Conclusion

Reducing algal blooms in water resources is a major issue in China as well as worldwide. Predicting algal densities, quantifying the drivers of algal growth, and revealing the interactions between water parameters are major steps in developing appropriate management strategies for algal blooms. However, the field work involved in lake monitoring is often labor-intensive. In order to maximize the cost-effectiveness of monitoring programs, it is important to determine which properties of the lake have the greatest predictive power for these blooms. In this regard, efforts are needed to embrace ML models and incorporate the most relevant and influential factors into these models to improve the accuracy of predictions and discover potential interactions. This study provides an ML workflow to improve the accuracy and interpretability of ML models for predicting algal concentrations.

Chapter 8
Spatiotemporal Environmental Health Modeling

The application of models in public health has profound implications. By accurately modeling the transmission patterns of infectious diseases, public health officials can better understand disease dynamics, enabling them to devise and implement effective control strategies. This approach not only helps to predict the demand for medical resources (such as beds, medications, and personnel) in advance and optimize resource allocation but also assesses the efficacy of various preventive and control measures, thereby enhancing the efficiency of public health emergency responses. Moreover, model analysis can reveal potential health risks and epidemiological characteristics, promoting personalized medicine and the development of precise public health policies, ultimately improving the overall health and resilience of society.

In the domain of environmental health, models play a crucial role as well. By analyzing environmental data (such as air pollution, water quality, and climate change), models can evaluate the impact of these factors on public health and facilitate the formulation of appropriate health protection measures. Through the modeling of environmental data, public health experts can identify and predict potential health threats, provide early warnings, and ensure timely intervention. This approach not only helps to mitigate the adverse health effects of environmental pollution on residents but also optimizes resource distribution, enhancing the efficiency and effectiveness of public health interventions. Ultimately, these model analysis results provide scientific evidence for policy-makers, supporting the development of more effective and sustainable public health policies.

J. Ren, Y. Tao, *Modeling in Life Sciences and Ecology*, Springer Asia Pacific Mathematics Series 4, https://doi.org/10.1007/978-981-95-1038-2_8

8.1 Mechanism-Based and Machine Learning Models for Epidemic Forecasting

Modeling pandemics presents numerous challenges. One significant difficulty in mathematical modeling is the selection of an appropriate model. The chosen model must be sufficiently simple to allow for the relatively easy prediction or estimation of certain parameters. Another challenge in modeling is the incorporation of preventive measures taken by the population, as these measures can alter the dynamics of disease transmission. Additional challenges include the difficulty to quantify the effects of isolation and social distancing accurately.

Mechanism-based mathematical models play an increasingly crucial role in public health prediction and control by interpreting data from epidemiological studies, forecasting disease trends, guiding intervention strategies, and optimizing resource allocation. These models are constructed by applying universal laws of nature within specific spatiotemporal contexts, employing deductive reasoning to provide meaningful representations of the underlying interactive processes. Compartmental models, such as susceptible-infected-removed (SIR) or susceptible-exposed-infected-removed (SEIR), exemplify this approach, relying on knowledge of physical principles governing infective parameters like incubation, recovery, and mortality [163–165]. The interpretability inherent in these models allows researchers to explore and understand the mathematical relationships between inputs and outputs, thereby enhancing insights into disease dynamics.

Complimentary to mathematical models that seek to explicitly incorporate the biology of the phenomena under investigation are data-driven models that seek to extract hidden patterns within large quantities of data. Without requiring specific knowledge about the underlying biology, machine learning approaches can be applied when only incomplete or limited knowledge is available for a study. They optimize and enhance existing models by employing algorithms such as neural networks, random forests, and support vector machines. These models compare predictions with actual observed data, adjusting parameters to minimize prediction errors [166]. Many efforts have been made to study epidemic forecasting applying either unsupervised or supervised learning on diverse data source [167–169].

This section aims to conduct a comparative analysis of compartmental model and machine learning models based on the criteria of accuracy and practicality and study the relative advantages and disadvantages of the two approaches to predictive modeling.

8.1.1 Data Source and Extraction

In early 2020, countries worldwide increasingly adopted sweeping measures, including full lockdowns, shutting down airports, imposing travel restrictions, and completely sealing their borders to contain COVID-19. All Canadians were urged to cancel or postpone nonessential travel in early March 2020. Once community transmission was confirmed, Canada strictly restricted its border access on March

18, 2020 [170]. Quebec and Ontario, the two most severely affected provinces, implemented lockdowns and set checkpoints at their borders to contain the emerging threat [171]. On June 8, 2020, a limited exemption at the US border was implemented, allowing immediate family members of citizens or permanent residents to return home. During this period, from March 28 to June 8, 2020, the Public Health Agency of Canada (PHAC) reported 96,233 confirmed cases and 1,930,141 tests conducted across the country. This data is utilized in our modeling for this section.

8.1.2 *Mechanism-Based Model*

8.1.2.1 Modeling and Simulation

In this subsection, an SIR model with diffusion terms and nonlocal delays is introduced, using data from the COVID-19 pandemic in Canada to demonstrate the predictive capabilities of a physics-based model. The modeling process includes data processing, parameter estimation and fitting, condition initialization, and model forecasting (Fig. 8.1).

The model considers three populations, susceptible $S(x, t)$, active $I(x, t)$, and removed $R(x, t)$ populations, at location x and time t, within a bounded domain Ω during the strict border shutdown. For simplicity, $\Omega = (0, l\pi)$ is considered for some $l > 0$, with diffusion rates for the three groups denoted as d_S, d_I, and d_R, respectively. In the context of COVID-19, the primary concern is the cumulative active cases, defined as

$$\text{Active cases} = \text{Confirmed cases} - \text{Death cases} - \text{Resolved cases}.$$

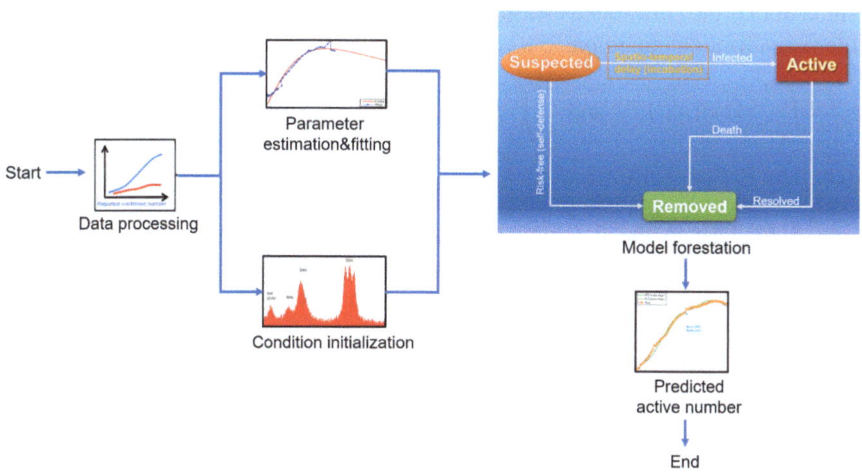

Fig. 8.1 Simplified schematic of the time-order expansion for the physics-based model

The confirmed group comprises patients with an infection resolved at rate $\gamma(I)$ over a specific period, and d represents the coronavirus-induced death rate. A delay of 1 to 2 weeks exists from virus contact to the onset of illness, seeking treatment, and reporting. Population movement, due to activities such as shopping, grabbing takeaway coffee, or walking dogs—even with social distancing measures in place—means that infected individuals could be present anywhere during the incubation period [172]. Therefore, an average population density distribution throughout the region is assumed, a concept first introduced by Gourley [33] and widely used in later studies [34, 50, 173, 174]. The spatial-average delay term is defined by

$$\begin{aligned}(M*u)(x,t) &= \int_{-\infty}^{t}\int_{0}^{l\pi} M(x,y,t-s)u(y,s)\mathrm{d}y\mathrm{d}s,\\ &= \int_{-\infty}^{t}\int_{0}^{l\pi} G(x,y,t-s)\tau(t-s)u(y,s)\mathrm{d}y\mathrm{d}s.\end{aligned} \tag{8.1}$$

This delay term calculates the probability of an individual having been at point y at time s, given that it is at point x at the current time t. The function $\tau : [0,\infty) \to [0,\infty)$ is a probability density function indicating the distribution of the delay, weighting the population t time units in the past. The function $G : \Omega \times \Omega \times (0,\infty) \to (0,\infty)$ is the Green's function of the diffusion equation with homogeneous Neumann boundary conditions and diffusion coefficient $D > 0$. $G_{ij}(x, y, t)$ satisfies

$$\begin{cases} G_t = DG_{yy}, & y \in (0, l\pi),\\ G_y = 0, & y = 0,\ l\pi,\\ G(x,y,0) = \delta(x-y), & x, y \in [0, l\pi]. \end{cases} \tag{8.2}$$

Here, $\delta(x)$ is the Dirac delta measure on Ω. The functions $\tau(t)$ and $G(x, y, t)$ satisfy the normalization conditions

$$\int_0^{\infty} \tau(t)\mathrm{d}t = 1, \quad \text{and} \quad \int_{\Omega} G(x,y,t)\mathrm{d}x = 1, \quad y \in \Omega, \quad t \geq 0. \tag{8.3}$$

The susceptible population $S(x, t)$ is exposed to active individuals at an infection rate $\beta(I)$. Due to the deterrent effect of increasing infections and government interventions, some individuals reduce their outdoor activities and practice self-isolation at a rate $\delta(I)$. The removed population $R(x, t)$ includes individuals who have died from the infection, recovered from the disease, or are in a risk-free environment. The functions $\beta(I)$, $\gamma(I)$, and $\delta(I)$ depend on I and their initial conditions, assumed to have the following properties:

$$\begin{aligned} &\beta(0) = 0, \quad \beta'(I) > 0, \quad \beta(I) \leq 1,\\ &\gamma(0) = 0, \quad \gamma'(I) > 0, \quad \gamma(I) \leq 1,\\ &\delta(0) = \delta_0 > 0, \quad \delta'(I) > 0, \quad \delta(I) \leq 1. \end{aligned}$$

With the border closed, no flux (Neumann) boundary conditions are imposed:

$$\frac{\partial S}{\partial x} = \frac{\partial I}{\partial x} = \frac{\partial R}{\partial x} = 0, \quad t > 0, \quad x = 0, l\pi. \tag{8.4}$$

The system is positive: All state variables take nonnegative values at any time t and position x within the domain. Thus, the initial conditions are chosen to ensure nonnegativity

$$\begin{aligned} S(x,t) &= S_0(x,t) \geq 0, \quad I(x,t) = I_0(x,t) \geq 0, \\ R(x,t) &= R_0(x,t) \geq 0, \quad t \in [-\infty, 0], \quad x \in (0, l\pi). \end{aligned} \tag{8.5}$$

The full model for $t > 0$ and $x \in (0, l\pi)$ is

$$\begin{cases} S_t(x,t) = d_S S_{xx} - S(x,t)(\beta(I(x,t)) + \delta(I(x,t))), \\ I_t(x,t) = d_I I_{xx} + M * \beta(I(x,t))S(x,t) - dI(x,t) - I(x,t)\gamma(I(x,t)), \\ R_t(x,t) = d_R R_{xx} + dI(x,t) + I(x,t)\gamma(I(x,t)) + S(x,t)\delta(I(x,t)), \\ \dfrac{\partial S}{\partial x} = \dfrac{\partial I}{\partial x} = \dfrac{\partial R}{\partial x} = 0, \\ S(x,t) = S_0(x,t) \geq 0, \quad I(x,t) = I_0(x,t) \geq 0, \quad t > 0, \quad x = 0, l\pi, \\ R(x,t) = R_0(x,t) \geq 0, \quad t \in [-\infty, 0], \quad x \in (0, l\pi). \end{cases} \tag{8.6}$$

with the boundary condition (8.4) and the initial condition (8.5). This model is referred to as the R-D model.

The confirmed rate $\beta(I)$ and resolved rate $\gamma(I)$ are estimated by

$$\beta(I) = \frac{\text{Confirmed cases}}{\text{Total tests}}, \quad \gamma(I) = \frac{\text{Resolved cases}}{\text{Confirmed cases}}.$$

MATLAB functions are used for fitting to determine the dynamics-dependent functional form. By fitting with real data, two nonlinear functions for active cases I describe the rates of confirmed and resolved cases

$$\beta(I) = \frac{kI}{bI^2 + c}, \quad \gamma(I) = \frac{I}{I + e}. \tag{8.7}$$

The fitting results are shown in Fig. 8.2. The confirmed rate $\beta(I)$ in stage 1 (from March 28 to May 3, 2020) maintained an increasing trend, decreasing since May 3, indicating that public health strategies, such as physical distancing and self-isolation, have reduced transmission probability. Public health guidelines protect people from infection; thus, the risk-free rate function is chosen as

$$\delta(I) = \frac{f(I + g)}{I + h}. \tag{8.8}$$

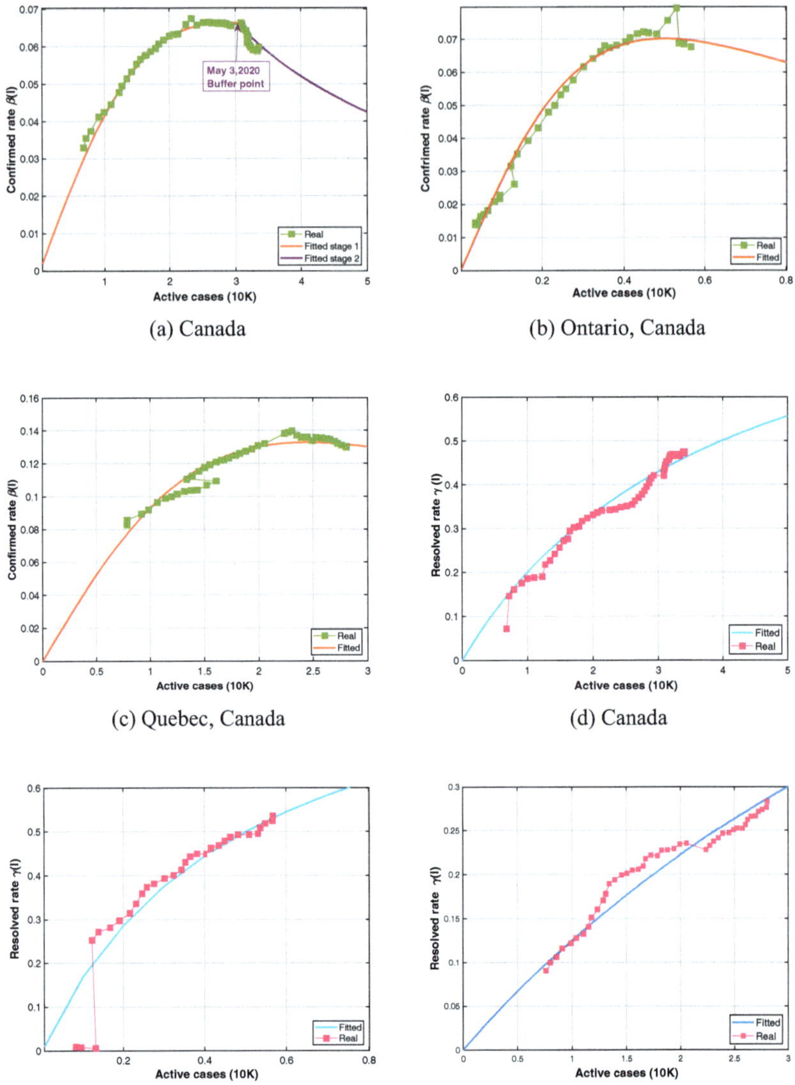

(a) Canada (b) Ontario, Canada

(c) Quebec, Canada (d) Canada

(e) Ontario, Canada (f) Quebec, Canada

Fig. 8.2 The fitting results of confirmed rate and resolved rate for Canada from March 28 to May 22, 2020; Ontario from March 17 to April 24, 2020; and Quebec from April 6 to May 19, 2020. (**a**) Canada. (**b**) Ontario, Canada. (**c**) Quebec, Canada. (**d**) Canada. (**e**) Ontario, Canada. (**f**) Quebec, Canada

This function increases from a small initial value to a constant value with the increasing active cases.

Two random scalars $norm(\mu, \sigma)$ and $randn$, along with three peak-shaped functions $h_1(x)$, $h_2(x)$, $h_3(x)$, describe the nonuniform Canadian population distri-

Fig. 8.3 The initial susceptible population. (**a**) Canada. (**b**) Ontario, Canada. (**c**) Quebec, Canada

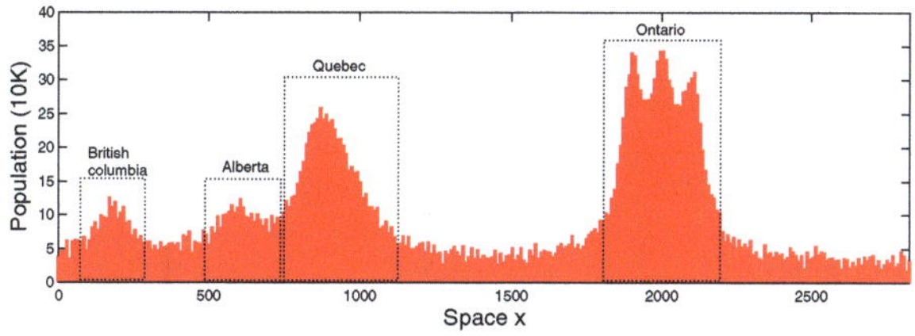

(a) Canada

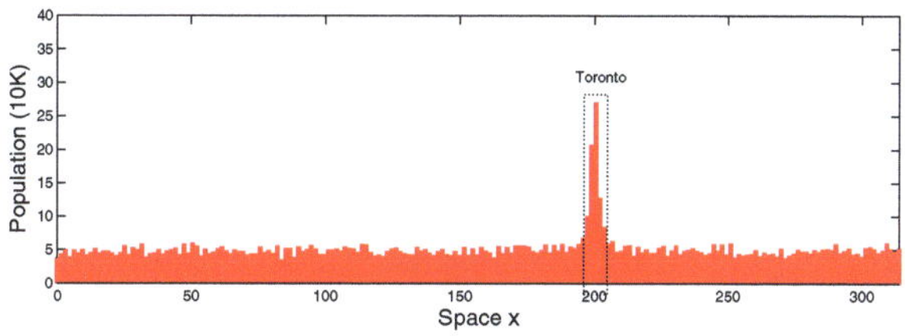

(b) Ontario, Canada

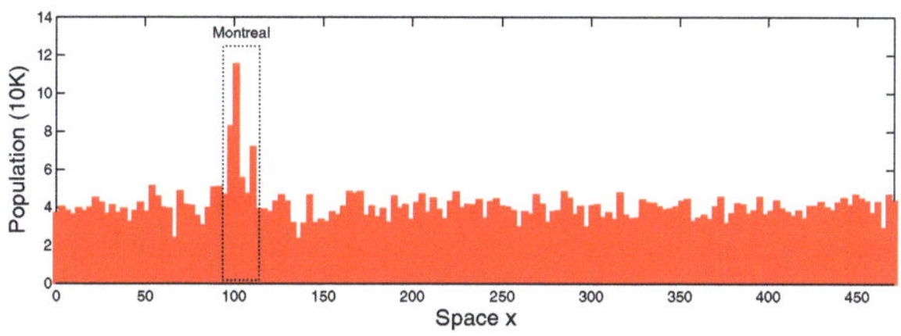

(c) Quebec, Canada

bution. The scalar $randn$ is drawn from the standard normal distribution with mean 0 and standard deviation 1. The scalar $norm$ is drawn from the normal distribution with mean parameter μ and standard deviation parameter σ. Four different peak susceptible populations in Canada are considered: Ontario, Quebec, Alberta, and British Columbia. For Ontario and Quebec, Toronto and Montreal are chosen as the peaks in the initial susceptible population (Fig. 8.3).

Further, the summary of the case number at all positions x in the region is calculated to show each group's dynamics in time order. The simulated total active cases and fitting results for Canada from March 28, 2020 to June 8; Ontario from March 18, 2020 to May 18; and Quebec from April 18, 2020 to June 8 are shown in time order in Fig. 8.4. The beginning time is chosen based on the reported data and time of closure of the border. The end times for Canada and Quebec are determined by the amendment on June 9 [170] and for Ontario by the reopening date on May 19 [175]. Additionally, there is a 1- to 2-week delay from onset to reporting to PHAC. The active cases with the reporting delay w_r are also presented by gray lines in Fig. 8.4.

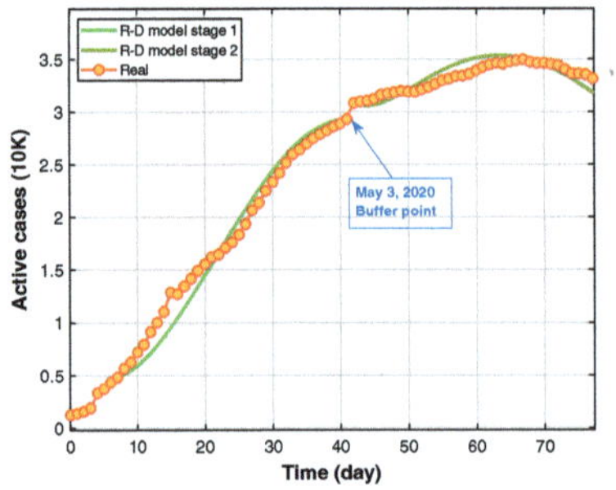

(a) Canada

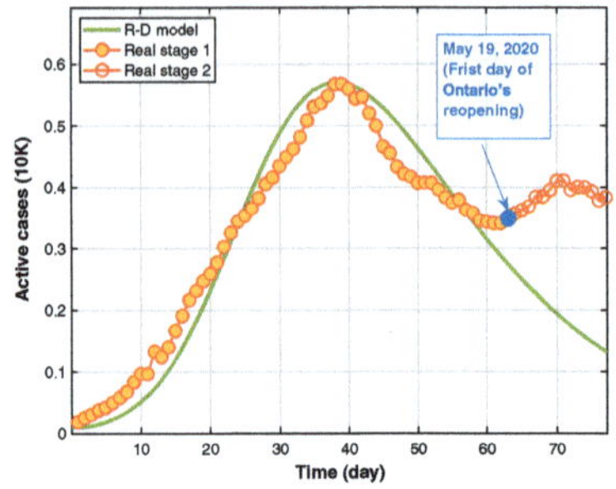

(b) Ontario, Canada

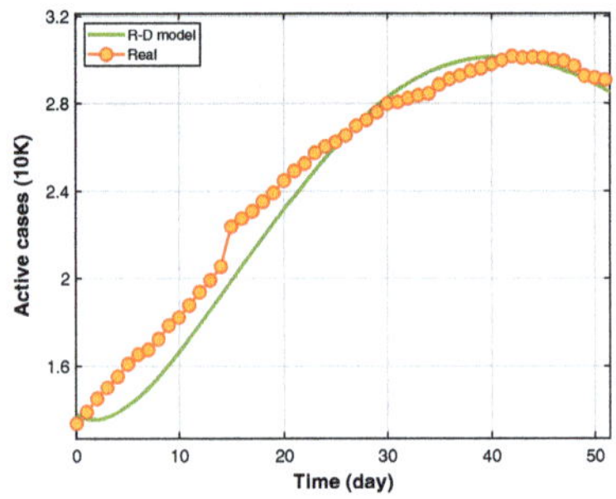

(c) Quebec, Canada

Fig. 8.4 Real and simulated active COVID-19 cases using R-D model. Simulated results (**a**) for Canada from March 28 to June 8, 2020; (**b**) for Ontario March 28 to May 19, 2020; (**c**) for Quebec from April 18 to June 8, 2020. (**a**) Canada. (**b**) Ontario, Canada. (**c**) Quebec, Canada

In Fig. 8.4a, a buffer point on May 3, 2020, with 30,873 active cases in Canadian cases is located. The inflection points of Ontario, Quebec, and Canada cases on April 24, 2020, May 31, 2020, and June 1, 2020, are estimated with peak values of 5,669, 30,100, and 35,001 cases, respectively. It is worth noting that in Fig. 8.4b, during the real stage 2, the growth of active cases was resurgent, since the gradually reopening plan was implemented in Ontario on May 19, 2020. The parameter values are shown in Tables 8.1 and 8.2. Furthermore, comparisons of the active numbers obtained through simulations using constant and data-dependent infection/resolved rates for Quebec are provided to demonstrate the improvement effect of data-dependent rates on the original model. The range of infection and recovery rate variations in Fig. 8.2c and f guided our selection of several constant infection and recovery rates. The results of these comparisons are presented in Fig. 8.5.

Table 8.1 Parameters values and initial conditions of R-D model for Canada

Parameter	Canada, stage 1	Canada, stage 2
k	0.38	0.21
b	1	1
c	8.3	1
d	0.075	0.08
e	4	3.5
f	0.05299	0.22
g	0.016	0.001
h	0.006	0.02
τ	8	24
τ_r	7	7
d_S	0.1	0.1
d_I	0.01	0.01
d_R	0.01	0.01
d_Q	0.1	0.1
S_0	$2.8 + 0.1\text{norm}(3, 8) + h_1(x)$	$0.78 + 0.1\text{norm}(0, 2) + 0.05h_1(x)$
I_0	$0.00165 + 0.0001\text{randn}$	$0.0095 + 0.0001\text{randn} + 0.00012h_1(x)$
R_0	$3.0 + 0.1\text{randn}$	$11.16 + 0.1\text{randn}$
Q_0	$0.0012 + 0.0001\text{randn}$	$0.018 + 0.0001\text{randn}$

Note: $h_1(x) = \frac{30500}{(x-190)^2+21x} + \frac{37500}{(x-590)^2+10x} + \frac{37500}{(x-990)^2+5} + \frac{47500}{(x-900)^2+5x} + \frac{47500}{(x-850)^2+5x} + \frac{47500}{(x-1900)^2+x} + \frac{47500}{(x-2000)^2+x} + \frac{47500}{(x-2100)^2+x}$.

8.1.2.2 Dynamics of R-D Model

In the following, we study the stability of model (8.6) to gain insight into disease dynamics. There is a conservation law due to the closure of model (8.6), which is obtained by adding the equations in (8.6) and integrate x on $(0, l\pi)$ to yield

$$\begin{aligned}&\int_0^{l\pi}(S_t(x,t)+I_t(x,t)+R_t(x,t))dx\\&=\frac{d}{dt}(\int_0^{l\pi}\int_0^{\infty}\int_{t-\eta}^{t}M(x,y,\eta)\beta(I(x,v))S(x,v)\mathrm{d}v\mathrm{d}\eta\mathrm{d}x).\end{aligned} \tag{8.9}$$

Then we obtain a value of the total population through

$$\begin{aligned}N_T=&\int_0^{l\pi}S(x,t)+I(x,t)+R(x,t)dx\\&+\int_0^{l\pi}\int_0^{\infty}\int_{t-\eta}^{t}M(x,y,\eta)\beta(I(x,v))S(x,v)\mathrm{d}v\mathrm{d}\eta\mathrm{d}x,\end{aligned} \tag{8.10}$$

Table 8.2 Parameters values and initial conditions of R-D model for Ontario and Quebec

Parameter	Ontario, Canada	Quebec, Canada
c	0.05	6
d	0.08	0.086
e	0.5	7
f	0.35	0.146
g	0.0026	0.024
h	0.008	0.099
τ	15	26
τ_r	10	10
d_S	0.1	0.1
d_I	0.01	0.01
d_R	0.01	0.01
d_Q	0.1	0.1
S_0	$4.7 + 0.1\text{norm}(0, 5) + h_2(x)$	$3.7 + 0.1\text{norm}(2, 5) + h_3(x)$
I_0	$0.0006 + 0.0001\text{randn}$	$0.0092 + 0.00001\text{randn}$
R_0	$2.2 + 0.1\text{randn}$	$1.55 + 0.1\text{randn}$
Q_0	$0.00001 + 0.000001\text{randn}$	$0.02 + 0.001\text{randn}$

Note: $h_2(x) = \frac{50}{(x-200)^2+2)}$, $h_3(x) = \frac{25}{(x-100)^2+2} + \frac{5}{(x-110)^2+2)}$

and this value will remain constant in time. Then, if a solution goes to an equilibrium (S^*, I^*, R^*), which should satisfy the conservation law as

$$N_T = \frac{\tilde{N_T}}{l\pi} = \beta(I^*)S^*T_f + S^* + I^* + R^*.$$

Here, $T_f = \int_0^\infty \eta\tau(\eta)d\eta$ is the mean delay. Equations (8.3) imply that the delay term (8.1) have no effect on the spatially uniform steady-state solutions. Setting $S = I = 0$, then from the conservation law, we get that $R = N_T$. This steady-state solution will be referred to as $(0, 0, N_T)$. Biologically, this represents the states where there is no susceptible and infected group in the ecosystem.

For the trivial solution $(0, 0, N_T)$, the Jacobi matrix is

$$J_1 = \begin{bmatrix} -\delta(0) - d_1\frac{n^2}{l^2} & 0 & 0 \\ 0 & -d - d_2\frac{n^2}{l^2} & 0 \\ \delta(0) & d & -d_3\frac{n^2}{l^2} \end{bmatrix}. \tag{8.11}$$

The characteristic equation around $(0, 0, N_T)$ is then

$$\Phi(\phi, \frac{n^2}{l^2}) = -(d_3\frac{n^2}{l^2} + \phi)(d_2\frac{n^2}{l^2} + d + \phi)(d_1\frac{n^2}{l^2} + \delta(0) + \phi), \tag{8.12}$$

Fig. 8.5 Real and simulated active cases using R-D model for Quebec from April 18 to June 8, 2020, with constant and data-dependent infection/resolved rate

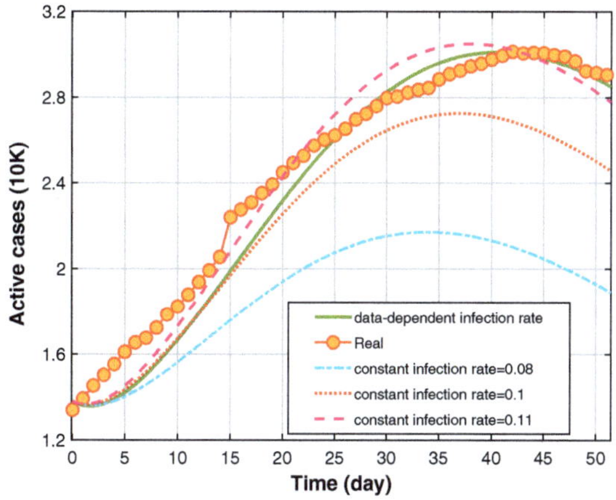

(a)

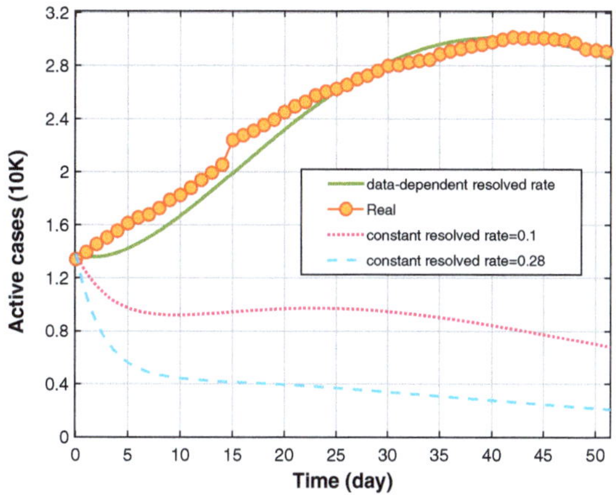

(b)

with

$$\phi_1 = -d_3\frac{n^2}{l^2} \le 0, \phi_2 = -(d_2\frac{n^2}{l^2} + d) \le 0, \phi_3 = -(d_1\frac{n^2}{l^2} + \delta(0)) \le 0.$$

Hence, the uniform equilibrium $(0, 0, N_T)$ of model (8.6) is locally asymptotically stable.

We investigate the global stability of the uniform equilibrium of model (8.6). Specifically, we consider the case where the delay function is given by

$$\tau(t) = \frac{1}{\tau}e^{-\frac{t}{\tau}}.$$

This choice of delay kernel allows us to reformulate model (8.6) as a reaction-diffusion model devoid of explicit delay terms, where the delay parameter manifests as a coefficient in the reformulated model.

Define

$$Q(x,t) = \int_{-\infty}^{t} \int_{0}^{l\pi} G(x,y,-s) e^{-\frac{t-s}{\tau}} \beta(I(x,s)) S(x,s)\, dy\, ds.$$

Model (8.6) is equivalent to the following system:

$$\begin{cases} S_t(x,t) = d_S S_{xx} - S(x,t)[\beta(I(x,t)) + \delta(I(x,t))], \\ I_t(x,t) = d_I I_{xx} + \dfrac{1}{\tau} Q(x,t) - \gamma(I(x,t)) I(x,t) - d_I I(x,t), \\ R_t(x,t) = d_R R_{xx} + \gamma(I(x,t)) I(x,t) + d_I I(x,t) + \delta I(x,t) S(x,t), \\ Q_t(x,t) = d_Q Q_{xx} - \dfrac{1}{\tau} Q(x,t) + S(x,t) \beta(I(x,t)). \end{cases} \tag{8.13}$$

Adding the equations in (8.13) and integrating over x from 0 to $l\pi$ yields

$$\begin{aligned} &\int_0^{l\pi} (S_t(x,t) + I_t(x,t) + R_t(x,t) + Q_t(x,t))\, dx \\ &= \frac{\partial}{\partial t} \int_0^{l\pi} (S(x,t) + I(x,t) + R(x,t) + Q(x,t))\, dx = 0, \end{aligned}$$

which implies the conservation law

$$\int_0^{l\pi} (S_t(x,t) + I_t(x,t) + R_t(x,t) + Q_t(x,t))\, dx = \tilde{N}_T. \tag{8.14}$$

Thus, the conservation law (8.14) can be expressed as

$$N_T' = \frac{\tilde{N}_T}{l\pi} = S^* + I^* + R^* + Q^*. \tag{8.15}$$

Following the conservation law (8.15), the equilibrium state is characterized by $(0, 0, N_T', 0)$. To demonstrate the existence of a classical solution to model (8.13), we employ the method of upper and lower solutions for a mixed quasi-monotone system.

Theorem 8.1 *Model (8.13) has a unique solution* $(S(x,t), I(x,t), R(x,t), Q(x,t))$ *such that* $S(x,t) > 0$*,* $I(x,t) > 0$*,* $R(x,t) > 0$*,* $Q(x,t) > 0$ *for* $t \in (0,\infty)$ *and* $x \in [0, l\pi]$.

Proof Define

$$g_1(x,t) = -S(\beta(I)+\delta(I)), \quad g_2(x,t) = \frac{1}{\tau}Q - \gamma(I)I - dI,$$

$$g_3(x,t) = \gamma(I)I + dI + \delta(I)S, \quad g_4(x,t) = -\frac{1}{\tau}Q + \beta(I)S.$$

Since in $\overline{R_+^4} = \{S \geq 0, I \geq 0, R \geq 0, Q \geq 0\}$,

$$\frac{\partial g_1}{\partial S} = -\,(\beta(I)+\delta(I)) < 0, \frac{\partial g_1}{\partial I} = -S(\beta'(I)+\delta'(I)) < 0, \frac{\partial g_1}{\partial R} = \frac{\partial g_1}{\partial Q} = 0,$$

$$\frac{\partial g_2}{\partial I} = -\,(\gamma(I)+d) < 0, \ \frac{\partial g_2}{\partial Q} = \frac{1}{\tau} > 0, \frac{\partial g_2}{\partial S} = \frac{\partial g_2}{\partial R} = 0,$$

$$\frac{\partial g_3}{\partial S} = \delta(I) > 0, \frac{\partial g_3}{\partial I} = \gamma(I) + d + \delta'(I)S > 0, \frac{\partial g_3}{\partial Q} = \frac{\partial g_3}{\partial R} = 0,$$

$$\frac{\partial g_4}{\partial S} = \beta(I) > 0, \frac{\partial g_4}{\partial Q} = -\frac{1}{\tau} < 0, \ \frac{\partial g_4}{\partial I} = \beta(I)'S > 0, \ \frac{\partial g_4}{\partial R} = 0,$$

then according to Definition 2.1 in [176], (8.13) is a mixed quasi-monotone system.

Let

$$(\hat{S}(x,t), \hat{I}(x,t), \hat{R}(x,t), \hat{Q}(x,t)) = (0,0,0,0)$$

and

$$(\tilde{S}(x,t), \tilde{I}(x,t), \tilde{R}(x,t), \tilde{Q}(x,t)) = (\tilde{S}^*(t), \tilde{I}^*(t), \tilde{R}^*(t), \tilde{Q}^*(t)),$$

where $(\tilde{S}^*(t), \tilde{I}^*(t), \tilde{R}^*(t), \tilde{Q}^*(t))$ is the unique solution of

$$\begin{cases} \dfrac{dS}{dt} = -S(\beta(I)+\delta(I)), \\ \dfrac{dI}{dt} = \dfrac{1}{\tau}Q, \\ \dfrac{dR}{dt} = \gamma(I)I + dI + \delta(I)S, \\ \dfrac{dQ}{dt} = \beta(I)S, \\ S(0) = \sup\limits_{[0,l\pi]} S_0(x), I(0) = \sup\limits_{[0,l\pi]} I_0(x), \\ R(0) = \sup\limits_{[0,l\pi]} R_0(x), Q(0) = \sup\limits_{[0,l\pi]} Q_0(x). \end{cases}$$

Then $(\hat{N}(x,t), \hat{P}(x,t), \hat{Z}(x,t), \hat{Q}(x,t))$ and $(\tilde{N}(x,t), \tilde{P}(x,t), \tilde{Z}(x,t), \tilde{Q}(x,t))$ are the lower-solution and upper-solution to (8.13), since

$$\begin{cases} \dfrac{\partial \tilde{S}(x,t)}{\partial t} - \triangle \tilde{S}(x,t) - g_1(\tilde{S}(x,t), \hat{I}(x,t), \hat{R}(x,t), \hat{Q}(x,t)) = 0 \\ \qquad = \dfrac{\partial \hat{S}(x,t)}{\partial t} - \triangle \hat{S}(x,t) - g_1(\hat{S}(x,t), \tilde{I}(x,t), \tilde{R}(x,t), \tilde{Q}(x,t)), \\ \dfrac{\partial \tilde{I}(x,t)}{\partial t} - \triangle \tilde{I}(x,t) - g_2(\tilde{I}(x,t), \hat{S}(x,t), \hat{R}(x,t), \hat{Q}(x,t)) = (\gamma(\tilde{I}) + d)\tilde{I} > 0 \\ \qquad = \dfrac{\partial \hat{I}(x,t)}{\partial t} - \triangle \hat{I}(x,t) - g_2(\hat{I}(x,t), \tilde{S}(x,t), \tilde{R}(x,t), \hat{Q}(x,t)), \\ \dfrac{\partial \tilde{R}(x,t)}{\partial t} - \triangle \tilde{R}(x,t) - g_3(\tilde{R}(x,t), \tilde{S}(x,t), \tilde{I}(x,t), \tilde{Q}(x,t)) = 0 \\ \qquad = \dfrac{\partial \hat{R}(x,t)}{\partial t} - \triangle \hat{R}(x,t) - g_3(\hat{R}(x,t), \hat{S}(x,t), \hat{I}(x,t), \hat{Q}(x,t)), \\ \dfrac{\partial \tilde{Q}(x,t)}{\partial t} - \triangle \tilde{Q}(x,t) - g_4(\tilde{Q}(x,t), \tilde{S}(x,t), \tilde{I}(x,t), \tilde{R}(x,t)) = \dfrac{1}{\tau}\tilde{Q} > 0 \\ \qquad = \dfrac{\partial \hat{Q}(x,t)}{\partial t} - \triangle \hat{Q}(x,t) - g_4(\hat{Q}(x,t), \hat{S}(x,t), \hat{I}(x,t), \hat{Q}(x,t)). \end{cases}$$

Also, the boundary conditions and the initial conditions are satisfied. Besides, for any (S_a, I_a, R_a, Q_a) and (S_b, I_b, R_b, Q_b) in X_T, the Lipschitz condition is clear, since

$$\begin{aligned} &|g_i(N_a, P_a, Z_a, Q_a) - g_i(N_b, P_b, Z_b, Q_b)| \\ &\quad \le K_i(|N_a - N_b| + |P_a - P_b| + |Z_a - Z_b| + |Q_a - Q_b|), \end{aligned}$$

where $K_i = K_i(x,t)$, $i = 1,2,3,4$. Therefore, Lemma 2.1 in [176] shows that (8.13) has a unique globally defined solution $(S(x,t), I(x,t), R(x,t), Q(x,t))$, which satisfies for $t \ge 0$,

$$\begin{aligned} 0 \le & S(x,t) \le \tilde{S}^*(t), 0 \le I(x,t) \le \tilde{I}^*(t), \\ 0 \le & R(x,t) \le \tilde{R}^*(t), 0 \le Q(x,t) \le \tilde{Q}^*(t). \end{aligned}$$

The strong maximum principle implies that $S(x,t), Q(x,t), I(x,t), R(x,t) > 0$ when $t > 0$ for all $x \in [0, l\pi]$. Immediately, from the conservation law (8.14), $S(t,x), I(t,x), R(t,x), Q(x,t)$ are bounded by the $\tilde{N_T'}$.

We now establish the global stability of the trivial equilibrium $(0, 0, N_T', 0)$.

Theorem 8.2 *The uniform equilibrium* $(0, 0, N_T', 0)$ *of model* (8.13) *is globally stable.*

Proof Considering the initial parabolic system (8.13), we observe that $S(x,t)$ satisfies

$$\begin{cases} S_t(x,t) \le d_S S_{xx} - \delta(I)S \le d_S S_{xx} - \delta_0 S, & t > 0, x \in (0, l\pi), \\ \dfrac{\partial S}{\partial x} = 0, & t > 0, x = 0, l\pi, \end{cases} \tag{8.16}$$

then from comparison principle of the parabolic equations, it is easy to verify that $S(x,t) < \check{S}(t)$. Here, $\check{S}(t)$ is the (spatial homogeneous) solutions of

$$\begin{cases} S_t(x,t) = d_S S_{xx} - \delta_0 S, & t > 0, x \in (0, l\pi), \\ \dfrac{\partial S}{\partial x} = 0, & t > 0, x = 0, l\pi, \\ S(0) = \sup S_0(x) & x \in [0, l\pi], \end{cases} \tag{8.17}$$

where $\check{S}(t) = S(0)e^{-\delta_0 t} > 0$. From Theorem 8.1, $S(x,t) > 0$ then we have

$$\lim_{t\to\infty, x\in[0,l\pi]} S(x,t) = 0.$$

It follows that there is a function, $\omega_1(t,x) = \beta(I)S$, such that

$$\lim_{t\to\infty, x\in[0,l\pi]} \omega_1(x,t) = 0,$$

and

$$\begin{cases} Q_t(x,t) < d_Q Q_{xx} + \omega_1 - \dfrac{Q}{\tau}, & t > 0, x \in (0, l\pi), \\ \dfrac{\partial Q}{\partial x} = 0, & t > 0, x = 0, l\pi, \end{cases} \tag{8.18}$$

then from the comparison principle, $Q(x,t) < \check{Q}(t)$. Here, $\check{Q}(t)$ is the solutions of

$$\begin{cases} Q_t(x,t) = d_Q Q_{xx} + \omega_1 - \dfrac{Q}{\tau}, & t > 0, x \in (0, l\pi), \\ \dfrac{\partial Z}{\partial x} = 0, & t > 0, x = 0, l\pi, \\ Q(0) = \sup Q_0(x) & x \in [0, l\pi], \end{cases} \tag{8.19}$$

where

$$\check{Q}(t) = e^{-\frac{1}{\tau}t} Q(0) + e^{-\frac{1}{\tau}t} \int_0^t e^{\frac{1}{\tau}u} \omega_1(u) du.$$

In the case where $\int_0^t e^{\frac{1}{\tau}u}\omega_1(u)du$ is obvious approaches zero as $t \to \infty$. Since $Q(x,t) > 0$, then we have

$$\lim_{t\to\infty, x\in[0,l\pi]} Q(x,t) = 0.$$

Similarly, there exists a function $\omega_2(x,t) = \dfrac{Q}{\tau}$, such that

$$\lim_{t\to\infty, x\in[0,l\pi]} \omega_2(x,t) = 0,$$

and

$$\begin{cases} I_t(x,t) < d_I I_{xx} + \omega_2 - dI, & t > 0, x \in (0, l\pi), \\ \dfrac{\partial I}{\partial x} = 0, & t > 0, x = 0, l\pi, \end{cases} \tag{8.20}$$

then from the comparison principle, $I(x,t) < \check{I}(t)$. Here, $\check{I}(t)$ is the (spatial homogeneous) solutions of

$$\begin{cases} I_t(x,t) = d_I I_{xx} + \omega_2 - dI, & t > 0, x \in (0, l\pi), \\ \dfrac{\partial I}{\partial x} = 0, & t > 0, x = 0, l\pi, \\ I(0) = \sup I_0(x) & x \in [0, l\pi], \end{cases} \tag{8.21}$$

where

$$\check{I}(t) = e^{-dt} I(0) + e^{-dt} \int_0^t e^{du}\omega_2(u)du.$$

As $t \to \infty$, we have

$$\lim_{t\to\infty, x\in[0,l\pi]} I(x,t) = 0.$$

Then from the conservation law (8.15), we have

$$\lim_{t\to\infty, x\in[0,l\pi]} R(x,t) = N_T'.$$

Therefore, $(0, 0, N_T', 0)$ is globally stable.

8.1.2.3 Shielding Mechanism and Its Synergistic Effect

In the following, we propose two shielding mechanisms based on a physics-based model. The concept of shielding mechanisms, initially introduced in [177], aims to increase interactions with recovered or risk-free individuals relative to those with individuals of unknown status. Our new shielding mechanisms differ from those in [177] by additionally considering individuals who have died from COVID-19. In this framework, we assume that recovered and risk-free individuals can safely interact with both susceptible and infectious individuals, effectively substituting interactions with susceptible and infectious individuals for interactions with removed individuals.

Let C_S, C_I, and C_R denote the contact rates of susceptible, active infectious, and removed (recovered and deceased) individuals, respectively. N_{cyc} represents the total population in the circulating baseline, defined as $N_{cyc} = S + I + R - D$, where D is the number of deaths caused by the disease. The force of infection is expected to be proportional to

$$C_S \frac{C_I}{C_S + C_I + C_R(R - D)},$$

where $C_0 = C_S + C_I + C_R(R - D)$ represents the weighted average of contacts. Particularly, in the absence of preferential interaction rates, this force simplifies to

$$C_S \frac{I}{N_{cyc}}.$$

Core Shielding Mechanism Under the framework of core shielding immunity, we introduce a relative preference $1 + \frac{aR}{R-D}$ indicating the likelihood that an individual will interact with a recovered person in what would otherwise be a potentially infectious encounter. Assume $C_S = C_I = C_0$ and $C_R = C_0\left(1 + \frac{aR}{R-D}\right)$. Consequently, the current infection force is proportional to $C_S \frac{I}{aR+N_{\text{cyc}}}$.

The effective infection force is attenuated by a factor of $\frac{N_{\text{cyc}}}{N_{\text{cyc}}+aR}$, where a represents the degree of shielding. For $a = 0$, this reduces to the conventional rate $\beta(I(x,t))$. Thus, the modified model incorporating core shielding is defined as follows:

$$\begin{cases} S_t(x,t) = d_S S_{xx} - \left(\dfrac{N_{cyc}\beta(I(x,t))}{N_{cyc} + aR} + \delta(I(x,t))\right) S(x,t), \\ I_t(x,t) = d_I I_{xx} + M * \dfrac{N_{cyc}\beta(I(x,t))}{N_{cyc} + aR} S(x,t) - I(x,t)\left(d + \gamma(I(x,t))\right), \\ R_t(x,t) = d_R R_{xx} + I(x,t)(d + \gamma(I(x,t))) + S(x,t)\delta(I(x,t)), \end{cases} \tag{8.22}$$

with the boundary condition (8.4) and the initial condition (8.5).

Flexible Shielding Mechanism In a flexible shielding mechanism, recovered and risk-free individuals are afforded a relative preference of $1 + \frac{aR}{R-D}$ in their contact rates compared to susceptible or active individuals.

Let C_B denote the baseline contact rate for susceptible and infectious individuals. Define $C_B = C_S = C_I$ and $C_R = C_B\left(1 + \frac{aR}{R-D}\right)$. The weighted average contact rate is then calculated as $C_0 = C_B(S + I) + C_B\left(1 + \frac{aR}{R-D}\right)(R - D)$. This formulation implies that $C_B = \frac{C_0}{N_{\text{cyc}}+aR}$. Consequently, the current infection force is proportional to $C_S\frac{N_{\text{cyc}}I}{(N_{\text{cyc}}+aR)^2}$. The effective infection force is reduced by a factor of $\frac{N_{\text{cyc}}^2}{(N_{\text{cyc}}+aR)^2}$. Therefore, the resulting model incorporating flexible shielding is defined as follows:

$$
\begin{cases}
S_t(x,t) = d_S S_{xx} - S(x,t)\left(\frac{N_{cyc}^2\beta(I(x,t))}{(N_{cyc}+aR)^2} + \delta(I(x,t))\right), \\
I_t(x,t) = d_I I_{xx} + M * \frac{N_{cyc}^2\beta(I(x,t))}{(N_{cyc}+aR)^2}S(x,t) - I(x,t)(d + \gamma(I(x,t))), \\
R_t(x,t) = d_R R_{xx} + I(x,t)(d + \gamma(I(x,t))) + S(x,t)\delta(I(x,t)),
\end{cases}
\tag{8.23}
$$

with the boundary condition (8.4) and the initial condition (8.5).

We use the Quebec scenario described above as our baseline. Within the core shielding mechanism, when $a = 1$, the peak number of active cases reaches 0.22% infected (18,730 cases out of a total population of 8,500,000), compared to approximately 0.35% (30,500 cases out of 8,500,000) in the baseline scenario without shielding. When $a = 2$, this peak value decreases to 0.17% (15,220 cases out of 8,500,000). Notably, when $a = 10$, there is no discernible peak value; active cases have continuously decreased, since the shielding began (Fig. 8.6).

Furthermore, Figs. 8.7 and 8.8 demonstrate that flexible shielding is robust in various scenarios and potentially more effective. Active cases are significantly suppressed once the shielding strategy is implemented (Fig. 8.6b). Figure 8.8a compares the time required to keep existing cases below 200 under different

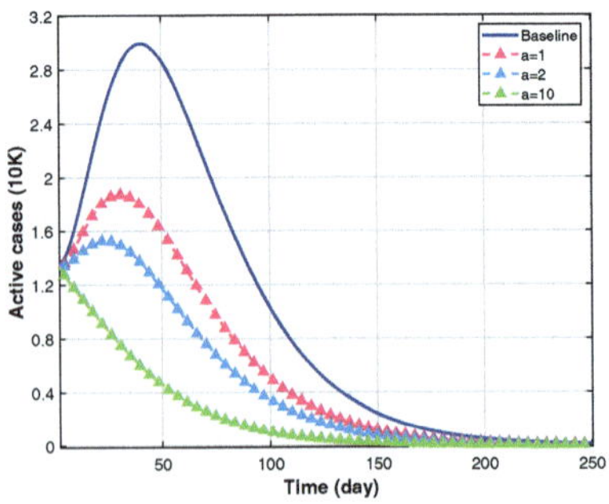

Fig. 8.6 Actives case under different levels of core shielding

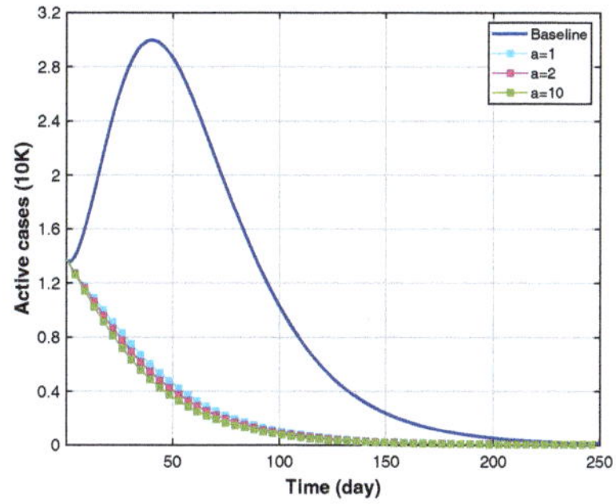

Fig. 8.7 Actives case under different levels of flexible shielding

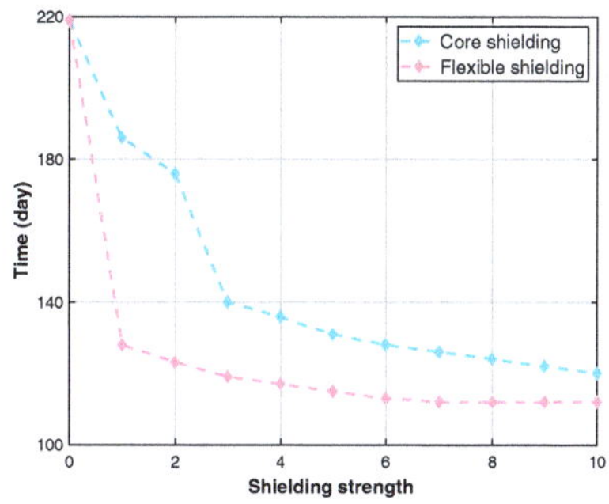

(a) Required time for less than 200 cases

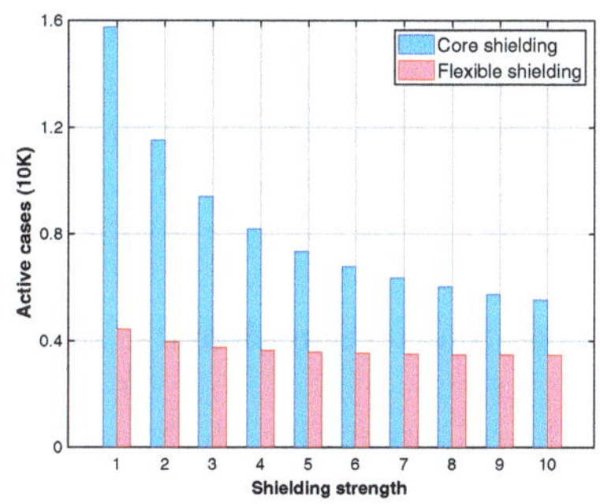

(b) Active cases on the last day

Fig. 8.8 Shield dynamics with different levels of shielding. (**a**) Required time for less than 200 cases. (**b**) Active cases on the last day

shielding degrees. Figure 8.8b compares the number of active cases under two shielding mechanisms when limited exemptions at the border were implemented on June 8, 2020. In the case where $a = 1$, the epidemic concludes with 15,740 infected individuals under the core shielding strategy, compared to 4,427 under the flexible scenario, representing reductions in transmission of approximately 45.85% and 84.77%, respectively, compared to scenarios without shield immunity.

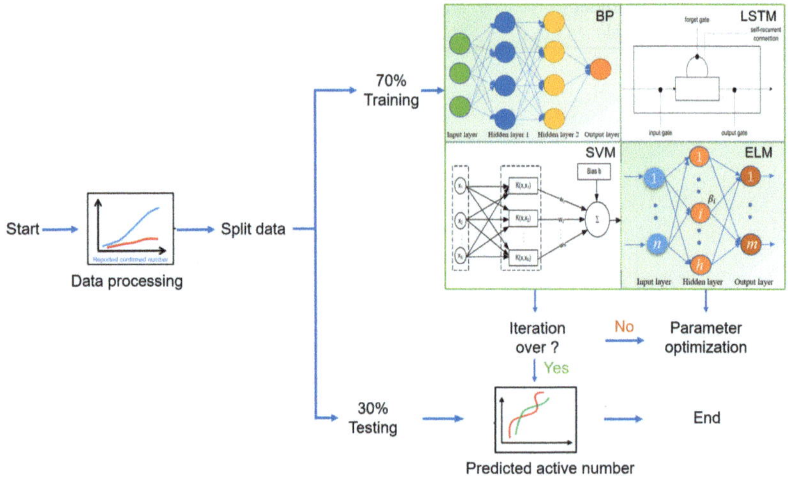

Fig. 8.9 Simplified schematic of the time-order expansion for BB models

8.1.3 Machine Learning Model

This subsection introduces methods for predicting the number of active cases using machine learning (ML) models. The workflow diagram is illustrated in Fig. 8.9, encompassing backpropagation (BP), long short-term memory (LSTM), support vector machine (SVM), and extreme learning machine (ELM). Earlier chapters have already covered BP and SVM; hence, we now focus on discussing the primary workflows of ELM and LSTM.

Long Short-Term Memory

When discussing the LSTM network, it is essential to delve into its key components and mathematical underpinnings systematically. LSTM is a variant of recurrent neural networks (RNNs) designed specifically to handle long sequence data by effectively capturing and utilizing long-term dependencies within sequences [178].

At the core of the LSTM architecture lies its intricate memory cell structure, comprising several crucial components:

1. Cell state (C_t): The cell state C_t carries information across different time steps t and is updated using the following formula:

$$C_t = f_t \odot C_{t-1} + i_t \odot \tilde{C}_t,$$

where C_{t-1} denotes the previous cell state, f_t represents the forget gate output, controlling what information to discard from the cell state, i_t denotes the input gate output, determining how new information should be integrated into the cell state, and $\tilde{C}_t$ is the candidate cell state for the current time step.

2. Forget gate: The forget gate output f_t, controlled by a sigmoid function, is defined as

$$f_t = \sigma(W_f \cdot [h_{t-1}, x_t] + b_f),$$

where W_f and b_f are the weights and biases of the forget gate, h_{t-1} is the previous hidden state, and x_t is the current input.

3. Input gate: The input gate, composed of sigmoid and tanh functions, determines the amount and content of new information $\tilde{C}_t$ to be updated:

$$i_t = \sigma(W_i \cdot [h_{t-1}, x_t] + b_i),$$

and

$$\tilde{C}_t = \tanh(W_C \cdot [h_{t-1}, x_t] + b_C),$$

where W_i, W_C, and respective biases b_i, b_C are the weights and biases of the input gate.

4. Output gate: The output gate controls the extraction of information from the cell state C_t to generate the current hidden state h_t:

$$o_t = \sigma(W_o \cdot [h_{t-1}, x_t] + b_o),$$

and

$$h_t = o_t \odot \tanh(C_t),$$

where W_o and b_o are the weights and biases of the output gate and o_t is the output of the output gate.

Through these components, LSTM networks effectively manage long-term dependencies in sequence data, making them suitable for tasks such as time series prediction and natural language processing [179, 180].

Extreme Learning Machine

The ELM stands out as a rapid learning algorithm designed for feedforward neural networks [181]. Unlike traditional gradient-based methods that adjust the weights iteratively, ELM adopts a unique approach focusing on the efficient training of neural networks. It achieves this through a straightforward three-step process: input weight initialization, hidden layer activation, and output weight computation.

In ELM, the input weights connecting the input layer to the hidden layer are randomly initialized. These weights are typically assigned values from a random distribution, such as uniform or Gaussian, and are not adjusted during training. Mathematically, for N input neurons and L hidden neurons, the input weight matrix $\mathbf{W}$ is defined as

$$\mathbf{W} = \left[w_{ij}\right]_{N\times L},$$

where w_{ij} represents the weight connecting the i-th input neuron to the j-th hidden neuron.

Once the input weights are initialized, the hidden layer neurons compute their activations using a nonlinear activation function, typically a sigmoid or radial basis function. The activation $\mathbf{H}$ of the hidden layer for M input samples can be expressed as

$$\mathbf{H} = \left[h(\mathbf{w}_1^\top \mathbf{x}_1 + b_1)\ h(\mathbf{w}_2^\top \mathbf{x}_2 + b_2)\ \cdots\ h(\mathbf{w}_L^\top \mathbf{x}_M + b_L)\right]^\top,$$

where $h(\cdot)$ is the chosen activation function, $\mathbf{w}_j$ denotes the j-th column of $\mathbf{W}$, $\mathbf{x}_i$ represents the input vector for the i-th sample, and b_j denotes the bias associated with the j-th hidden neuron.

In the final step, the output weights $\mathbf{B}$ connecting the hidden layer to the output layer are computed in a least-squares sense. This involves solving the linear system

$$\mathbf{HB} = \mathbf{T},$$

where $\mathbf{T}$ is the target output matrix for the training samples.

ELM offers several advantages over traditional neural network training methods. By initializing input weights randomly and fixing them, ELM skips the iterative weight adjustment process, leading to significantly faster training times. ELM is scalable to large datasets and can handle high-dimensional input spaces efficiently. Despite its simplicity, ELM often generalizes well and achieves competitive performance compared to more complex models. In summary, ELM's rapid learning capability and efficient training procedure make it a popular choice for various applications, including regression, classification, and time series prediction tasks [182].

The optimal parameter values for the forecasting process of ML models are determined using the grid search technique with tenfold cross-validation. The recorded data is divided into two parts: 70% for training and 30% for testing. The forecasting results produced by four ML models for Canada, and its provinces Ontario and Quebec are sequentially presented in Fig. 8.10.

These forecasting models are evaluated using the following metrics: coefficient of determination (R^2), relative error (RE), and root mean square error (RMSE). The relevant results are displayed in Table 8.3.

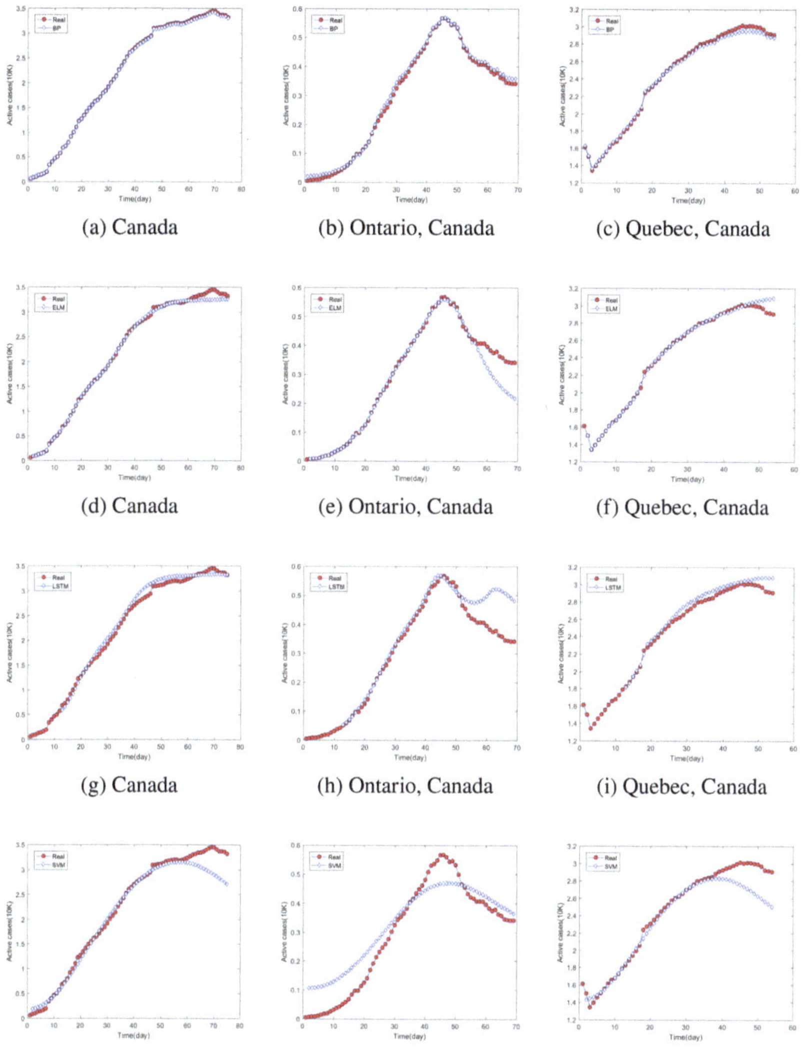

(a) Canada (b) Ontario, Canada (c) Quebec, Canada

(d) Canada (e) Ontario, Canada (f) Quebec, Canada

(g) Canada (h) Ontario, Canada (i) Quebec, Canada

(j) Canada (k) Ontario, Canada (l) Quebec, Canada

Fig. 8.10 Real and forecasted active cases in Canada and its province Ontario and Quebec using BP, ELM, LSTM and SVM models. (**a**) Canada. (**b**) Ontario, Canada. (**c**) Quebec, Canada. (**d**) Canada. (**e**) Ontario, Canada. (**f**) Quebec, Canada. (**g**) Canada. (**h**) Ontario, Canada. (**i**) Quebec, Canada. (**j**) Canada. (**k**) Ontario, Canada. (**l**) Quebec, Canada

Table 8.3 Validation metrics for R-D, BP, ELM, LSTM, and SVM models

Scenario	Model	R^2	MAE	$RMSE$
Canada	R-D	0.9729	3850	5070
	BP	0.9998	80	1051
	ELM	0.9971	352	4983
	LSTM	0.9863	142	2359
	SVM	0.9656	1236	17255
Ontario,Canada	R-D	0.9029	5850	4991
	BP	0.9965	346	512
	ELM	0.9514	2094	3295
	LSTM	0.8026	3222	4844
	SVM	0.8689	816	1201
Quebec,Canada	R-D	0.9584	1144	1380
	BP	0.9968	168	1982
	ELM	0.9933	192	3163
	LSTM	0.9602	252	3445
	SVM	0.9158	926	4269

8.1.4 Relative Advantages and Disadvantages of the Two Approaches

Compared to physics-based models, ML models offer greater flexibility due to their ability to derive patterns directly from data, reducing reliance on predefined relationships between variables. This adaptability enables ML models to deliver more robust and timely forecasts. In contrast, physics-based models rely heavily on assumptions about underlying mechanisms, the accuracy of fitting procedures, and the precise identification of model parameters. In our research, the complexities associated with estimating confirmed and resolved rate functions based on intricate intrinsic relationships led to extended estimation processes and significant uncertainty in predictive outcomes. Physics-based models require ongoing uncertainty analysis and real-time updates to maintain the reliability and timeliness of forecasts. Despite efforts to integrate dynamic-dependent rates into our diffusion-reaction model, achieving high predictive accuracy remains challenging. In contrast, models like BP, LSTM, and ELM demonstrate superior empirical performance and operational efficiency through their automated learning processes.

Physics-based models excel in providing insightful representations of interactive processes by leveraging fundamental laws of nature and deductive reasoning within specific spatiotemporal contexts. For instance, the R-D model incorporation of Neumann boundary conditions and its consideration of low diffusion rates effectively capture the ongoing dynamics within communities following the implementation of lockdown measures. Moreover, the model's formulation of the risk-free rate function reflects human responses to changing epidemiological conditions and varying infection risks. Dynamical analyses, including stability and sensitivity studies, further enhance our understanding of disease transmission mechanisms.

In contrast, ML algorithms often operate as black boxes, lacking the ability to offer interpretable representations of these complex interactive processes and causal relationships. Researchers frequently lack insight into how final predictions are generated, hindering their understanding of the underlying mechanisms at play. This opacity stems from the inherent nature of ML models, where predictions are derived through intricate patterns learned from data rather than through explicit rules or principles. The interpretability issue also presents a significant challenge in fields where understanding the decision-making process is crucial for developing effective interventions and strategies. In contrast, physics-based models play a pivotal role in informing intervention strategies by elucidating the interdependencies between variables and simulating the potential outcomes of various control measures. For instance, this section evaluates the effectiveness of two shielding mechanisms, highlighting their potential in mitigating epidemic peaks and reducing the duration of spread.

8.1.4.1 Conclusion

Physics-based models have a long-standing history in scientific research, providing high interpretability and insights into underlying mechanisms. In contrast, ML models have rapidly evolved, offering reliable predictions based on data-driven approaches. Over the past decade, the gap between these two methodologies has significantly decreased. ML models excel in generating accurate predictions but often lack the ability to elucidate the transmission mechanisms underlying epidemic phenomena. Physics-based models, on the other hand, offer robust interpretability by leveraging established physical laws and principles. However, they are highly dependent on model assumptions and the fitting process, which can sometimes limit their applicability.

To bridge the gap between interpretability and predictive accuracy, future research aims to integrate physics-based models with advanced ML techniques. One promising direction is the development of physics-informed neural networks (PINNs). PINNs combine the strengths of physics-based models and neural networks by incorporating physical laws into the training process of neural networks. This hybrid approach ensures that the model adheres to known physical principles while benefiting from the flexibility and predictive power of ML. By integrating PINNs, researchers can create models that not only predict epidemic trends accurately but also provide meaningful insights into the underlying transmission dynamics. For example, traditional epidemic models, such as the SIR (susceptible-infected-recovered) model, rely on differential equations to describe the spread of diseases. While these models offer valuable insights, they can be limited by their assumptions and the need for precise parameter estimation. PINNs address these limitations by embedding the principles of these differential equations into the neural network architecture, allowing for more flexible and accurate modeling.

The convergence of physics-based and ML models signifies a collaborative advancement in the field of epidemic modeling. This synergy is expected to lead to more robust and interpretable models, capable of addressing the complexities of public health challenges. Future research will continue to explore and refine these hybrid models, leveraging the strengths of both approaches to improve epidemic predictions and control measures.

8.2 Analyzing Pulmonary Tuberculosis Incidence and Its Contributing Factor

Pulmonary tuberculosis (PTB), caused by *Mycobacterium tuberculosis*, is a chronic infectious disease transmitted through the respiratory system, posing a significant threat to public health and human life security [183]. The primary mode of transmission is through respiratory droplets released during coughing, sneezing, or speaking by infected individuals [184]. During the "13th Five-Year Plan" period (2016–2020), nationwide efforts including tuberculosis screening, BCG vaccination, and non-pharmacological interventions contributed to a continuous decline in tuberculosis incidence, decreasing from 65 cases per 100,000 population in 2016 to 58 cases per 100,000 population in 2020 [169]. This underscores the importance of identifying preventable and modifiable risk factors in effectively managing and alleviating the burden of PTB.

Beyond suboptimal antituberculosis treatment, evidence indicates that socioeconomic and environmental factors such as population density, per capita GDP (pGDP), temperature, altitude, wind speed, and precipitation play crucial roles in driving PTB incidence [169]. Furthermore, long-term or short-term exposure to air pollution significantly increases the burden of PTB through immune system impairment, oxidative stress induction [185], and inflammatory responses [186]. However, inconsistencies in previous research regarding the nonlinear response between pollutant concentration levels and PTB incidence can be attributed to studies predominantly focusing on isolated environmental factors rather than capturing the complex interactions within the human population's actual exposure to a highly intricate socioeconomic and natural environment. Addressing this complexity, machine learning (ML) methods offer significant advantages in analyzing and adapting to such intricate environments.

Despite demonstrating impressive predictive capabilities, ML models face challenges in interpretability, limiting comprehensive insights into the spatiotemporal dynamics of PTB incidence and its relationships with various environmental factors [187]. To address this issue, recent developments in explainable artificial intelligence (XAI) aim to enhance the interpretative capacity of ML models. Accordingly, this chapter primarily utilizes ML and XAI to construct a national-scale ML workflow aimed at systematically quantifying the contributions of socioeconomic and environmental factors to PTB incidence rates and identifying critical thresholds associated with factors influencing PTB incidence.

8.2.1 Data Extraction and Preprocessing

8.2.1.1 Incidence Data

The tuberculosis incidence rate data used were obtained from the Public Health Science Data Center (PHSDC) (https://www.phsciencedata.cn/Share/index.jsp). This repository systematically collects reported tuberculosis cases through the Infectious Disease Network Reporting System from 2004 onward. Our analysis focused on monthly tuberculosis incidence rates spanning 2013 to 2019 across the 31 provinces of mainland China, including municipalities directly governed by the central government and autonomous regions while excluding Hong Kong, Macau, and Taiwan. It is important to note that the data utilized in this research comprises monthly provincial-level incidence rates during the specified period.

8.2.1.2 Natural and Pollution Data

Air quality data were sourced from the China High-Resolution and High-Quality Near-Surface Air Pollution Dataset (CHAP). This dataset integrates diverse data sources such as ground-based observations, atmospheric reanalysis, emission inventories, and model simulations, utilizing artificial intelligence techniques for high spatial and temporal resolution and precision. The selected CHAP dataset covers the period from 2013 to 2019 and includes six near-surface air pollutants: $PM_{2.5}$, PM_{10}, O_3, NO_2, SO_2, and CO, collected contemporaneously with the tuberculosis incidence data. Thus, this section focuses on acute rather than chronic exposure to pollutants.

Meteorological data were obtained from the European Centre for Medium-Range Weather Forecasts (ECMWF) ERA5 dataset, offering monthly averaged data on temperature, wind speed, and relative humidity. The precipitation data used were derived from the China 1 km Monthly Precipitation Dataset, obtained from the National Tibetan Plateau Data Center, ensuring spatial consistency through a spatial downscaling scheme.

8.2.1.3 Socioeconomic and Medical Data

Data on socioeconomic factors such as population density and per capita GDP (pGDP), as well as healthcare resources including the number of healthcare technical personnel and practicing physicians per thousand population, were sourced from the WorldPop dataset (https://www.worldpop.org/) and the China Statistical Yearbook (http://www.stats.gov.cn/sj/ndsj/), respectively. To address any missing data, a cubic spline interpolation method was applied, ensuring the continuity and comprehensiveness of the dataset used in this section. The category, variable, abbreviation, and unit of the data used are consolidated and presented in Table 8.4.

Table 8.4 Summary of predictor variables

Category	Variable (Unit)	Abbreviation
Incidence	Monthly incidence rate (per 100,000)	–
Pollution	Monthly mean concentration of $PM_{2.5}$ (μ/m^3)	$PM_{2.5}$
	Monthly mean concentration of PM_{10} (μ/m^3)	PM_{10}
	Monthly mean concentration of SO_2 (μ/m^3)	SO_2
	Monthly mean concentration of NO_2 (μ/m^3)	NO_2
	Monthly mean concentration of CO (mg/m^3)	CO
	Monthly mean concentration of O3 (μ/m^3)	O3
Climate	Monthly mean perception (mm)	PRE
	Monthly mean relative humidity (%)	RH
	Monthly mean wind speed (m/s)	WS
	Monthly mean temperature (□)	TEMP
Socioeconomic	Resident population (million)	POP
	Monthly GDP per capita (RMB)	pGDP
	Number of doctors per 100,000 people	Doctor
	Number of healthcare technicians per 100,000 people	Techni

8.2.1.4 Data Preprocessing

In addressing occasional data gaps due to sensor interruptions, inaccessible sampling sites, or deliberate confidentiality measures, a methodological approach using the ArcGIS raster calculator is employed in this section. This tool computes missing values by averaging surrounding pixel values, ensuring continuity and reliability in the dataset.

This research utilizes ArcGIS to establish an iterative model for processing raster data to derive monthly averages of pollution and climate variables across diverse provinces. The workflow is delineated as follows:

1. Grid creation: The Create Fishnet tool, adapted as the create fishnet feature class tool, constructs a grid based on national administrative planning data. This tool generates a feature class with predefined specifications including grid cell size, row count, and column count. For computational accuracy and efficiency, parameters were set to 350 rows and 600 columns, resulting in the creation of a point feature layer (fishnet_label) and corresponding polygon features.
2. Value extraction: Utilizing the value to points tool, raster data values are extracted based on the point feature layer (fishnet_label), thereby generating a new point feature layer (extract_fishnet_label).
3. Geospatial intersection: The identity tool is employed to compute geometric intersections between input features and administrative planning data for China, with the point feature layer (extract_fishnet_label) serving as input features. This process assigns regional identifiers to raster point values in extract_fishnet_label. Null values outside national borders are automatically filled as -999.

4. Data filtering: To ensure data validity, the Filter tool is applied with a condition set to values $>=0$, effectively removing invalid entries.
5. Statistical summarization: The summary statistics tool computes mean values for feature points within each province, yielding provincial averages for the variables under study.
6. National aggregation: A weighted average of provincial means is calculated to derive the national average, ensuring comprehensive coverage and representation of spatial data throughout the analytical process.

This methodological framework ensures a systematic and rigorous treatment of spatial data, facilitating accurate analysis and interpretation of pollution and climate variables across different provinces in China.

8.2.2 Predictive Model of PTB Incidence

8.2.2.1 Seasonality Text

Due to China's sustained commitment to public health and increased resource allocation, the PTB incidence has shown a consistent year-on-year decline (refer to Fig. 8.11). By 2019, the reported PTB incidence rate had significantly decreased to an average of 58 cases per 100,000 population, marking the lowest in recent years. Among the most affected regions in China are the northwestern provinces of Xinjiang, Tibet, and Qinghai. A comparison of PTB incidence rates between 2013 and 2019 across China's 31 provinces and directly administered municipalities reveals notable declines, particularly in the northeastern provinces of Heilongjiang, Jilin, Inner Mongolia, as well as in the eastern provinces of Shandong and Zhejiang. The Augmented Dickey-Fuller (ADF) test [188] was employed to examine whether PTB incidence exhibits seasonality. This statistical test is crucial for assessing the presence of a unit root in time series data, which indicates non-stationarity. The ADF test extends the Dickey-Fuller test by incorporating lagged differences of the dependent variable to account for higher-order autoregressive processes. Its formula is expressed as

$$\Delta y_t = \alpha + \beta t + \gamma y_{t-1} + \sum_{i=1}^{p} \delta_i \Delta y_{t-i} + \epsilon_t,$$

where Δy_t represents the first difference of the series at time t, defined as $y_t - y_{t-1}$; y_{t-1} denotes the lagged level of the series; t signifies the time trend; and $\alpha, \beta, \gamma, \delta_i$, and ϵ_t are parameters and error terms involved in the model.

The ADF test results identified seasonal patterns in PTB incidence rates across various provinces, notably higher rates observed in winter and spring compared to summer and autumn. These provinces exhibiting seasonality are visually indicated in red in Fig. 8.12.

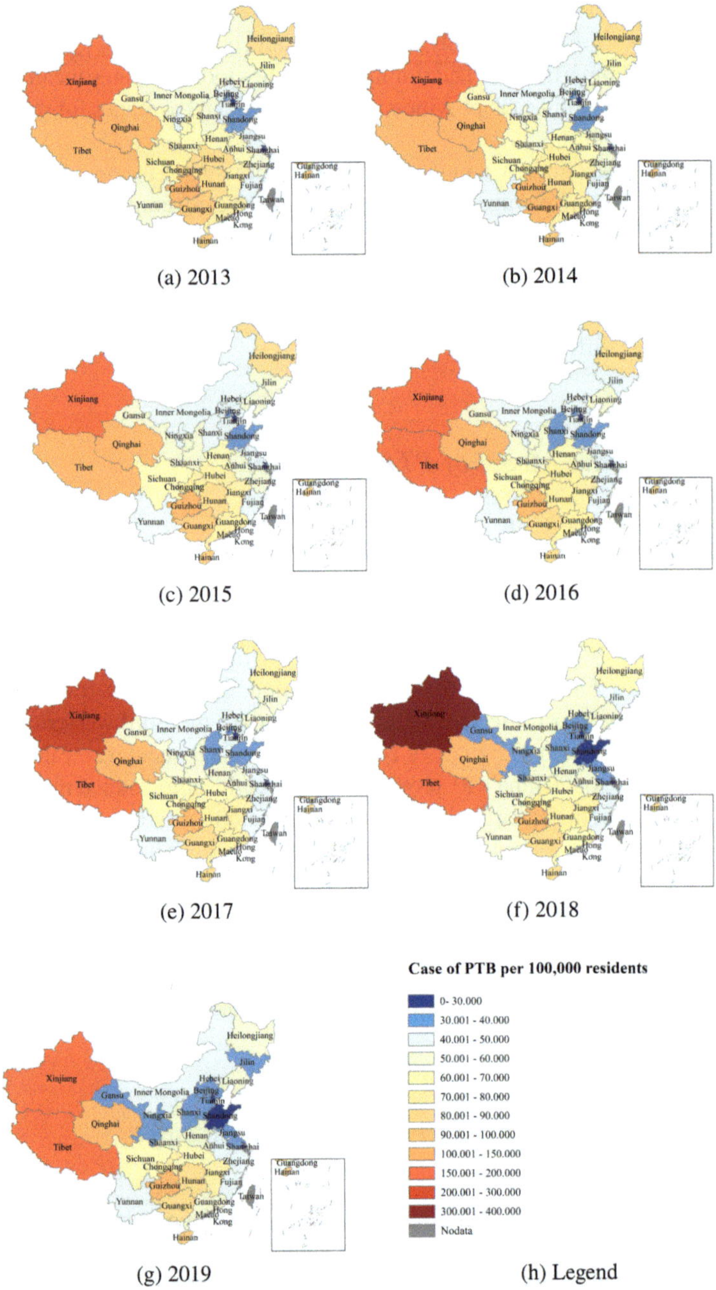

Fig. 8.11 The spatiotemporal distribution of PTB incidence rate in mainland China during 2013 and 2019. (**a**) 2013. (**b**) 2014. (**c**) 2015. (**d**) 2016. (**e**) 2017. (**f**) 2018. (**g**) 2019. (**h**) Legend

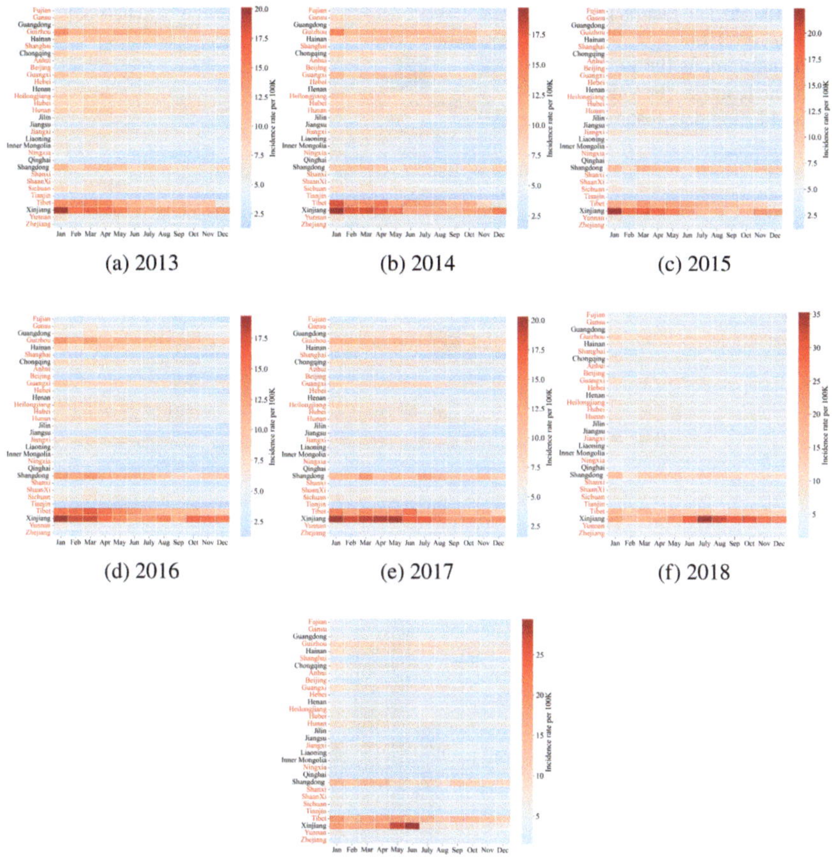

(a) 2013 (b) 2014 (c) 2015

(d) 2016 (e) 2017 (f) 2018

(g) 2019

Fig. 8.12 Monthly PTB incidence rates in mainland China from 2013 to 2019. Provinces highlighted in red denote regions identified with seasonality by ADF test. (**a**) 2013. (**b**) 2014. (**c**) 2015. (**d**) 2016. (**e**) 2017. (**f**) 2018. (**g**) 2019

8.2.2.2 Feature Selection

Before these variables were integrated into ML models, a feature screening process was conducted to evaluate their Pearson correlations. As depicted in Fig. 8.13a, the correlation coefficients between the numbers of healthcare technicians and doctors were found to exceed an absolute value of 0.75. Additionally, the variance inflation factors (VIFs) for 14 candidate features were assessed, as illustrated in Fig. 8.13b. Notably, VIFs for "Techni," "Doctor," and "$PM_{2.5}$" exceeded 10, indicating significant collinearity. $PM_{2.5}$ serves as a critical air quality indicator and is essential to retain in the models. Given that both "Techni" and "Doctor" represent indicators of medical resources within target provinces, and their Pearson

Fig. 8.13 (**a**) The heatmap of Pearson correlation coefficients among the 14 candidate features. (**b**) VIFs for the 14 candidate features

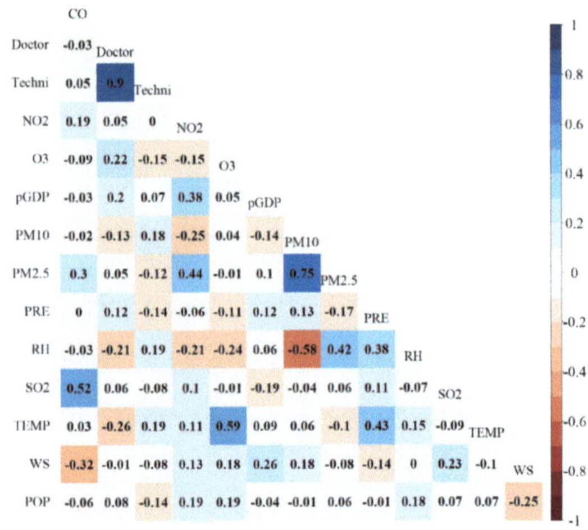

(a)

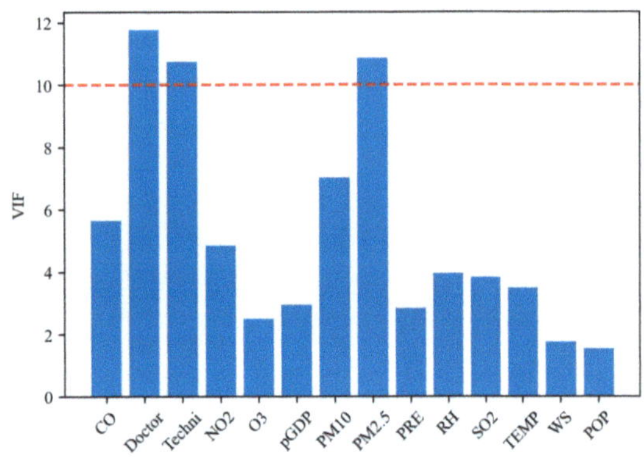

(b)

correlation coefficient is 0.9, the decision was made to eliminate the healthcare technician variable to mitigate potential issues of multicollinearity.

8.2.2.3 Model Construction and Evaluation

To forecast the monthly incidence rate of PTB, a diverse set of machine learning models was employed, including categorical boosting (Catboost), gradient boosting decision tree (GBDT), gradient boosting machine (LightGBM), extreme gradient boosting (XGBoost), K-nearest neighbor (KNN), random forest (RF), and classification and regression tree (CART). These algorithms have been previously introduced, and further elaboration is not provided here.

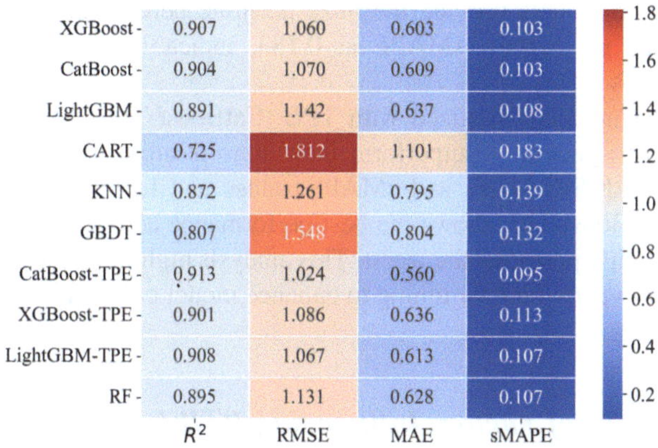

Fig. 8.14 Model performance metrics (R^2, MAE, RMSE, and sMAPE) of ten candidate models under tenfold cross-validation

To enhance the predictive accuracy of the models, the tree-structured Parzen estimator (TPE) optimization method was employed. TPE is a probabilistic model utilized in Bayesian optimization and hyperparameter tuning [144]. It efficiently explores hyperparameter spaces by modeling the probability of improvement across various configurations. Unlike traditional grid search or random search methods, TPE dynamically adjusts its search strategy based on the observed performance of previous configurations. In this section, parameter optimization using TPE was applied to fine-tune the parameters of LightGBM, Catboost, and XGBoost models, thereby refining their performance.

The heatmap (Fig. 8.14) provides a comparative evaluation of different machine learning models based on four metrics: the coefficient of determination (R^2), root mean square error (RMSE), mean absolute error (MAE), and symmetric mean absolute percentage error (sMAPE).

The R^2 values for most models are high, indicating good predictive performance. XGBoost and CatBoost-TPE exhibit the highest R^2 values at 0.907 and 0.913, respectively, suggesting these models explain a large proportion of the variance in the data. CART shows the lowest R^2 value at 0.725, indicating poorer performance relative to other models. Lower RMSE values indicate better model performance by showing smaller differences between predicted and observed values. CatBoost-TPE and XGBoost-TPE achieve the lowest RMSE values at 1.024 and 1.086, respectively, reflecting high accuracy in predictions. CART has the highest RMSE at 1.812, indicating it performs the worst among the evaluated models. Similar to RMSE, lower MAE values suggest better model performance by showing smaller absolute errors in predictions. CatBoost-TPE and XGBoost again perform best with the lowest MAE values at 0.560 and 0.613, respectively. CART shows the highest MAE at 1.101, reaffirming its relatively poor performance. Lower sMAPE values indicate better predictive accuracy in terms of percentage errors. CatBoost-TPE achieves the lowest sMAPE at 0.095, followed by XGBoost-TPE at 0.103, suggesting these models provide the most accurate percentage error predictions. CART exhibits the highest sMAPE at 0.183, indicating the least accuracy in percentage terms.

Overall, the models optimized with TPE (CatBoost-TPE, XGBoost-TPE, and LightGBM-TPE) generally outperform their non-optimized counterparts, demonstrating lower RMSE, MAE, and sMAPE values and higher R^2 values. Among all models, CatBoost-TPE shows the best performance across most metrics, while CART consistently performs the worst. This analysis highlights the effectiveness of using TPE for hyperparameter tuning to enhance model performance.

8.2.3 Analysis of Factors Influencing PTB Incidence

Figure 8.15 presents the SHAP summary plots illustrating the impact of 13 features on the incidence rate of PTB, as determined by the Catboost-TPE models. The SHAP values for population show both positive and negative impacts on the model's output, indicating that the influence of population size is variable and context-dependent. Population size, pGDP, and PM_{10} concentration are identified as the primary determinants influencing the incidence rate of PTB. High pGDP values exhibit a significant positive correlation with the predicted incidence rate, as shown by the left-tailed distribution of SHAP values. A similar trend is observed in the SHAP values associated with the number of doctors. Additionally, the SHAP

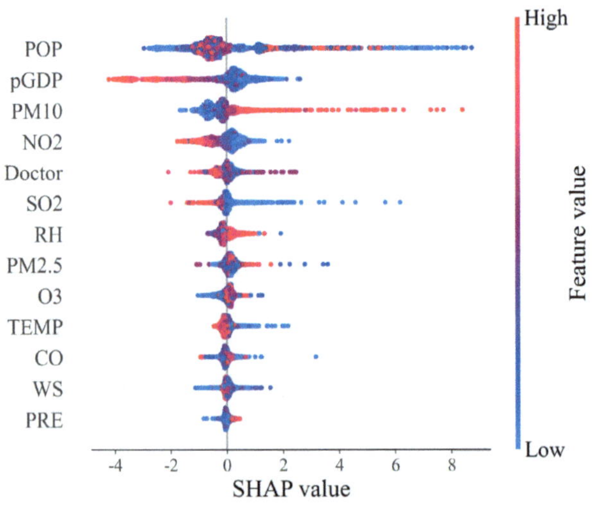

Fig. 8.15 SHAP summary plots for PTB incidence rate. The position of the dot along the x-axis shows the magnitude and direction of the feature's impact on the prediction. The y-axis represents feature importance, ranked in descending order. The color from blue to red represents the feature's value from low to high

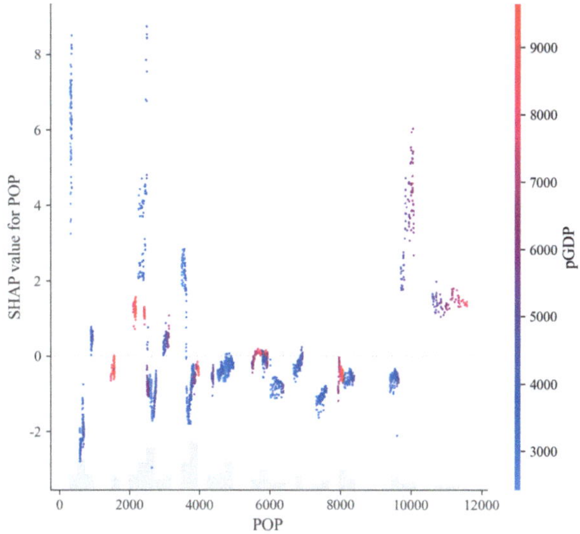

Fig. 8.16 The impact of POP on the incidence rate of PTB. The x-axis represents population size, the y-axis indicates the SHAP value for population, and the color gradient shows corresponding pGDP values

values indicate that elevated concentrations of $PM_{2.5}$ and PM_{10} have a right-tailed distribution, correlating with an increased incidence rate. Conversely, lower concentrations of NO_2 and SO_2, along with higher temperatures, are associated with a decreased incidence rate.

Figure 8.16 illustrates the impact of POP on the incidence rate of PTB, with pGDP as the secondary feature. The SHAP values for population display a mix of positive and negative impacts, indicating variable influence on PTB incidence. Distinct clusters at lower population levels suggest consistent impacts within certain ranges, while populations above 10,000 show increased SHAP values, highlighting higher positive impacts on PTB incidence. Higher pGDP values correlate with increased SHAP values, suggesting more pronounced PTB incidence in economically developed regions. Critical thresholds, such as populations around 10,000, show significant SHAP value transitions, marking tipping points where population impact intensifies. Lower population values, especially in economically underdeveloped areas, exhibit negative SHAP values, indicating a mitigating effect on PTB incidence. This analysis underscores the complex interactions between population size and economic factors, informing targeted public health strategies.

Population size and pGDP have emerged as pivotal factors affecting PTB incidence rates, exerting significant influence on disease transmission and control. Population size directly influences PTB transmission, with higher population density regions experiencing greater transmission probabilities due to increased contact rates [189]. As population size rises, the contact rates increase, consequently escalating the risk of PTB transmission [190]. There is a noticeable declining trend

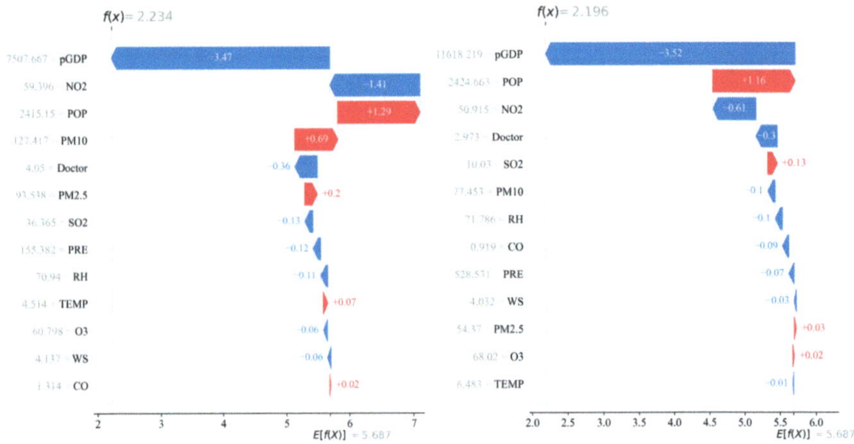

(a) Shanghai, January 2013 (b) Shanghai, January 2019

Fig. 8.17 The SHAP breakdown plots for PTB incidence in Shanghai in January 2013 and January 2019. (**a**) Shanghai, January 2013. (**b**) Shanghai, January 2019

in overall PTB incidence rates, although the rate of decline varies across different regions. This variation is potentially associated with factors such as differing rates of economic development. Economic status significantly impacts PTB incidence; lower pGDP is often accompanied by insufficient healthcare facilities, limited medical resources, and lower health awareness. These conditions can contribute to delays in the diagnosis and treatment of tuberculosis, thereby increasing the risk of transmission [191]. Additionally, lower economic levels may lead to social issues such as poverty and unfavorable living conditions, potentially rendering individuals more susceptible to infection with Mycobacterium tuberculosis [192].

To further elucidate the relationship between economic development and PTB incidence, Shanghai was chosen as an exemplar to highlight the contributions of pGDP and population size to the incidence of PTB in January 2013 and January 2019 (Fig. 8.17). The analysis revealed consistent trends: pGDP exhibited a negative correlation with PTB incidence, indicating its favorable influence on reducing PTB rates. Conversely, population size showed a positive correlation with PTB incidence, suggesting its propensity to increase PTB rates. These findings underscore the complex interplay between socioeconomic factors and PTB incidence. Higher economic development, as indicated by higher pGDP, is associated with better healthcare infrastructure and greater health awareness, which contribute to lower PTB incidence rates. On the other hand, higher population densities increase the likelihood of disease transmission, emphasizing the need for targeted public health interventions in densely populated regions. Therefore, in formulating comprehensive TB prevention strategies, it is crucial to consider both environmental risks and economic development factors to effectively mitigate the incidence of PTB.

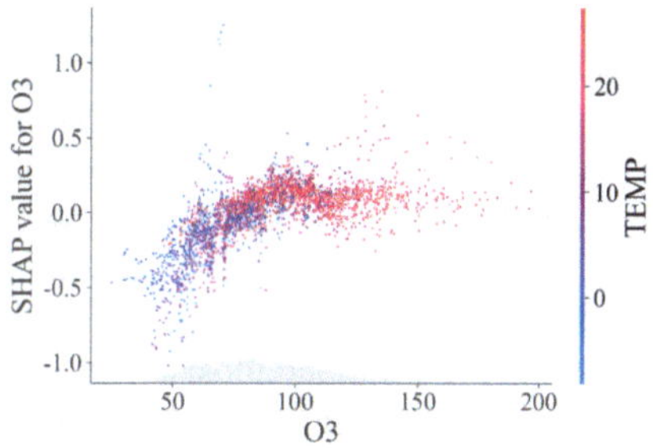

(a)

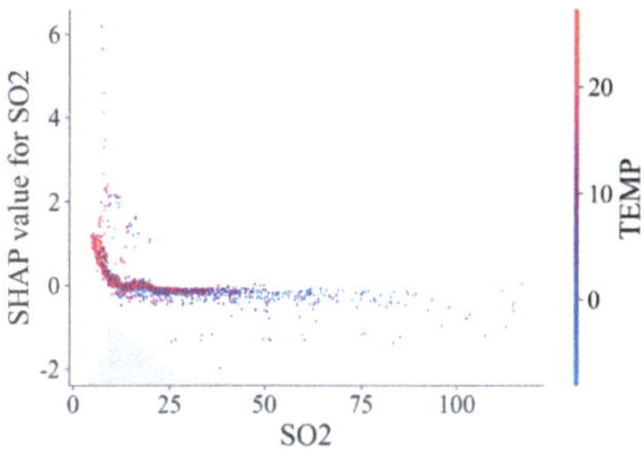

(b)

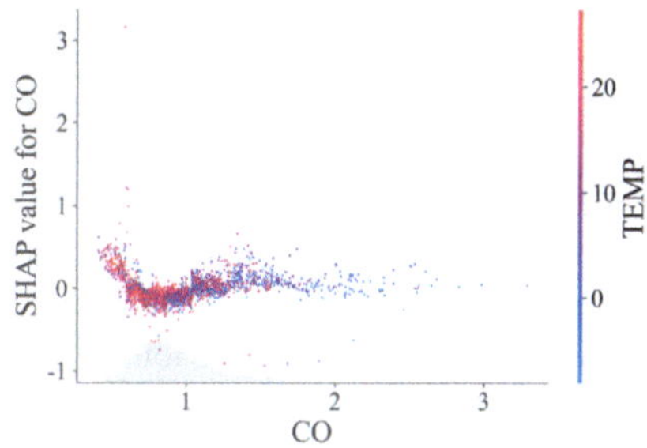

(c)

Fig. 8.18 SHAP dependence plot for PTB incidence rate, with the *x*-axis showing the primary driver value, the *y*-axis the SHAP value, and point colors indicating the temperature value

Figure 8.18a shows that as the concentration of O_3 increases, the SHAP values generally increase, indicating a positive impact on the model's output. Higher O_3 levels are associated with an increase in PTB incidence. The color gradient, indicating temperature, shows that higher temperatures (red dots) tend to correspond with higher SHAP values, suggesting that the positive impact of O_3 on PTB incidence is more pronounced in warmer conditions. Figure 8.18b and c reveal that at lower concentrations of SO_2 and CO, the SHAP values are positive, indicating an increased impact on PTB incidence. As the concentrations of SO_2 and CO increase, the SHAP values decrease and become more neutral. The color gradient shows that higher temperatures (red dots) correspond with more positive SHAP values, suggesting that the mitigating effect of high SO_2 and CO levels on PTB incidence is more significant in colder conditions.

The literature on the relationship between air pollutants and PTB incidence is contentious. Dimala and Kadia suggested that air pollutants such as SO_2 are associated with an increased incidence of PTB, while exposure to CO, NO_2, and O3 is not (Dimala and Kadia, 2022). Yang et al. propose that all air pollutants ($PM_{2.5}$, PM_{10}, SO_2, NO_2, CO, and O3) in Urumqi, China, might elevate the risk of tuberculosis [193]. In a nationwide survey, Zhu et al. found that higher levels of

SO_2 were linked to a lower PTB incidence, but the impact of ozone was precisely the opposite [194]. In the realm of ozone's antimicrobial properties, research findings suggest that ozone may exert a pronounced effect against microorganisms through mechanisms such as the disruption of microbial cell membrane structures, inhibition of their metabolic activity, and induction of oxidative stress reactions [195]. These distinctive characteristic positions ozone as a potential immunomodulatory agent, offering promise in bolstering defenses against the intrusion of Mycobacterium tuberculosis, the causative agent of tuberculosis. However, it is imperative to acknowledge the close correlation between ozone and emissions from traffic-related sources. The principal mechanisms governing ozone generation involve the interaction of nitrogen oxides and volatile organic compounds with sunlight. Due to the inherent instability of ozone in the environment, it readily reacts with other air pollutants, giving rise to novel and intricate compounds. Therefore, a judicious interpretation of the effects of ozone exposure is warranted, recognizing that the reaction by-products may introduce supplementary considerations for public health [196].

These varied findings underscore the complexity of the relationship between air pollution and PTB incidence. While certain pollutants may directly influence the incidence of PTB through various biological mechanisms, their interactions and the resulting by-products also play a significant role. This complexity necessitates a comprehensive approach to understanding and mitigating the effects of air pollution on public health, particularly in relation to PTB.

Figure 8.19 shows an increase in SHAP values with rising $PM_{2.5}$ and PM_{10} concentrations, indicating a positive impact on PTB incidence. The color gradient reveals that higher temperatures (red dots) are associated with lower SHAP values, suggesting that the adverse effects of $PM_{2.5}$ and PM_{10} on PTB incidence are amplified in lower temperatures. These results revealed a significant association between exposure to ambient PM and an increased risk of monthly incidence of PTB, consistent with findings from [193, 194, 197].

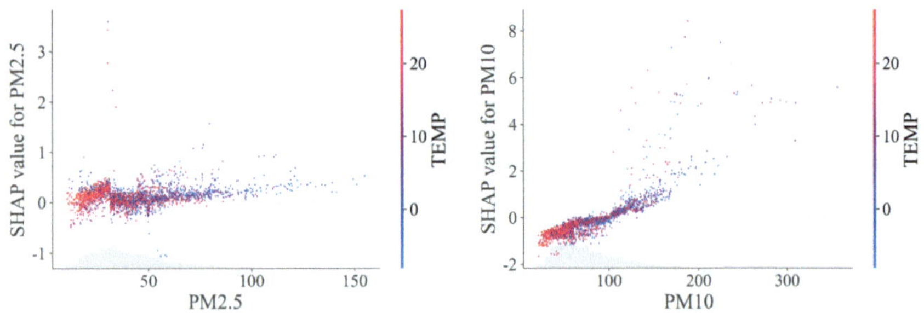

(a) (b)

Fig. 8.19 SHAP dependence plot for PTB incidence rate, with the x-axis showing the primary driver value, the y-axis the SHAP value, and point colors indicating the temperature value

Compelling evidence suggests that exposure to air pollutants, particularly $PM_{2.5}$ particles, may inhibit crucial immune defense mechanisms, thereby increasing susceptibility to active tuberculosis and PTB-related mortality [198]. For instance, particulate matter has been found to inhibit the activity of antimicrobial peptides against Mycobacterium tuberculosis within respiratory epithelial cells, altering the phagocytic capability of macrophages [185]. This implies that elevated concentrations of $PM_{2.5}$ and PM_{10} may adversely affect pulmonary immune responses, enhancing vulnerability to Mycobacterium tuberculosis, the causative agent of PTB. Heightened levels of particulate matter may induce pulmonary inflammation and oxidative stress, potentially compromising the immune response and disease resistance, thereby increasing the risk of PTB infection [199]. Additionally, air pollutants may contribute to the dissemination of Mycobacterium tuberculosis through non-immunological mechanisms. Certain airway toxins, such as particulate matter, irritant gases, environmental tobacco smoke, and mixed pollutants, could escalate the frequency and intensity of coughing in tuberculosis patients [200], potentially leading to increased transmission of Mycobacterium tuberculosis to uninfected individuals [201]. Consequently, elevated concentrations of $PM_{2.5}$ and PM_{10} may not only impair the immune system's defense against tubercle bacilli but also potentially increase the risk of infection by promoting the transmission process. These intricate interactions likely contribute to the observed association between high concentrations of particulate matter and the incidence of PTB.

Figure 8.20a illustrates the impact of the number of doctors per 1,000 population on the PTB incidence, with pGDP as the secondary feature indicated by color. The SHAP values mostly cluster around zero, suggesting a moderate and variable impact of doctor density on PTB incidence. At lower numbers of doctors (1 to 2 per 1,000), the SHAP values show significant variability with both positive and negative impacts, indicating that the effect on PTB incidence is highly context-dependent. In regions with lower pGDP (blue dots), an increase in doctor numbers is associated with more negative SHAP values, indicating a reduction in PTB incidence. Conversely, in higher pGDP areas (red dots), the SHAP values are closer to zero or slightly positive, suggesting a less pronounced impact. A transition is observed around 2 to 3 doctors per 1,000 population, where SHAP values cluster tightly around zero, indicating a consistent but neutral effect. Beyond this range, the number of doctors shows a predominantly neutral or slightly positive effect on PTB incidence. Overall, the plot highlights the nuanced and context-dependent influence of doctor density on PTB incidence, varying with economic development levels.

Figure 8.20b shows a slight increase in SHAP values with rising precipitation, indicating a weak positive impact on PTB incidence. Figure 8.20c indicates that as RH increases, SHAP values initially decrease, suggesting a negative impact on PTB incidence at lower humidity levels. However, beyond a certain point, SHAP values increase, indicating a positive impact at higher humidity levels. These trends are align with previous literature findings [202, 203]. The color gradient shows that higher temperatures (red dots) are associated with higher SHAP values, suggesting that the positive impact of high RH or PRE on PTB incidence is more significant in warmer temperatures.

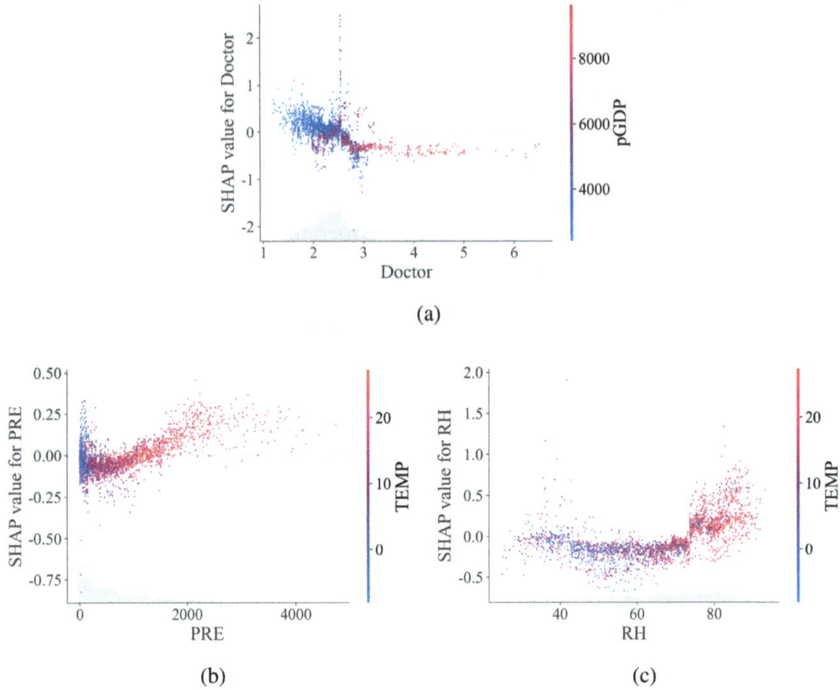

Fig. 8.20 SHAP dependence plot for PTB incidence rate, with the x-axis showing the primary driver value, the y-axis the SHAP value, and point colors indicating the secondary driver value

Mycobacterium tuberculosis is known to exhibit sensitivity to environmental conditions, particularly temperature and humidity. Higher temperatures and lower humidity are likely to be unfavorable for the survival and transmission of Mycobacterium tuberculosis in the external environment. Elevated temperatures may impose thermal stress on the bacteria, thereby diminishing their viability outside the host. Concurrently, reduced humidity levels may lead to decreased moisture in the air, making it more challenging for Mycobacterium tuberculosis to effectively disseminate. These findings suggest that environmental conditions play a crucial role in the transmission dynamics of PTB. In warmer and more humid environments, the survival and transmission of Mycobacterium tuberculosis may be facilitated, leading to higher incidence rates of PTB. Conversely, in hotter and drier conditions, the bacterium's viability and transmission potential are likely reduced, resulting in lower PTB incidence rates. Understanding these environmental influences is essential for developing effective public health strategies to control PTB, particularly in regions experiencing significant variations in temperature and humidity.

8.2.4 Conclusion

This section utilizes data from 13 natural and socioeconomic factors across 31 provinces and municipalities in China to construct a predictive model for the monthly incidence rate of PTB from 2013 to 2019 using machine learning algorithms. By applying interpretable machine learning techniques, the study found significant correlations between environmental risk factors, particularly air pollutants, and PTB incidence. Additionally, factors such as population size and socioeconomic development are also identified as crucial determinants of the monthly PTB incidence rate. Given the significant influence of environmental and socioeconomic factors on PTB incidence, it is crucial to integrate these considerations into national TB prevention strategies. Emphasizing these factors can enhance the effectiveness of public health interventions and policies aimed at reducing TB incidence.

This workflow, combining ML and XAI, demonstrates the potential applications in public health. As dynamical systems, machine learning, and XAI continue to evolve, integrating these approaches may offer even more robust tools for research and policy development in public health. Future studies should explore this integration further to improve predictive accuracy and interpretability of health-related models.

References

1. Perko, L.: Differential Equations and Dynamical Systems. Springer Science & Business Media, Berlin (2013)
2. Chen, S., Shi, J., Shuai, Z., Wu, Y.: Two novel proofs of spectral monotonicity of perturbed essentially nonnegative matrices with applications in population dynamics. SIAM J. Appl. Math. **82**(2), 654–676 (2022)
3. Sasmal, S.K., Dong, Y., Takeuchi, Y.: Mathematical modeling on t-cell mediated adaptive immunity in primary dengue infections. J. Theor. Biol. **429**, 229–240 (2017)
4. Shan, C., Zhu, H.: Bifurcations and complex dynamics of an SIR model with the impact of the number of hospital beds. J. Differ. Equations **257**(5), 1662–1688 (2014)
5. Zhu, H., Campbell, S.A., Wolkowicz, G.S.K.: Bifurcation analysis of a predator-prey system with nonmonotonic functional response. SIAM J. Appl. Math. **63**(2), 636–682 (2003)
6. Tang, B., Bragazzi, N.L., Li, Q., Tang, S., Xiao, Y., Wu, J.: An updated estimation of the risk of transmission of the novel coronavirus (2019-nCov). Infectious Disease Modelling **5**, 248–255 (2020)
7. Liu, S., Beretta, E.: A stage-structured predator-prey model of beddington-deangelis type. SIAM J. Appl. Math. **66**(4), 1101–1129 (2006)
8. Sun, G., Jusup, M., Jin, Z., Wang, Y., Wang, Z.: Pattern transitions in spatial epidemics: Mechanisms and emergent properties. Phys. Life Rev. **19**, 43–73 (2016)
9. Liu, S., Wang, X., Wang, L., Song, H.: Competitive exclusion in delayed chemostat models with differential removal rates. SIAM J. Appl. Math. **74**(3), 634–648 (2014)
10. Du, Y., Lin, Z.: Spreading-vanishing dichotomy in the diffusive logistic model with a free boundary. SIAM J. Math. Anal. **42**(1), 377–405 (2010)
11. Zhong, S., Zhang, K., Bagheri, M., Burken, J.G., Gu, A., Li, B., et al.: Machine learning: new ideas and tools in environmental science and engineering. Environ. Sci. Technol. **55**(19), 12741–12754 (2021)
12. Myles, A.J., Feudale, R.N., Liu, Y., Woody, N.A., Brown, S.D.: An introduction to decision tree modeling. J. Chemometrics: J. Chemometrics Soc. **18**(6), 275–285 (2004)
13. Quinlan, J.R.: Induction of decision trees. Mach. Learn. **1**, 81–106 (1986)
14. Freund, Y., Schapire, R.E.: A decision-theoretic generalization of on-line learning and an application to boosting. J. Comput. Syst. Sci. **55**(1), 119–139 (1997)
15. Breiman, L.: Random forests. Mach. Learn. **45**(1), 5–32 (2001)
16. Geurts, P., Ernst, D., Wehenkel, L.: Extremely randomized trees. Mach. Learn. **63**, 3–42 (2006)

J. Ren, Y. Tao, *Modeling in Life Sciences and Ecology*, Springer Asia Pacific Mathematics Series 4, https://doi.org/10.1007/978-981-95-1038-2

17. Zhang, S., Li, X., Zong, M., Zhu, X., Wang, R.: Efficient kNN classification with different numbers of nearest neighbors. IEEE Trans. Neural Networks Learn. Syst. **29**(5), 1774–1785 (2017)
18. Goodfellow, I., Bengio, Y., Courville, A.: Deep Learning. MIT Press, New York (2016)
19. Bishop, C.M.: Pattern recognition and machine learning. Springer Google Scholar **2**, 1122–1128 (2006)
20. Swets, J.A.: Signal Detection Theory and ROC Analysis in Psychology and Diagnostics: Collected Papers. Psychology Press, Milton Park (2014)
21. Harrell, F.E. Jr, Lee, K.L., Mark, D.B.: Multivariable prognostic models: issues in developing models, evaluating assumptions and adequacy, and measuring and reducing errors. Stat. Med. **15**(4), 361–387 (1996)
22. Gerds, T.A., Schumacher, M.: Consistent estimation of the expected brier score in general survival models with right-censored event times. Biom. J. **48**(6), 1029–1040 (2006)
23. Khalid, S., Khalil, T., Nasreen, S.: A survey of feature selection and feature extraction techniques in machine learning. In: 2014 Science and Information Conference, pp. 372–378 (IEEE, New York, 2014)
24. Nelsen, R.: Kendall tau Metric. Encyclopaedia of Mathematics. Springer ISBN, New York (2001)
25. Combettes, P.L., Wajs, V.R.: Signal recovery by proximal forward-backward splitting. Multiscale Model. Simul. **4**(4), 1168–1200 (2005)
26. Thabtah, F., Hammoud, S., Kamalov, F., Gonsalves, A.: Data imbalance in classification: experimental evaluation. Inf. Sci. **513**, 429–441 (2020)
27. Lin, T.-Y., Goyal, P., Girshick, R., He, K., Dollár, P.: Focal loss for dense object detection. In: Proceedings of the IEEE International Conference on Computer Vision, pp. 2980–2988 (2017)
28. Fan, W., Stolfo, S.J., Zhang, J., Chan, P.K.: AdaCost: misclassification cost-sensitive boosting. Proc. 16th Int. Conf. Mach. Learn. **99**, 97–105 (1999)
29. Friedman, J.H.: Greedy function approximation: a gradient boosting machine. Ann. Stat. **29**(5), 1189–1232 (2001)
30. Lundberg, S.M., Lee, S.-I.: A unified approach to interpreting model predictions. Adv. Neural Inf. Proces. Syst. **30** (2017)
31. Shapley, L.: A value for n-person games. In: Contributions to Theory Games (AM-28). Princeton University, Princeton, p. 2 (1953)
32. Weinan, E.: The dawning of a new era in applied mathematics. Not. Am. Math. Soc. **68**(4), 565–571 (2021)
33. Gourley, S.A., Britton, N.F.: A predator-prey reaction-diffusion system with nonlocal effects. J. Math. Biol. **34**(3), 297–333 (1996)
34. Tao, Y., Campbell, S.A., Poulin, F.J.: Dynamics of a diffusive nutrient-phytoplankton-zooplankton model with spatio-temporal delay. SIAM J. Appl. Math. **81**(6), 2405–2432 (2021)
35. Franks, P.J.S.: NPZ models of plankton dynamics: their construction, coupling to physics, and application. J. Oceanogr. **58**(2), 379–387 (2002)
36. Kloosterman, M., Campbell, S.A., Poulin, F.J.: A closed NPZ model with delayed nutrient recycling. J. Math. Biol. **68**, 815–850 (2014)
37. Gourley, S.A., Bartuccelli, M.: Parameter domains for instability of uniform states in systems with many delays. J. Math. Biol. **35**, 843–867 (1997)
38. Pao, C.-V.: Nonlinear Parabolic and Elliptic Equations. Springer Science & Business Media, New York (2012)
39. Kielhöfer, H.: Bifurcation Theory: An Introduction with Applications to PDEs, vol. 156. Springer Science & Business Media, New York (2006)
40. Yi, F., Wei, J., Shi, J.: Bifurcation and spatiotemporal patterns in a homogeneous diffusive predator–prey system. J. Differ. Equations **246**(5), 1944–1977 (2009)
41. May, R.M.: Limit cycles in predator-prey communities. Science **177**(4052), 900–902 (1972)

42. Rosenzweig, M.L., MacArthur, R.H.: Graphical representation and stability conditions of predator-prey interactions. Am. Nat. **97**(895), 209–223 (1963)
43. Holling, C.S.: The components of predation as revealed by a study of small-mammal predation of the european pine sawfly1. Can. Entomol. **91**(5), 293–320 (1959)
44. Holling, C.S.: The functional response of predators to prey density and its role in mimicry and population regulation. Memoirs Entomol. Soc. Canada **97**(S45), 5–60 (1965)
45. Beddington, J.R.: Mutual interference between parasites or predators and its effect on searching efficiency. J. Anim. Ecol. **44**(1), 331–340 (1975)
46. DeAngelis, D.L. Goldstein, R.A., O'Neill, R.V.: A model for tropic interaction. Ecology **56**(4), 881–892 (1975)
47. Crowley, P.H., Martin, E.K.: Functional responses and interference within and between year classes of a dragonfly population. J. N. Am. Benthol. Soc. **8**(3), 211–221 (1989)
48. Arditi, R.,Ginzburg, L.R.: Coupling in predator-prey dynamics: ratio-dependence. J. Theor. Biol. **139**(3), 311–326 (1989)
49. Song, P., Lou, Y., Xiao, Y.: A spatial seirs reaction-diffusion model in heterogeneous environment. J. Differ. Equations **267**(9), 5084–5114 (2019)
50. Tao, Y., Ren, J.: The stability and bifurcation of homogeneous diffusive predator–prey systems with spatio–temporal delays. Discrete Contin. Dyn. Syst. - Ser. B **27**(1), (2022)
51. Wang, J., Wei, J., Shi, J.: Global bifurcation analysis and pattern formation in homogeneous diffusive predator–prey systems. J. Differ. Equations **260**(4), 3495–3523 (2016)
52. García-Jiménez, C., Gutiérrez-Salmerón, M., Chocarro-Calvo, A., García-Martinez, J.M., Castaño, A., De la Vieja, A.: From obesity to diabetes and cancer: epidemiological links and role of therapies. Br. J. Cancer **114**(7), 716–722 (2016)
53. Lean, M.E.J., Leslie, W.S., Barnes, A.C., Brosnahan, N., Thom, G., McCombie, L., Peters, C., Zhyzhneuskaya, S., Al-Mrabeh, A., Hollingsworth, K.G., et al.: Primary care-led weight management for remission of type 2 diabetes (direct): an open-label, cluster-randomised trial. The Lancet **391**(10120), 541–551 (2018)
54. Tsilidis, K.K., Kasimis, J.C., Lopez, D.S., Ntzani, E.E., Ioannidis, J.P.A.: Type 2 diabetes and cancer: umbrella review of meta-analyses of observational studies. BMJ **350** (2015)
55. Kim, D.W., Zavala, E., Kim, J.K.: Wearable technology and systems modeling for personalized chronotherapy. Curr. Opin. Syst. Biol. **21**, 9–15 (2020)
56. Ha, J., Sherman, A.: Type 2 diabetes: one disease, many pathways. Am. J. Physiol. Endocrinol. Metab. **319**(2), E410–E426 (2020)
57. Grant, A.D., Lewis, D.M., Kriegsfeld, L.J.: Multi-timescale rhythmicity of blood glucose and insulin delivery reveals key advantages of hybrid closed loop therapy. J. Diabetes Sci. Technol. **16**(4), 912–920 (2022)
58. Tao, Y., Sun, Y., Zhu, H., Lyu, J., Ren, J.: Nilpotent singularities and periodic perturbation of a gi β model: a pathway to glucose disorder. J. Nonlinear Sci. **33**(3), 49 (2023)
59. Topp, B., Promislow, K., Devries, G., Miura, R.M., Finegood, D.T.: A model of β-cell mass, insulin, and glucose kinetics: pathways to diabetes. J. Theor. Biol. **206**(4), 605–619 (2000)
60. Stein, M.B., McCarthy, M.J., Chen, C.-Y., Jain, S., Gelernter, J., He, F., Heeringa, S.G., Kessler, R.C., Nock, M.K., Ripke, S., et al.: Genome-wide analysis of insomnia disorder. Mol. Psychiatry **23**(11), 2238–2250 (2018)
61. Brenachot, X., Ramadori, G., Ioris, R.M., Veyrat-Durebex, C., Altirriba, J., Aras, E., Ljubicic, S., Kohno, D., Fabbiano, S., Clement, S., et al.: Hepatic protein tyrosine phosphatase receptor gamma links obesity-induced inflammation to insulin resistance. Nat. Commun. **8**(1), 1820 (2017)
62. Aulinas, A., Colom, C., García Patterson, A., Ubeda, J., María, M.A., Orellana, I., Adelantado, J.M., De Leiva, A., Corcoy, R.: Smoking affects the oral glucose tolerance test profile and the relationship between glucose and hba1c in gestational diabetes mellitus. Diabet. Med. **33**(9), 1240–1244 (2016)
63. Micha, R., Peñalvo, J.L., Cudhea, F., Imamura, F., Rehm, C.D., Mozaffarian, D.: Association between dietary factors and mortality from heart disease, stroke, and type 2 diabetes in the united states. JAMA **317**(9), 912–924 (2017)

64. Depner, C.M., Melanson, E.L., McHill, A.W., Wright, K.P. Jr.: Mistimed food intake and sleep alters 24-hour time-of-day patterns of the human plasma proteome. Proc. Natl. Acad. Sci. **115**(23), E5390–E5399 (2018)
65. Ando, H., Ushijima, K., Shimba, S., Fujimura, A.: Daily fasting blood glucose rhythm in male mice: a role of the circadian clock in the liver. Endocrinology **157**(2), 463–469 (2016)
66. Marijon, E., Mocumbi, A., Narayanan, K., Jouven, X., Celermajer, D.S.: Persisting burden and challenges of rheumatic heart disease. Eur. Heart J. **42**(34), 3338–3348 (2021)
67. Ordunez, P., Martinez, R., Soliz, P., Giraldo, G., Mujica, O.J., Nordet, P.: Rheumatic heart disease burden, trends, and inequalities in the americas, 1990–2017: a population-based study. Lancet Glob. Health **7**(10), e1388–e1397 (2019)
68. Watkins, D.A., Johnson, C.O., Colquhoun, S.M., Karthikeyan, G., Beaton, A., Bukhman, G., Forouzanfar, M.H., Longenecker, C.T., Mayosi, B.M., Mensah, G.A., et al.: Global, regional, and national burden of rheumatic heart disease, 1990–2015. N. Engl. J. Med. **377**(8), 713–722 (2017)
69. Guttapadu, R., Prakash, N., Chatterjee, R., Sastry, U.M.K., Subramanyam, J.R., Chakravortty, D., Chandra, N.: Profiling system-wide variations and similarities between rheumatic heart disease and acute rheumatic fever–a pilot analysis. PLoS Negl. Trop. Dis. **17**(4), e0011263 (2023)
70. Dooley, L.M., Ahmad, T.B., Pandey, M., Good, M.F., Kotiw, M.: Rheumatic heart disease: a review of the current status of global research activity. Autoimmun. Rev. **20**(2), 102740 (2021)
71. Arvind, B., Ramakrishnan, S.: Rheumatic fever and rheumatic heart disease in children. Indian J. Pediatr. **87**, 305–311 (2020)
72. Rauschert, S., Raubenheimer, K., Melton, P.E., Huang, R.C.: Machine learning and clinical epigenetics: a review of challenges for diagnosis and classification. Clin. Epigenetics **12**(1), 1–11 (2020)
73. Ngiam, K.Y., Khor, W.: Big data and machine learning algorithms for health-care delivery. Lancet Oncol. **20**(5), e262–e273 (2019)
74. Lee, Y.W., Choi, J.W., Shin, E.-H.: Machine learning model for predicting malaria using clinical information. Comput. Biol. Med. **129**, 104151 (2021)
75. Ghorbani, M., Kazi, A., Baghshah, M.S., Rabiee, H.R., Navab, N.: RA-GCN: graph convolutional network for disease prediction problems with imbalanced data. Med. Image Anal. **75**, 102272 (2022)
76. Akter, S., Das, D., Haque, R.U., Tonmoy, M.I.Q., Hasan, M.R., Mahjabeen, S., Ahmed, M.: Ad-covnet: an exploratory analysis using a hybrid deep learning model to handle data imbalance, predict fatality, and risk factors in alzheimer's patients with COVID-19. Comput. Biol. Med. **146**, 105657 (2022)
77. Razzaghi, T., Safro, I., Ewing, J., Sadrfaridpour, E., Scott, J.D.: Predictive models for bariatric surgery risks with imbalanced medical datasets. Ann. Oper. Res. **280**, 1–18 (2019)
78. Zhang, M.,Zhou, Z.: ML-KNN: a lazy learning approach to multi-label learning. Pattern Recogn. **40**(7), 2038–2048 (2007)
79. Qiao, Y., Yang, X., Wu, E.: The research of bp neural network based on one-hot encoding and principle component analysis in determining the therapeutic effect of diabetes mellitus. In: IOP Conference Series: Earth and Environmental Science, vol. 267, pp. 042178. IOP Publishing, New York (2019)
80. Centofanti, F., Fontana, M., Lepore, A., Vantini, S.: Smooth lasso estimator for the function-on-function linear regression model. Comput. Stat. Data Anal. **176**, 107556 (2022)
81. Zhou, Z., Feng, J.: Deep forest. Natl. Sci. Rev. **6**(1), 74–86 (2019)
82. Chen, T., Guestrin, C.: XGboosta Scalable Tree Boosting System, pp. 785–794 (2016)
83. Chawla, N.V.: Data mining for imbalanced datasets: an overview. In: Data Mining and Knowledge Discovery Handbook, pp. 875–886 (2010)
84. Kotsiantis, S., Kanellopoulos, D., Pintelas, P.: Handling imbalanced datasets: A review. GESTS Int. Trans. Comput. Sci. Eng. **30**(1), 25–36 (2006)

85. He, V.Y.F., Condon, J.R., Ralph, A.P., Zhao, Y., Roberts, K., de Dassel, J.L., Currie, B.J., Fittock, M., Edwards, K.N., Carapetis, J.R.: Long-term outcomes from acute rheumatic fever and rheumatic heart disease: a data-linkage and survival analysis approach. Circulation **134**(3), 222–232 (2016)
86. Lawrence, J.G., Carapetis, J.R., Griffiths, K., Edwards, K., Condon, J.R.: Acute rheumatic fever and rheumatic heart disease: incidence and progression in the northern territory of australia, 1997 to 2010. Circulation **128**(5), 492–501 (2013)
87. Janssens, U., Dujardin, R., Graf, J., Lepper, W., Ortlepp, J., Merx, M., Zarse, M., Reffelmann, T., Hanrath, P.: Value of SOFA (sequential organ failure assessment) score and total maximum SOFA score in 812 patients with acute cardiovascular disorders. Critical Care **5**, 1–1 (2001)
88. Liu, J., Wang, D., Xiong, Y., Liu, B., Lin, J., Zhang, S., Wu, B., Wei, C., Liu, M.: Association between coagulation function and cerebral microbleeds in ischemic stroke patients with atrial fibrillation and/or rheumatic heart disease. Aging Dis. **8**(2), 131 (2017)
89. Pradhan, R.R., Jha, A., Nepal, G., Sharma, M.: Rheumatic heart disease with multiple systemic emboli: a rare occurrence in a single subject. Cureus **10**(7) (2018)
90. DeBakey, M.E.: Left ventricular bypass pump for cardiac assistance: clinical experience. Am. J. Cardiol. **27**(1), 3–11 (1971)
91. Mickerson, J.N., Swale, J.: Diuretic effect of steroid therapy in obstinate heart failure. Br. Med. J. **1**(5126), 876 (1959)
92. McClave, S.A., Snider, H.L., Spain, D.A.: Preoperative issues in clinical nutrition. Chest **115**(5), 64S–70S (1999)
93. Evans, A.S., Hosseinian, L., Mohabir, T., Kurtis, S., Mechanick, J.I.: Nutrition and the cardiac surgery intensive care unit patient–an update. J. Cardiothorac. Vasc. Anesth. **29**(4), 1044–1050 (2015)
94. Horwich, T.B., Kalantar-Zadeh, K., Mac Lellan, R.W., Fonarow, G.C.: Albumin levels predict survival in patients with systolic heart failure. Am. Heart J. **155**(5), 883–889 (2008)
95. Uthamalingam, S., Kandala, J., Daley, M., Patvardhan, E., Capodilupo, R., Moore, S.A., Januzzi, J.L. Jr.: Serum albumin and mortality in acutely decompensated heart failure. Am. Heart J. **160**(6), 1149–1155 (2010)
96. Don, B.R., Kaysen, G.: Poor nutritional status and inflammation: serum albumin: relationship to inflammation and nutrition. Semin. Dial. **17**(6), 432–437 (2004)
97. Arrigo, M., Jessup, M., Mullens, W., Reza, N., Shah, A.M., Sliwa, K., Mebazaa, A.: Acute heart failure. Nat. Rev. Dis. Prim. **6**(1), 16 (2020)
98. Breast Cancer. https://www.who.int/news-room/fact-sheets/detail/breast-cancer
99. Coles, C.E., Anderson, B.O., Cameron, D., Cardoso, F., Horton, R., Knaul, F.M., Mutebi, M., Lee, N., Abraham, J.E., André, F., et al.: The lancet breast cancer commission: tackling a global health, gender, and equity challenge. Lancet **399**(10330), 1101–1103 (2022)
100. Houssein, E.H., Emam, M.M., Ali, A.A., Suganthan, P.N.: Deep and machine learning techniques for medical imaging-based breast cancer: a comprehensive review. Expert Syst. Appl. **167**, 114161 (2021)
101. Zizaan, A., Idri, A.: Machine learning based breast cancer screening: Trends, challenges, and opportunities. Comput. Methods Biomech. Biomed. Eng.: Imaging Visualization **11**(3), 976–996 (2023)
102. Cox, D.R.: The analysis of multivariate binary data. Appl. Stat. **21**(2), 113–120 (1972)
103. Gordon, L., Olshen, R.A.: Tree-structured survival analysis. Cancer Treat. Rep. **69**(10), 1065–1069 (1985)
104. Ishwaran, H., Kogalur, U.B., Blackstone, E.H., Lauer, M.S.: Random Survival Forests (2008)
105. Bai, M., Zheng, Y., Shen, Y.: Gradient boosting survival tree with applications in credit scoring. J. Oper. Res. Soc. **73**(1), 39–55 (2022)
106. Katzman, J.L., Shaham, U., Cloninger, A., Bates, J., Jiang, T., Kluger, Y.: Deepsurv: personalized treatment recommender system using a cox proportional hazards deep neural network. BMC Med. Res. Methodol. **18**(1), 1–12 (2018)
107. Pillai, N., Dasgupta, A., Sudsakorn, S., Fretland, J., Mavroudis, P.D.: Machine learning guided early drug discovery of small molecules. Drug Discov. Today **27**(8), 2209–2215 (2022)

108. Vamathevan, J., Clark, D., Czodrowski, P., Dunham, I., Ferran, E., Lee, G., Li, B., Madabhushi, A., Shah, P., Spitzer, M., et al.: Applications of machine learning in drug discovery and development. Nat. Rev. Drug Discov. **18**(6), 463–477 (2019)
109. Siebenmorgen, T., Menezes, F., Benassou, S., Merdivan, E., Didi, K., Mourão, A.S.D., Kitel, R. Liò, P., Kesselheim, S., Piraud, M., et al.: Misato: machine learning dataset of protein–ligand complexes for structure-based drug discovery. Nat. Comput. Sci. **4**(5), 1–12 (2024)
110. Lin, H., Zhang, Q., Kong, X., Zhang, L., Zhang, Y., Tang, Y., Xu, H.: Machine learning prediction of antiviral-hpv protein interactions for anti-hpv pharmacotherapy. Sci. Rep. **11**(1), 24367 (2021)
111. Jaccard, P.: Nouvelles recherches sur la distribution florale. Bull. Soc. Vaud. Sci. Nat. **44**, 223–270 (1908)
112. Fernández-de Gortari, E., García-Jacas, C.R., Martinez-Mayorga, K., Medina-Franco, J.L.: Database fingerprint (DFP): an approach to represent molecular databases. J. Cheminf. **9**(1), 1–9 (2017)
113. Yap, C.W.: PaDEL-descriptor: an open source software to calculate molecular descriptors and fingerprints. J. Comput. Chem. **32**(7), 1466–1474 (2011)
114. Luo, L., Chen, Q.: Research on amino acid sequence feature extraction methods. Comput. Technol. Dev. **2**(2), 206–208 (2010)
115. Yegnanarayana, B.: Artificial Neural Networks. PHI Learning Pvt. Ltd., New York (2009)
116. Rodriguez, J.J., Kuncheva, L.I., Alonso, C.J.: Rotation forest: a new classifier ensemble method. IEEE Trans. Pattern Anal. Mach. Intell. **28**(10), 1619–1630 (2006)
117. Zheng, R., Zhang, S., Zeng, H., Wang, S., Sun, K., Chen, R., Li, L., Wei, W., He, J.: Cancer incidence and mortality in China, 2016. J. Natl. Cancer Center **2**(1), 1–9 (2022)
118. Delaney, G., Jacob, S., Featherstone, C., Barton, M.: The role of radiotherapy in cancer treatment: estimating optimal utilization from a review of evidence-based clinical guidelines. Cancer: Interdisciplinary Int. J. Am. Cancer Soc. **104**(6), 1129–1137 (2005)
119. Colli, L.M., Machiela, M.J., Zhang, H., Myers, T.A., Jessop, L., Delattre, O., Yu, K., Chanock, S.J.: Landscape of combination immunotherapy and targeted therapy to improve cancer management. Cancer Res. **77**(13), 3666–3671 (2017)
120. Zhang, Y., Zhang, Z.: The history and advances in cancer immunotherapy: understanding the characteristics of tumor-infiltrating immune cells and their therapeutic implications. Cell. Mol. Immunol. **17**(8), 807–821 (2020)
121. Wang, F.: Cancer cell culture–methods and protocols. In Vitro Cell. Dev. Biol. Anim. **39**(10), 476–476 (2003)
122. Lamb, J., Crawford, E.D., Peck, D., Modell, J.W., Blat, I.C., Wrobel, M.J., Lerner, J., Brunet, J.-P., Subramanian, A., Ross, K.N., et al.: The connectivity map: using gene-expression signatures to connect small molecules, genes, and disease. Science **313**(5795), 1929–1935 (2006)
123. Yang, W., Soares, J., Greninger, P., Edelman, E.J., Lightfoot, H., Forbes, S., Bindal, N., Beare, D., Smith, J.A., Thompson, I.R., et al.: Genomics of drug sensitivity in cancer (GDSC): a resource for therapeutic biomarker discovery in cancer cells. Nucleic Acids Res. **41**(D1), D955–D961 (2012)
124. Kim, S., Chen, J., Cheng, T., Gindulyte, A., He, J., He, S., Li, Q., Shoemaker, B.A., Thiessen, P.A., Yu, B., et al.: Pubchem in 2021: new data content and improved web interfaces. Nucleic Acids Res. **49**(D1), D1388–D1395 (2021)
125. Knijnenburg, T.A., Klau, G.W., Iorio, F., Garnett, M.J., McDermott, U., Shmulevich, I., Wessels, L.F.A.: Logic models to predict continuous outputs based on binary inputs with an application to personalized cancer therapy. Sci. Rep. **6**(1), 1–14 (2016)
126. Schneider, S., Thurnher, D., Kadletz, L., Seemann, R., Brunner, M., Kotowski, U., Schmid, R., Lill, C., Heiduschka, G.: Effects of neratinib and combination with irradiation and chemotherapy in head and neck cancer cells. Oral Dis. **22**(8), 797–804 (2016)
127. Ihnen, M., Zu Eulenburg, C., Kolarova, T., Qi, J.W., Manivong, K., Chalukya, M., Dering, J., Anderson, L., Ginther, C., Meuter, A., et al.: Therapeutic potential of the poly (adp-ribose) polymerase inhibitor rucaparib for the treatment of sporadic human ovarian cancerparp inhibitor rucaparib in ovarian cancer. Mol. Cancer Ther. **12**(6), 1002–1015 (2013)

128. Ji, D., Li, L., Wang, Y., Zhang, J., Cheng, M., Sun, Y., et al.: The heaviest particulate air-pollution episodes occurred in northern China in January, 2013: insights gained from observation. Atmos. Environ. **92**, 546–556 (2014)
129. Chameides, W.L., Yu, H., Liu, S.C., Bergin, M., Zhou, X., Mearns, L., et al.: Case study of the effects of atmospheric aerosols and regional haze on agriculture: an opportunity to enhance crop yields in China through emission controls? Proc. Natl. Acad. Sci. **96**(24), 13626–13633 (1999)
130. Kan, H., Chen, R., Tong, S.: Ambient air pollution, climate change, and population health in China. Environ. Int. **42**, 10–19 (2012)
131. Yin, P., Brauer, M., Cohen, A.J., Wang, H., Li, J., Burnett, R.T., et al.: The effect of air pollution on deaths, disease burden, and life expectancy across China and its provinces, 1990–2017: an analysis for the Global Burden of Disease Study 2017. Lancet Planet. Heath **4**(9), e386–e398 (2020)
132. Zhang, X., Chen, X., Zhang, X.: The impact of exposure to air pollution on cognitive performance. Proc. Natl. Acad. Sci. **115**(37), 9193–9197 (2018)
133. Geng, G., Xiao, Q., Zheng, Y., Tong, D., Zhang, Y., Zhang, X., Zhang, Q., He, K., Liu, Y.: Impact of China's air pollution prevention and control action plan on pm2. 5 chemical composition over eastern China. Sci. China Earth Sci. **62**(12), 1872–1884 (2019)
134. Li, K., Jacob, D.J., Liao, H., Zhu, J., Shah, V., Shen, L., et al.: A two-pollutant strategy for improving ozone and particulate air quality in China. Nat. Geosci. **12**(11), 906–910 (2019)
135. Zhang, Q., Geng, G.: Impact of clean air action on $pm_{2.5}$ pollution in China. Sci. China Earth Sci. **62**(12), 1845–1846 (2019)
136. Cui, H., Li, J., Sun, Y., Milne, R., Tao, Y., Ren, J.: A novel framework for quantitative attribution of particulate matter pollution mitigation to natural and socioeconomic drivers. Sci. Total Environ. **926**, 171910 (2024)
137. Jin, L., Apte, J.S., Miller, S.L., Tao, S., Wang, S., Jiang, G., et al.: Global endeavors to address the health effects of urban air pollution. Environ. Sci. Technol. **56**(11), 6793–6798 (2022)
138. Baba, K., Shibata, R., Sibuya, M.: Partial correlation and conditional correlation as measures of conditional independence. Aust. N. Z. J. Stat. **46**(4), 657–664 (2004)
139. Lei, W., Jiao, L., Xu, G., Zhou, Z.: Urban scaling in rapidly urbanising China. Urban Stud. **59**(9), 1889–1908 (2022)
140. Le Vine, S., Wu, C., Polak, J.: A nationwide study of factors associated with household car ownership in China. IATSS Res. **42**(3), 128–137 (2018)
141. Luo, K., Li, G., Fang, C., Sun, S.: $Pm_{2.5}$ mitigation in China: Socioeconomic determinants of concentrations and differential control policies. J. Environ. Manag. **213**, 47–55 (2018)
142. Bengio, Y.: Learning deep architectures for AI. Found. Trends Mach. Learn. **2**(1), 1–127 (2009).
143. Fausett, L.V.: Fundamentals of Neural Networks: Architectures, Algorithms and Applications. Pearson Education India, Noida (2006)
144. Bergstra, J., Bardenet, R., Bengio, Y., Kégl, B.: Algorithms for hyper-parameter optimization. Adv. Neural Inf. Proces. Syst. **24** (2011)
145. Lundberg, S.M., Erion, G.G., Lee, S.-I.: Consistent individualized feature attribution for tree ensembles. arXiv preprint arXiv:1802.03888 (2018)
146. Unger, N., Shindell, D.T., Koch, D.M., Streets, D.G.: Cross influences of ozone and sulfate precursor emissions changes on air quality and climate. Proc. Natl. Acad. Sci. **103**(12), 4377–4380 (2006)
147. Gao, M., Liu, Z., Zheng, B., Ji, D., Sherman, P., Song, S., et al.: China's emission control strategies have suppressed unfavorable influences of climate on wintertime $pm_{2.5}$ concentrations in Beijing since 2002. Atmos. Chem. Phys. **20**(3), 1497–1505 (2020)
148. Li, T., Li, Y., An, D., Han, Y., Xu, S., Lu, Z., et al.: Mining of the association rules between industrialization level and air quality to inform high-quality development in China. J. Environ. Manag. **246**, 564–574 (2019)
149. Wang, F., Wang, R., Wang, J.: Measurement of China's green GDP and its dynamic variation based on industrial perspective. Environ. Sci. Pollut. Res. **27**(35), 43813–43828 (2020)

150. Liu, D., Xu, C., Yu, Y., Rong, K., Zhang, J.: Economic growth target, distortion of public expenditure and business cycle in China. China Econ. Rev. **63**, 101373 (2020)
151. Zhu, J., Lin, B.: Economic growth pressure and energy efficiency improvement: empirical evidence from chinese cities. Appl. Energy **307**, 118275 (2022)
152. Shen, F., Liu, B., Luo, F., Wu, C., Chen, H., Wei, W.: The effect of economic growth target constraints on green technology innovation. J. Environ. Manag. **292**, 112765 (2021)
153. Wang, Q., Zhao, Z., Shen, N., Liu, T.: Have chinese cities achieved the win–win between environmental protection and economic development? from the perspective of environmental efficiency. Ecol. Indic. **51**, 151–158 (2015)
154. Dai, Y., Yang, S., Zhao, D., Hu, C., Xu, W., Anderson, D.M., Li, Y., Song, X.-P., Boyce, D.G., Gibson, L., et al.: Coastal phytoplankton blooms expand and intensify in the 21st century. Nature **615**(7951), 280–284 (2023)
155. Loewen, C.J.G., Wyatt, F.R., Mortimer, C.A., Vinebrooke, R.D., Zurawell, R.W.: Multiscale drivers of phytoplankton communities in north-temperate lakes. Ecol. Appl. **30**(5), e02102 (2020)
156. Zhang, Q., Dong, X., Yang, X., Odgaard, B.V., Jeppesen, E.: Hydrologic and anthropogenic influences on aquatic macrophyte development in a large, shallow lake in China. Freshw. Biol. **64**(4), 799–812 (2019)
157. Rao, K., Zhang, X., Yi, X., Li, Z., Wang, P., Huang, G., Guo, X.: Interactive effects of environmental factors on phytoplankton communities and benthic nutrient interactions in a shallow lake and adjoining rivers in China. Sci. Total Environ. **619**, 1661–1672 (2018)
158. Lürling, M., Eshetu, F., Faassen, E.J., Kosten, S., Huszar, V.L.M.: Comparison of cyanobacterial and green algal growth rates at different temperatures. Freshw. Biol. **58**(3), 552–559 (2013)
159. Ma, S., Lv, X., Geng, S., Wang, G., Yang, S., Gao, Y.: Canonical correspondence analysis of relationship between characteristics of phytoplankton community and environmental factors in wolong lake. In: IOP Conference Series: Earth and Environmental Science, vol. 621, pp. 012107. IOP Publishing, New York (2021)
160. Misra, A.K., Chandra, P., Raghavendra, V.: Modeling the depletion of dissolved oxygen in a lake due to algal bloom: effect of time delay. Adv. Water Resour. **34**(10), 1232–1238 (2011)
161. Smith, V.H.: Low nitrogen to phosphorus ratios favor dominance by blue-green algae in lake phytoplankton. Science **221**(4611), 669–671 (1983)
162. Smith, V.H.: Light and nutrient effects on the relative biomass of blue-green algae in lake phytoplankton. Can. J. Fish. Aquat. Sci. **43**(1), 148–153 (1986)
163. He, S., Peng, Y., Sun, K.: SEIR modeling of the COVID-19 and its dynamics. Nonlinear Dyn. **101**, 1667–1680 (2020)
164. Tao, Y., Zhu, H., Ren, J.: Comparative study of physics-based model and machine learning model for epidemic forecasting and countermeasure. Comput. Appl. Math. **43**(3), 148 (2024)
165. Engbert, R., Rabe, M.M., Kliegl, R., Reich, S.: Sequential data assimilation of the stochastic SEIR epidemic model for regional COVID-19 dynamics. Bull. Math. Biol. **83**(1), 1 (2021)
166. Rudin, C.: Stop explaining black box machine learning models for high stakes decisions and use interpretable models instead. Nat. Mach. Intell. **1**(5), 206–215 (2019)
167. Evangelou, N., Cui, T., Bello-Rivas, J.M., Makeev, A., Kevrekidis, I.G.: Tipping points of evolving epidemiological networks: machine learning-assisted, data-driven effective modeling. Chaos: An Interdisciplinary J. Nonlinear Sci. **34**(6) (2024)
168. Watson, G.L., Xiong, D., Zhang, L., Zoller, J.A., Shamshoian, J., Sundin, P., Bufford, T., Rimoin, A.W., Suchard, M.A., Ramirez, C.M.: Pandemic velocity: Forecasting COVID-19 in the us with a machine learning and Bayesian time series compartmental model. PLoS Comput. Biol. **17**(3), e1008837 (2021)
169. Tao, Y., Zhao, J., Cui, H., Liu, L., He, L.: Exploring the impact of socioeconomic and natural factors on pulmonary tuberculosis incidence in China (2013–2019) using explainable machine learning: a nationwide study. Acta Tropica **253**, 107176 (2024)
170. Canada to allow family of citizens across U.S. border amid COVID-19 shutdown. https://globalnews.ca/news/7039072/coronavirus-canada-us-border-exception/

171. Many provinces, territories enforcing border checkpoints and travel restrictions. https://www.cbc.ca/news/canada/coronavirus-checkpoints-travel-restrictions-border-province-1.5522467
172. Clement, J.C., Ponnusamy, V.K., Sriharipriya, K.C., Nandakumar, R.: A survey on mathematical, machine learning and deep learning models for COVID-19 transmission and diagnosis. IEEE Rev. Biomed. Eng. **15**, 325–340 (2021)
173. Chen, S.,Yu, J.: Stability analysis of a reaction–diffusion equation with spatiotemporal delay and Dirichlet boundary condition. J. Dyn. Diff. Equat. **28**, 857–866 (2016)
174. Shi, Q., Shi, J., Wang, H.: Spatial movement with distributed memory. J. Math. Biol. **82**(4), 1–32 (2021)
175. Reopening Ontario. https://www.ontario.ca/page/reopening-ontario
176. Pao, C.V., Ruan, W.H.: Quasilinear parabolic and elliptic systems with mixed quasimonotone functions. J. Differ. Equ. **255**(7), 1515–1553 (2013)
177. Weitz, J.S., Beckett, S.J., Coenen, A.R., Demory, D., Dominguez-Mirazo, M., Dushoff, J., Leung, C., Li, G., Măgălie, A., Park, S.W., et al.: Modeling shield immunity to reduce COVID-19 epidemic spread. Nat. Med. **26**, 849–854 (2020)
178. Hochreiter, S., Schmidhuber, J.: Long short-term memory. Neural Comput. **9**(8), 1735–1780 (1997)
179. Schmidhuber, J.: Deep learning in neural networks: An overview. Neural Netw. **61**, 85–117 (2015)
180. Graves, A., Schmidhuber, J.: Framewise phoneme classification with bidirectional LSTM and other neural network architectures. Neural Netw. **18**(5–6), 602–610 (2005)
181. Huang, G.-B., Zhu, Q.-Y., Siew, C.-K.: Extreme learning machine: theory and applications. Neurocomputing **70**(1–3), 489–501 (2006)
182. Huang, G.-B., Zhou, H., Ding, X., Zhang, R.: Extreme learning machine for regression and multiclass classification. IEEE Trans. Syst. Man Cybern. B Cybern. **42**(2), 513–529 (2011)
183. Menzies, N.A., Quaife, M., Allwood, B.W., Byrne, A.L., Coussens, A.K., Harries, A.D., Marx, F.M., Meghji, J., Pedrazzoli, D., Salomon, J.A., et al.: Lifetime burden of disease due to incident tuberculosis: a global reappraisal including post-tuberculosis sequelae. Lancet Glob. Health **9**(12), e1679–e1687 (2021)
184. Kyu, H.H., Maddison, E.R., Henry, N.J., Ledesma, J.R., Wiens, K.E., Reiner, R., Biehl, M.H., Shields, C., Osgood-Zimmerman, A., Ross, J.M., et al.: Global, regional, and national burden of tuberculosis, 1990–2016: results from the global burden of diseases, injuries, and risk factors 2016 study. Lancet Infect. Dis. **18**(12), 1329–1349 (2018)
185. Trauer, J.M., Dodd, P.J., Gomes, M.G.M., Gomez, G.B., Houben, R.M.G.J., McBryde, E.S., Melsew, Y.A., Menzies, N.A., Arinaminpathy, N., Shrestha, S., et al.: The importance of heterogeneity to the epidemiology of tuberculosis. Clin. Infect. Dis. **69**(1), 159–166 (2019)
186. Hahad, O., Lelieveld, J., Birklein, F., Lieb, K., Daiber, A., Münzel, T.: Ambient air pollution increases the risk of cerebrovascular and neuropsychiatric disorders through induction of inflammation and oxidative stress. Int. J. Mol. Sci. **21**(12), 4306 (2020)
187. Sibandze, D.B., Magazi, B.T., Malinga, L.A., Maningi, N.E., Shey, B.-A., Pasipanodya, J.G., Mbelle, N.N.: Machine learning reveals that mycobacterium tuberculosis genotypes and anatomic disease site impacts drug resistance and disease transmission among patients with proven extra-pulmonary tuberculosis. BMC Infect. Dis. **20**, 1–15 (2020)
188. Harris, R.I.D.: Testing for unit roots using the augmented dickey-fuller test: some issues relating to the size, power and the lag structure of the test. Econ. Lett. **38**(4), 381–386 (1992)
189. Dye, C., Williams, B.G.: The population dynamics and control of tuberculosis. Science **328**(5980), 856–861 (2010)
190. van Leth, F., Van der Werf, M.J., Borgdorff, M.W.: Prevalence of tuberculous infection and incidence of tuberculosis: a re-assessment of the styblo rule. Bull. World Health Organ. **86**(1), 20–26 (2008)
191. Dye, C., Lönnroth, K., Jaramillo, E., Williams, B.G., Raviglione, M.: Trends in tuberculosis incidence and their determinants in 134 countries. Bull. World Health Organ. **87**(9), 683–691 (2009)

192. Kirigia, J.M., Muthuri, D.K.R.: Productivity losses associated with tuberculosis deaths in the world health organization african region. Infect. Dis. Poverty **5**(03), 28–39 (2016)
193. Yang, J., Zhang, M., Chen, Y., Ma, L., Yadikaer, R., Lu, Y., Lou, P., Pu, Y., Xiang, R., Rui, B.: A study on the relationship between air pollution and pulmonary tuberculosis based on the general additive model in Wulumuqi, China. Int. J. Infect. Dis. **96**, 42–47 (2020)
194. Zhu, S., Wu, Y., Wang, Q., Gao, L., Chen, L., Zeng, F., Yang, P., Gao, Y., Yang, J.: Long-term exposure to ambient air pollution and greenness in relation to pulmonary tuberculosis in China: a nationwide modelling study. Environ. Res. **214**, 114100 (2022)
195. Borges, G.Á., Elias, S.T., da Silva, S.M.M., Magalhães, P.O., Macedo, S.B., Ribeiro, A.P.D., Guerra, E.N.S.: In vitro evaluation of wound healing and antimicrobial potential of ozone therapy. J. Craniofac. Surg. **45**(3), 364–370 (2017)
196. Smith, G.S., Van Eeden, S.K.D., Garcia, C., Shan, J., Baxter, R., Herring, A.H., Richardson, D.B., Van Rie, A., Emch, M., Gammon, M.D.: Air pollution and pulmonary tuberculosis: a nested case–control study among members of a northern California health plan. Environ. Health Perspect. **124**(6), 761–768 (2016)
197. Li, J.-X., Luan, Q., Li, B., Dharmage, S.C., Heinrich, J., Bloom, M.S., Knibbs, L.D., Popovic, I., Li, L., Zhong, X., et al.: Outdoor environmental exposome and the burden of tuberculosis: Findings from nearly two million adults in northwestern China. J. Hazard. Mater. **459**, 132222 (2023)
198. Popovic, I., Magalhaes, R.J.S., Ge, E., Marks, G.B., Dong, G.-H., Wei, X., Knibbs, L.D.: A systematic literature review and critical appraisal of epidemiological studies on outdoor air pollution and tuberculosis outcomes. Environ. Res. **170**, 33–45 (2019)
199. Liu, F., Zhang, Z., Chen, H., Nie, S.: Associations of ambient air pollutants with regional pulmonary tuberculosis incidence in the central chinese province of Hubei: a Bayesian spatial-temporal analysis. Environ. Health **19**, 1–10 (2020)
200. Patterson, B., Wood, R.: Is cough really necessary for tb transmission? Tuberculosis **117**, 31–35 (2019)
201. Dinkele, R., Gessner, S., McKerry, A., Leonard, B., Leukes, J., Seldon, R., Warner, D.F., Wood, R.: Aerosolization of mycobacterium tuberculosis by tidal breathing. Am. J. Respir. Crit. Care Med. **206**(2), 206–216 (2022)
202. Huang, K., Hu, C.-Y., Yang, X.-Y., Zhang, Y., Wang, X.-Q., Zhang, K.-D., Li, Y.-Q., Wang, J., Yu, W.-J., Cheng, X., et al.: Contributions of ambient temperature and relative humidity to the risk of tuberculosis admissions: A multicity study in central China. Sci. Total Environ. **838**, 156272 (2022)
203. Wang, Q., Li, Y.-L., Yin, Y.-L., Hu, B., Yu, C.-C., Wang, Z.-D., Li, Y.-H., Xu, C.-J., Wang, Y.-B.: Association of air pollutants and meteorological factors with tuberculosis: a national multicenter ecological study in China. Int. J. Biometeorol. **67**(10), 1629–1641 (2023)

FSC
www.fsc.org
MIX
Papier aus verantwortungsvollen Quellen
Paper from responsible sources
FSC® C105338

If you have any concerns about our products,
you can contact us on
ProductSafety@springernature.com

In case Publisher is established outside the EU,
the EU authorized representative is:
Springer Nature Customer Service Center GmbH
Europaplatz 3, 69115 Heidelberg, Germany

Printed by Libri Plureos GmbH
in Hamburg, Germany